Cell Communication

WILEY SERIES IN THE DYNAMICS OF CELL BIOLOGY

Russell J. Barrnett, Series Editor

Cell Communication, **Edited by Rody P. Cox**

Electron Microscopic Concepts of Secretion: Ultrastructure of Endocrine and Reproductive Organs, **Edited by Melvin Hess**

CELL COMMUNICATION

Edited by

RODY P. COX, M.D.

Division of Human Genetics
Departments of Medicine and Pharmacology
New York University School of Medicine

A WILEY BIOMEDICAL-HEALTH PUBLICATION

JOHN WILEY & SONS,   New York • London • Sydney • Toronto

Library of Congress Cataloging in Publication Data:

Cox, Rody P   1926-
  Cell communication.

  (Wiley series in the dynamics of cell biology) (A Wiley biomedical-health publication)
  Includes bibliographical references.

  1. Cell interaction.   I. Title.   [DNLM: 1. Cells—Physiology. QH631 C877c]

QH604.2.C69   574.8′76   74-10547
ISBN 0-471-18135-8

Printed in the United States of America

10 9 8 7 6 5 4 3 2 1

Authors

M. E. Balis
The Sloan-Kettering Cancer Center
New York, New York

A. Basten
Immunology Unit
Department of Bacteriology
University of Sydney
Sydney, Australia

Rody P. Cox
Division of Human Genetics
Departments of Medicine and Pharmacology
New York University School of Medicine
New York, New York

C. N. D. Cruickshank
M.R.C. Unit on Experimental Pathology of Skin
The Medical School
The University of Birmingham
Birmingham, England

Joseph Dancis
Department of Pediatrics
New York University School of Medicine
New York, New York

Gerald D. Fischbach
Department of Pharmacology
Harvard Medical School
Boston, Massachusetts

Norton B. Gilula
The Rockefeller University
New York, New York

Albert Harris
Department of Zoology
University of North Carolina
Chapel Hill, North Carolina

Gerald M. Kolodny
Radiology Research Laboratory
Harvard Medical School
Boston, Massachusetts

Marjorie R. Krauss
Division of Human Genetics
Department of Medicine
New York University School of Medicine
New York, New York

J. F. A. P. Miller
Walter and Eliza Hall Institute of Medical Research
Melbourne, Australia

Elizabeth F. Neufeld
National Institute of Arthritis, Metabolism and Digestive Diseases
National Institutes of Health
Bethesda, Maryland

Elena Ottolenghi-Nightingale
National Academy of Sciences
Washington, D. C.

v

Harry Rubin   Judson D. Sheridan
Department of Molecular Biology   Department of Zoology
University of California   University of Minnesota
Berkeley, California   Minneapolis, Minnesota

Preface

Cell communication is a general phenomenon of importance in understanding many biological processes. It interfaces with nearly all disciplines of biology and medicine. From an evolutionary point of view cell interactions provide multicellular organisms with mechanisms for homeostasis and for regulating cellular activities. Some forms of communication require cell contact, whereas others involve transfer through the external milieu.

It is the purpose of this monograph to review a number of examples of cell interactions in several biological systems. The processes described appear different, but the fundamental mechanisms responsible for communication may be quite similar. Although specialized techniques are required for the study of each of the systems, the concepts and approaches to understanding the cellular and molecular bases of these interactions are again similar. Each of the presentations describes experimental evidence on which the current concepts of cell communication are based. It is unavoidable, and probably even desirable, that the special bias of the individual authors is evident in their contributions. Many of the presentations also include new ideas and challenging hypotheses for which compelling evidence is at present lacking. This monograph critically appraises the current status and limitations of our understanding of cell communication and indicates future directions for investigation. We hope it will stimulate workers in other areas of biology to develop new systems and models for exploring cell interactions.

RODY P. COX

New York, New York
March 1974

Contents

CHAPTER ONE

Junctions Between Cells

NORTON B. GILULA

The Rockefeller University
New York, New York

Cell-to-cell communication, as a general biological phenomenon, consists of both long-range and short-range interactions between cells in both excitable and non-excitable tissues. The short-range interactions require direct physical contact between cells. These interactions are associated with a variety of communication phenomena, such as the immune response, cell fusion, electrical (ionic) coupling, metabolic cooperation, synaptic transmission (chemical), and intercellular adhesion.

Short-range cellular interactions are frequently accompanied by distinct intercellular membrane specializations. These membrane specializations or intercellular contacts can be characterized with physiological and ultrastructural probes into several groups (Table 1). The different groups of intercellular contacts can, in turn, reflect different functional capabilities of various cellular interactions.

The most extensive intercellular contacts that can be distinguished as plasma membrane and intercellular matrix modifications are those that are involved in intercellular adhesion, electrical coupling, metabolic coupling, the chemical synapse, and transepithelial permeability regulation. In this chapter emphasis will be placed on (1) a description of the basic features of the major intercellular contacts; (2) a discussion of their physiological significance; and (3) a careful ex-

Table 1. Summary of Specialized Intercellular Contacts

Contact	Origin	Function	Form	Type
Gap junction	Vertebrate and invertebrate	Intercellular communication; electrical or low-resistance synapse	Plaque	1. Gap junction A (nexus) 2. Gap junction B—arthropods
Tight junction	Vertebrate	Epithelial transcellular permeability regulation; sealing element	Belt or band	1. Zonula and fascia occludens (belt or band) 2. Endothelial cell junction (discontinuous belt or band) 3. Sertoli-Sertoli cell junction (belt)
Septate junction	Invertebrate	Partial permeability barrier; adhesion	Belt	1. Septate junction 2. Zonula continua—arthropods
Synaptic junction	Vertebrate and invertebrate	Chemical synaptic transmission	Plaque	1. Chemical synapse-excitatory and inhibitory 2. Neuromuscular junction
Desmosome	Vertebrate and invertebrate	Mechanical support; adhesion	Plaque, belt, or band	1. Desmosome and hemidesmosome (macula adhaerens) 2. Intermediate desmosome (zonula adhaerens) 3. Fascia adhaerens 4. Septatelike contact

amination of the structural and physiological evidence that has implicated one of the intercellular contacts, the gap junction, in the phenomenon of intercellular communication (ionic and metabolic cell coupling).

DEFINITION OF A CELL JUNCTION

A cell junction is a specialized region of short-range contact between two cells which is associated with a differentiation of the contributing cell surface membranes and/or the intervening intercellular matrix. Cells may maintain a reasonable intercellular distance (up to 200 Å) at these sites, or they may eliminate the intercellular space by a true fusion of their surface membranes. Cell junctions have now been found throughout the metazoan animal kingdom, including both invertebrates and vertebrates, and they are expressed both *in vivo* and in culture.

In both plants and animals the cytoplasmic bridges between cells are the extreme example of cytoplasmic continuity between cells. However, since they are not frequent elements in mammalian organisms (with the exception of germ cell interactions), these structures will not be considered in this chapter.

METHODS OF CHARACTERIZING CELL JUNCTIONS

Cell junctions usually occur as localized sites of interaction involving small regions of the interacting cell surfaces. Due to their small size, with a few exceptions, cell junctions must be characterized by electron-microscopic examination. With the electron microscope, junctions may be examined in (1) conventional thin-section preparations, (2) material that is treated with an intercellular "stain," or "tracer" substance, (3) freeze-fractured specimens that expose the internal aspects of the junctional membranes, and (4) "negative" stain or shadowed material of isolated (*in vitro*) cell junctions. Physiologically, some cell junctions between intact cells can be studied with microelectrode recording techniques, together with dye injections, and biochemical information can be obtained by studying isolated cell junctions from plasma membrane-enriched subcellular fractions.

TYPES OF CELL JUNCTIONS

Gap Junction

CONVENTIONAL THIN SECTIONS. The gap junction was described in thin sections as early as 1958 (1), and since then there have been varying descriptions of this junction (2–6) due to different tissue sources and preparation (fixation and staining) procedures. From these studies, some of the following names have been applied to this structure: longitudinal connecting surfaces (1), quintuple-layered interconnection (3), external compound membranes (2), nexus (4), and gap junction (6). A detailed discussion of this history has recently appeared (7).

The gap junction was first clearly resolved in its present form by Revel and Karnovsky in 1967. They utilized a preservation procedure (glutaraldehyde, osmium, *en bloc* uranyl acetate staining) that resolved the gap junction (in cross section) into a seven-layered (septilaminar) structure. A seven-layered structure has now been demonstrated in all of the structures described earlier. The seven layers result from the parallel apposition of two 75 Å thick unit membranes (trilaminar in appearance) separated by a 20–40 Å electron-lucent space, or "gap"

(Fig. 1). This structure is 150–190 Å, or a maximum of 40 Å greater than the combined thickness of two unit membranes. The appearance of this arrangement of two membranes separated by a "gap" led to the use of the term "gap junction" to describe this structure (6). The term "nexus," which was initially used to describe a pentalaminar (five-layered) structure (4), is now often used interchangeably with the term "gap junction." It is now clear that the structure originally described as a nexus (pentalaminar) can be preserved as a gap junction (septilaminar) when treated with *en bloc* uranyl acetate staining (6,8–10).

The gap junction with the septilaminar appearance has now been described between cells in both vertebrates (6,8–11) and invertebrates (12–16) as well as between cells in culture (17–20). Due to its presence between nonexcitable as well as excitable cells, this structure is possibly the most frequent cell junction found in animal organisms. Instead of listing the many locations where this structure has been identified, perhaps it is more interesting to consider some places where the structure has not yet been found. This list includes notably mature skeletal muscle cells (myotubes) and circulating blood cells.

TRACERS. An electron-opaque material, colloidal lanthanum hydroxide, was successfully used to demonstrate a penetrable region in the gap junction by Revel and Karnovsky in 1967 (6). This material penetrates or traces a central region of the junction (about 55 Å in thickness) that is slightly larger than the electron-lucent gap. This observation clearly indicates that there is an extracellular continuity (pathway) through the gap region of the junction. Lanthanum penetration has become a characteristic feature of gap junctions that

clearly distinguishes them from the occluding pentalaminar structures, which are truly "tight" and represent membrane-to-membrane fusions (6,8–11,13, 16,21). Other substances, such as pyroantimonate and ruthenium red, can also be used to demonstrate the penetrability of the gap junction (11,22). It is also interesting to note that certain high-molecular-weight substances, such as horseradish peroxidase, are excluded from the "gap" region (23); this is presumably due to a size limitation in the extracellular region of the gap.

Revel and Karnovsky also observed that in *en face* view a lanthanum-impregnated gap junction is comprised of a polygonal lattice of 70–80 Å subunits (6). The electron-dense lanthanum outlines the subunits, which have a 90–100 Å center-to-center spacing. The lanthanum also is frequently present as a 15–20 Å dense dot occupying the central region of the 80 Å subunits. These *en face* characteristics have now been found in studies of gap junctions from a variety of different sources (6,8–10,17). A similar membrane polygonal lattice had also been observed prior to the observations of Revel and Karnovsky in 1967. Robertson (2) described a polygonal lattice with 90 Å center-to-center spacing at the electrotonic membrane synapse (goldfish medulla), and Benedetti and Emmelot (24, 25) found a similar polygonal lattice when examining a plasma membrane fraction from rat liver with negative staining. Since 1967 both of these earlier observations have been substantiated as characteristic gap junctions (8,9).

NEGATIVE STAIN. Negative staining with heavy metal salts, such as sodium phosphotungstate, ammonium molybdate, and uranyl acetate, has provided some useful complementary information about the gap junction. Since gap junctions are portions of subcellular plasma mem-

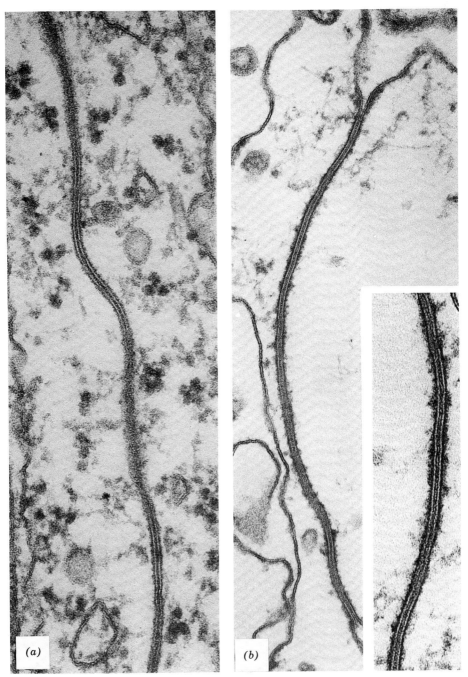

Fig. 1. Thin-section appearance of the gap junction between intact cells and in plasma-membrane-enriched subcellular fractions. (*A*) Extensive gap junction between Don hamster fibroblasts in cell culture. The gap junction is present in both transverse and slightly oblique planes. ×180,000. (*B*) Gap junction in an enriched plasma membrane subfraction from rat liver. The junction can be further purified from the nonjunctional plasma membranes present in this fraction. ×105,600. During the isolation procedure, the structural integrity of the gap junction is maintained as indicated by its appearance in thin sections (inset): two closely apposed junctional membranes are separated by an electron-translucent 20–40 Å space, or "gap." The isolated liver gap junction frequently has a discontinuous layer of electron-dense material associated with the cytoplasmic surfaces of the junctional membranes. ×160,000.

brane preparations, they can be observed in these samples with the negative-staining techniquie. Benedetti and Emmelot originally observed a polygonal lattice of subunits in a membrane preparation from rat liver (24). They initially felt that this lattice was a component of the general plasma membrane; however, they later (25) equated this array with the lattice that was described by Revel and Karnovsky. Later Goodenough and Revel (9) conclusively demonstrated that the negative-stain polygonal lattice of subunits was directly related to the thin-section image of a septilaminar gap junction. The negative-stain technique is now an important probe for assaying gap junctions in isolated membrane preparations (Fig. 2). In this regard it is interesting to note that the polygonal lattice is more easily visualized with negative staining after the membranes have been treated with detergent, such as deoxycholate (9,25) or sarkosyl (26). In the case of deoxycholate treatment, the lattice is still present even though the thin-section appearance of the gap junction may be altered from a septilaminar structure to a pentalaminar one (disappearance of the gap) (9,25).

FREEZE-FRACTURE. Studies on gap junctions that are frozen and then fractured to expose internal membrane components have provided important information about the structural characteristics of the gap junctional membranes. These studies have also been instrumental in providing a more comprehensive view of the form and distribution of gap junctions between cells in tissue and in culture.

The freeze-fracture process exposes two complementary internal membrane components or fracture faces (26,28–31). With the cell surface plasma membrane, these two fracture faces can be distinguished on the basis of their relationship to the cytoplasm and the extracellular space. They are commonly referred to as (1) an inner membrane fracture face (face A), which is adjacent to the cytoplasm; and (2) an outer membrane fracture face (face B), which is adjacent to the extracellular space. Therefore a single unit membrane is split into two components (faces A and B) when fractured. In the last few years the terms "face A" and "face B" have emerged as the popular convention for referring to the two junctional membrane fracture faces (7,32).

Freeze-fracturing dramatically demonstrates that the gap junction is a uniquely specialized region of the plasma membrane (9–11,14,18,20,26,31). As seen in Fig. 3, gap junctions are generally present as segregated domains or localized regions in plaques which are distinctly differentiated from regions of nonjunctional membrane (see also Fig. 9). Gap junctions possess a characteristic polygonal arrangement of homogeneous 70–80 Å particles on face A with a 90–100 Å center-to-center spacing. A 20–25 Å central dot or depression is frequently present in the center of these junctionl particles (10,14). Face B of gap junctional membranes contains a polygonal arrangement of pits, or depressions, which have a similar packing. Freeze-fracture observations have indicated that gap junctions can exist as large plaques (9–11,26, 33), as small plaques (11,18,20,33,34) or as a single band or strand of particles (35). At present there is no documented information that precisely defines the minimal size of a detectable gap junction. Theoretically there should exist a gap junction containing only a single particle that is matched by a similar particle in the adjacent cell membrane. The packing of face A particles within a gap junctional plaque can vary significantly. These variations usually include a homogeneous polygonal packing (liver and

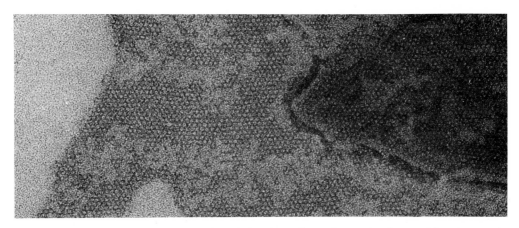

Fig. 2. Negative-stained gap junction from isolated rat liver plasma membranes (phosphotungstic acid at pH 7.0). The polygonal lattice of the gap junction is frequently discontinuous when "stained" at a low temperature (4°C). An electron-dense dot is present in the center of some of these polygonal subunits. ×160,000.

pancreas) (Fig. 3), an interrupted packing where there are small particle aggregates separated by smooth regions within a plaque (myocardium, certain cells in culture, ciliary epithelium of the eye, and adrenal cortex; see Fig. 9), and a single strand or two of aggregated particles (between photoreceptor cells in the vertebrate retina) (35). In general, the gap junctions that are present between cells in any specialized region of an organ are strikingly similar with regards to size, distribution, and packing characteristics. Gap junctions usually exist as individual (isolated) junction elements, but they can also be associated with another junction element, the tight junction (11,26,33).

CHEMICAL AND PHYSICAL PROPERTIES. There have been a limited number of observations on the physicochemical properties of gap junctions. These observations have been made on (1) junctions in intact tissue and (2) more recently on isolated junctions *in vitro*.

Studies on intact tissue have revealed that the gap junctions are generally insensitive to treatment with proteolytic enzymes or divalent-cation chelators, such as ethylenediaminetetraacetic acid (EDTA) (36–39), but they are affected by osmotic or tonicity changes (36,40,41). The gap junctions can be opened, or "unzipped," by hypertonic (sucrose) treatment (40–42), and the junctional membrane fracture faces are not detectably altered during this process (42). Tissue dissociation into single cell populations using proteolytic treatments disrupts the normal interactions of cells without affecting the integrity of the gap junctions (38). In this case the gap junctions are present as the entire 150–200 Å thick complex, which is attached to only one cell. In general, gap junctions display a remarkable resistance to physical or mechanical stress.

Enzymatically, no endogenous activity has been found to be consistently associated with gap junctions either *in vivo* or *in vitro*. However, there have been two separate reports that an ATPase reaction product can be localized cytochemically at the gap junction in intact myocardium (43) and in isolated gap junctions from rat liver (44). Due to the

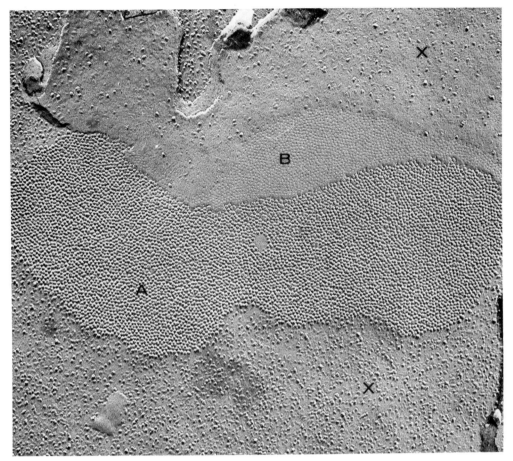

Fig. 3. Freeze-fractured gap junction from intact mouse liver. The freeze-fractured gap junction possesses a unique membrane differentiation that is characterized by a polygonal arrangement of membrane particles on fracture face A (indicated by the letter "A") and a polygonal array of complementary pits or depressions on fracture face B indicated by the letter "B"). Fracture face A is synonymous with the term "inner membrane fracture face," which corresponds to the cytoplasmic leaflet of a plasma membrane; fracture face B is synonymous with the term "outer membrane fracture face," which corresponds to the extracellular leaflet of a plasma membrane. Note that the gap junction is present as a plaquelike region that is segregated from regions of nonjunctional plasma membrane (X). The nonjunctional membrane fracture faces are characterized by a random distribution of a heterogeneous (size) particle population. ×96,000.

rather spurious results with cytochemical ATPase localizations, these observations must await further substantiation with other methods. Other cytochemical probes on gap junctions have produced negative results so far. These include the absence of colloidal iron hydroxide binding at low pH (45), the absence of concanavalin A binding (45), and the absence of cationic derivatized ferritin binding (45).

Gap junctions are normal components of subcellular plasma membrane fractions (Fig. 1*B*), and this fact has led to several structural and biochemical observations on *in vitro* gap junctions. Three different studies have been focused on gap junctions from rat and mouse liver.

Benedetti and Emmelot were the first to identify the gap junctional hexagonal lattice in plasma membrane preparations from rat liver (24,25). They used these preparations to study a variety of effects on the gap junctional lattice observed with negative staining. In summary, they found that the lattice image was temperature dependent (enhanced at 37°C), neuraminidase treatment did not affect the lattice spacing, and papain or trypsin caused a reduction in the lattice spacing (46). Benedetti and Emmelot also utilized a brief detergent treatment (1% deoxycholate) to isolate an enriched gap junctional fraction (25). Along with the gap junctions, this fraction contains a significant amount of amorphous material, thus hindering biochemical analysis.

Goodenough and Revel have used a variety of chemical probes to determine the biochemical content of the isolated mouse liver gap junctions, using the presence of the 20–40 Å gap as an indicator of structural integrity or intactness (9,39). In two separate studies they reported that the 20–40 Å gap can disappear after (1) extraction with 60% acetone; (2) treatment with 1% deoxycholate; or (3) treatment with phospholipase C (from *Clostridium welchii*) at 1 mg/ml. They also found that the 20–40 Å gap had a remarkable resistance to treatment with (1) $0.02M$ EDTA; (2) 6 M urea; or (3) Pronase (1 mg/ml).

In a recent report Goodenough and Stockenius have isolated a preparation of mouse liver gap junctions that has an exceptionally high degree of purity (26). This purification is based on a collagenase digestion, a brief treatment with the detergent sarkosyl NL-97, a brief ultrasonication, and finally a sucrose gradient. This junction preparation has only three detectable protein components on sodium dodecyl sulfate–polyacrylamide gels; the most prominent

component has a molecular weight in the range of 20,000. A thin-layer chromatography profile of this material indicates the presence of some neutral lipid and some phospholipid (tentatively phosphatidylcholine and phosphatidylethanolamine). A hexagonal lattice with 86 Å center-to-center spacing was also reported from low-angle X-ray diffraction studies on both wet and dried junctional preparations.

Evans and Gurd have recently isolated an enriched gap junctional preparation from mouse liver based on a resistance to the detergent N-laurylsarcosinate (47). Although they have obtained a substantial amount of chemical information on this preparation, it is difficult to determine what portion of their information directly applies to the gap junction, since there is a significant amount of amorphous material within the preparation. In a separate study they have also reported that the detergent-resistant fraction has a slow degradation rate in relation to other membrane components; this information was obtained by administering radioactive leucine (double-label technique) to intact mice and then following the fate of the labels in the membrane fractions (48).

The most recent study on isolated gap junctions is focused on the "synaptic discs" or electrical synapses in the goldfish medulla (49). In this study Zampighi and Robertson found that divalent-cation chelators (EDTA and EGTA) produce discontinuities in the polygonal lattice. This effect is also accompanied by fragmentation of the junction in some instances. Also the fragmentation effect can be enhanced by applying 0.3% deoxycholate in the presence of a divalent-cation chelator.

A VARIATION—GAP JUNCTION B. A structural pleomorphism has recently been established for the gap junctions in a variety of arthropod tissues (32,50–54). As seen in Fig. 4, in thin sections the arthropod gap junctions appear quite similar to the "conventional" gap junction (12,13,54–58), even though the intercellular "gap" is slightly larger than normal (about 30–40 Å). In lanthanum-treated preparations the polygonal lattice of subunits has slightly larger dimensions that the conventional gap junction (13,54,55,58).

In freeze-fractured tissue three basic structural differences can be observed in the arthropod gap junction (Figs. 5 and 6):

1. The gap junctional membranes contain two complementary fracture faces: the A face (inner or juxtacytoplasmic membrane fracture face), which contains pits or depressions; and the B face (outer membrane fracture face), which contains junctional membrane particles.

2. The gap junctional particles on the B fracture face are large and often heterogeneous in size; the particles are 110 Å or larger in diameter, and they are frequently present as fused aggregates of two or more particles.

3. The gap junctional particles are generally present in an irregular, nonpolygonal, packing.

These freeze-fracture characteristics clearly distinguish the arthropod gap junction from those described in other organisms so far. This arthropod junction has been termed gap junction B due to the disposition of the junctional membrane particles on fracture face B (32,50, 51).

The gap junction B pleomorphism is now clearly established for arthropods; however, too little information is available on other invertebrate gap junctions to be able to extend this pleomorphism at present. It is interesting to note that molluscan gap junctions are similar to the conventional structure, or gap junc-

tion A (14,59). Therefore it will be of evolutionary interest to characterize the gap junctions of annelids and other closely related phyla in the future. Other cell junctions, such as the tight and septate, already provide taxonomic distinction between invertebrate and vertebrate tissues; hence it may also be possible that a pleomorphism may exist in the gap junction that is associated with an evolutionary divergence.

The arthropod junctional pleomorphism is perhaps most significant with regards to the physiology of intercellular communication. The original observation of electrotonic coupling (60), as well as a large body of subsequent information concerning dye and macromolecular intercellular transfer (12,55,61,62), has been made on arthropod tissues. The junctional pleomorphism may indicate that the physiological properties of these junctions may be uniquely or qualitatively different from those associated with gap junction A. At any rate it is important to note this difference when one attempts to apply the physiological phenomena from arthropod tissues to other systems.

Tight Junction

CONVENTIONAL THIN SECTIONS. The tight junction is practically a ubiquitous structure between vertebrate epithelial cells (5,7–9,11,33). The structure is characterized by a true fusion or union of the membranes of adjacent cells. At the site of fusion the membranes are usually 140–150 Å thick. A tight junction may exist as a beltlike structure or as an isolated band. In thin sections it has been demonstrated that the tight junctions are capable of excluding or occluding the diffusion of large molecules between cells (5,9,63). In 1963 Farquhar and Palade

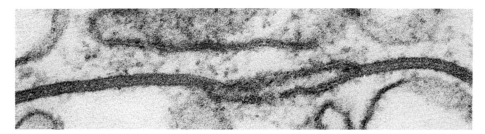

Fig. 4. Thin-section appearance of the gap junction B found in arthropod tissues. These gap junctions (from the crayfish hepatopancreas) are generally not distinguishable from the conventional gap junction in thin-section images. ×180,000.

Fig. 5. Freeze-fractured gap junction B. In the crayfish hepatopancreas the gap junctions are present as plaquelike structures varying in size and distribution between cells. The gap junctional plaque is comprised of a loose packing of heterogeneous particles that reside on fracture face B (B) (outer membrane fracture face). The complementary fracture face (A) contains an array of depressions. In this image nine identifiable gap junctions are present. Note the asymmetric particle distribution in the nonjunctional membrane regions (X): the A face contains more particles than the B face. ×100,000.

11

Fig. 6. Freeze-fractured gap junction B. The junctional particles located on the B fracture face (B) are heterogeneous in size (they are 110 Å in diameter when single) and are not frequently packed in polygonal arrays. The complementary depressions are found on fracture face A (A), and some large particles (arrows), presumably from the B fracture face, are often associated with the complementary element. ×120,000.

described tight junctions in a variety of mammalian epithelial tissues that were present in beltlike (zonula) or bandlike (fascia) arrangements (5). Thus they termed these structures zonula occludens and fascia occludens, respectively. Further, they demonstrated that a charac- teristic series of cell contacts is present in the apical region of interaction between many epithelial cells. This series of cell contacts has been described as the "junc- tional complex" (Fig. 7), and it includes the zonula occludens (tight junction) plus two desmosomal structures, the

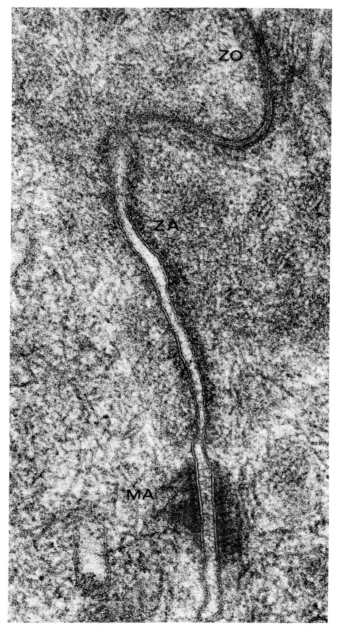

Fig. 7. Junctional complex between epithelial cells of the rat small intestine. The lateral plasma membranes fuse in the apical region to form the tight junction, or zonula occludens (ZO). The desmosomal elements of the junctional complex, the zonula adhaerens (ZA) and the macula adhaerens (MA), or desmosome, are present below the tight junction. ×119,000.

zonula adhaerens and macula adhaerens
(5). Although another type of tight junction, the focal or macula occludens, has
been frequently described in thin sections, it has generally been characterized
as a band or isolated segment of tight
junction (fascia occludens) when examined with freeze-fracturing.

TRACERS. Tracer studies, using electron-
dense exogenous materials, have demonstrated that the tight junction (zonula
occludens) is an effective occluding barrier to the intercellular diffusion of large
molecules (5,8,9,11,63,64). Recent studies
have indicated that small ionic species
may penetrate the zonula occludens in
certain epithelia (65,66).

FREEZE-FRACTURE. The freeze-fractured
tight junction is comprised of two complementary fracture face components arranged in a continuous meshwork
around the borders of epithelial cells
(Fig. 8). The inner membrane fracture
face (face A) is characterized by the
presence of short segments of anastomosing ridges, and the outer membrane fracture face (face B) contains a complementary arrangement of grooves (9,11,31,33,
67–69). These fracture face components
(ridges and grooves) represent the internal membrane features at the sites of
membrane fusion. In freeze-fractured
specimens the ridges are frequently observed as discontinuous elements consisting of particulate components (80–85 Å
in diameter) (11,69).

The tight junctional ridges are usually
linked to other ridges to form the zonula
occludens. However, isolated bands or
segments of ridges also exist; these isolated bands correspond to the fascia occludens structure (5,11). Tight junctional
ridges are frequently observed in close
association with gap junctional particles
(Fig. 9), and in some instances these
ridges completely surround or sequester
the particles of the gap junction (11,26).

FUNCTION. Several studies in the past
have identified a "tight" junction as the
low-resistance structure between electrically coupled cells. It is now clear
that the "tight" junctions in those studies were probably "gap" junctions, and
not the true zonula occludens. In all
epithelia that have recently been carefully examined both tight junctions (zonula occludens) and gap junctions are
present. Thus it has been virtually impossible to demonstrate a possible role
of the tight junction (by itself) in ionic
coupling between cells.

Studies with macromolecular and ionic
tracers have indicated that the zonula
occludens in certain epithelia is an effective barrier to the transepithelial diffusion of material via the intercellular
spaces (5,8,9,11,63). Similar tracer studies
performed on a variety of epithelia with
different transepithelial permeabilities
(from "leaky" to "tight") indicate that
under certain conditions the tight junctions may be penetrated in "leaky" epithelia (65,66). Other observations have
indicated that the tight junction is sensitive to osmotic changes (70–72) and that
this sensitivity may be related to transepithelial permeability regulation. In a
recent study Claude and Goodenough
(69) have related the freeze-fracture morphology of tight junctions to the transepithelial permeability properties of a
variety of epithelia. They have found
that the junctions in "very leaky" epithelia consist of one junctional strand,
the junctions in "very tight" epithelia
consist of five or more strands, and the
junctions in epithelia with intermediate
permeabilities contain an intermediate
junctional arrangement.

Septate Junction

CONVENTIONAL THIN SECTIONS. The septate junction was first described in

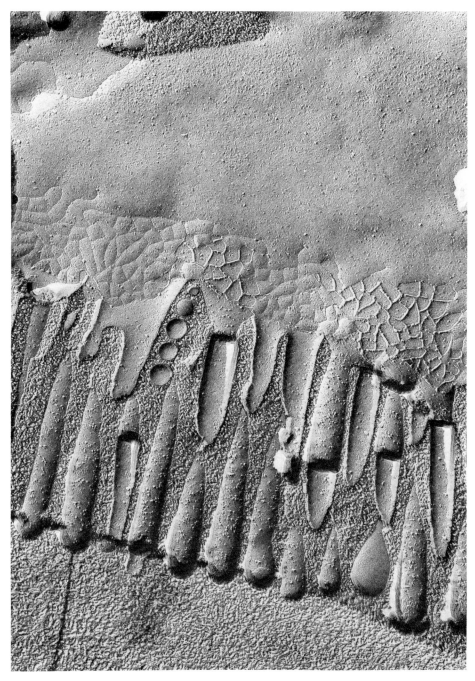

Fig. 8. Freeze-fractured zonula occludens (tight junction) between epithelial cells of the rat small intestine. The zonula occludens is comprised of a meshwork arrangement of ridges (A fracture face) and grooves (B fracture face). The ridges and grooves represent the actual sites of true membrane fusion. The arrangement of anastomosing ridges and grooves is responsible for the occlusion property of the zonula occludens. This belt, or zonula, of membrane fusion provides a permeability barrier for the diffusion of large macromolecules from the luminal to the basal aspects of these cells. ×70,000. From D. S. Friend and N. B. Gilula *J. Cell Biol.* **52,** 758 (1972), Fig. 21. Reprinted with permission from the Rockefeller University Press.

Fig. 9. Freeze-fracture image of the cell junctions between the epithelial layers in the ciliary epithelium of the rabbit eye. In this tissue the apical or luminal surfaces of the two epithelial cell sheets are joined by the arrangement of tight and gap junctions present in this image. The gap junctions (GJ) are present as plaques with polygonal arrays of particles. These particle arrays are segregated into subgroups within a plaque by the presence of particle-free zones or aisles. The tight junctions (arrows) are present as single or double ridges (fasciae occludens) that are closely apposed to the gap junctional particles. In this tissue the tight junctional ridges do not circumscribe or surround the gap junctions. ×90,000. From unpublished data of N. B. Gilula, E. Raviola, and G. Raviola.

Hydra (73), and since then it has been observed between a variety of cells in many invertebrate organisms (13,14,16, 32,55,74–76). The true septate junction has only been found in invertebrate tissues, and septatelike structures that are clearly different have been reported in vertebrate tissues (77–79).

The septate junction is the most extensive cell junction observed. It often extends for microns in length, and it occurs as a zonula, or belt, surrounding the apical or basal regions of cells. In transverse view the junction is comprised of two adjacent cell membranes joined by a periodic arrangement of electron-dense bars, or septa (Fig. 10*A*). The septa are nearly constant in width, but the spacing between them may vary (74,80). In tangential views (Fig. 10*B*) the septa are recognizable as parallel, pleated or corrugated sheets. The interaction of the adjacent pleated sheets often produces a series of hexagonal units. Studies on septate junctions from different sources have indicated that there may be considerable variations in the three-dimensional appearance of the intercellular septa (16,74–76).

TRACER STUDIES. Tracer substances or "stains" (e.g., lanthanum and ruthenium red) definitely penetrate the septate junction and produce a negative contrast image of the intercellular septal sheets (13,16,74,76). In negative image the septa appear as transparent 40–50 Å wide bars, whereas the interseptal spaces are electron dense. In these preparations the septa form a definite continuum between the adjacent cell membranes (74). The ruthenium red treatment also indicates that the septa are chemically different from the content of the interseptal spaces (74).

FREEZE-FRACTURE. The septate junctional membranes contain two complementary internal fracture faces (Fig. 11) (74). Fracture face A (inner membrane face) contains rows of 85 Å particles, and fracture face B (outer membrane face) contains interrupted rows of depressions. The particles within a row are 210 Å apart (center to center), and the particles between rows are not in register. The particle rows are frequently parallel; however, their distribution and arrangement are strikingly similar to those of the intercellular septa observed in thin section (14,59,74,76).

An integrated optical diffraction analysis of thin-section, tracer, and freeze-fracture information on the septate junction from mussel gill epithelium demonstrates a precise correspondence between the intramembrane junctional particles and the intercellular septa (74). In essence, this correspondence provides a structural basis for cytoplasmic continuity between two adjacent cells.

FUNCTION. The septate junction has been proposed as a site of intercellular adhesion (73), a permeability barrier (81), and a pathway for low-resistance ionic coupling (74,82,83).

Studies with tracers indicate that the septate junction, at best, is only a partially occluding structure (13,16,74,76). In some cases the tracer material can penetrate the entire length of the septate junction (13).

The septate junction is definitely present between cells that are ionically coupled (55,61,74,82,83); however, recent studies have indicated that gap junctions coexist with septate junctions between these cells (13,14,16,55). Therefore it will be impossible to accurately determine the involvement of the septate junction in ionic coupling until a cellular system is studied in which the septate junction is present but the gap junction is definitely absent.

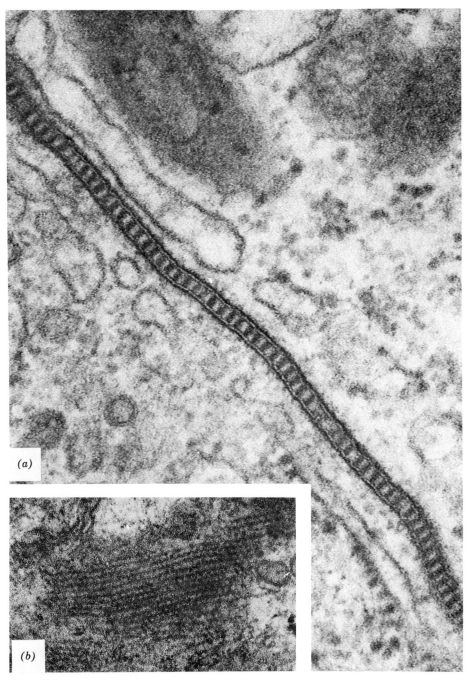

Fig. 10. Thin-section appearance of the invertebrate septate junction. (*A*) Transverse image of the septate junction between two molluskan ciliated epithelial cells. The two lateral plasma membranes are joined by a periodic arrangement of electron-dense bars, or septa, which are present within the intercellular space. ×160,000. (*B*) A tangential view of the septate junction, revealing the polygonal arrangement of the parallel intercellular septal sheets. These septal sheets form a continuous belt, or zonula, around the cells. ×80,000.

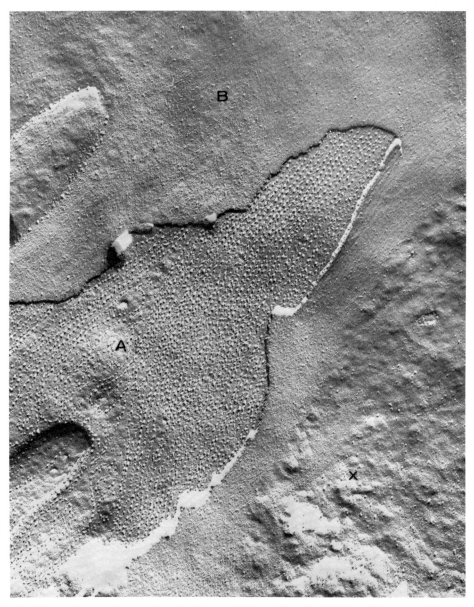

Fig. 11. Freeze-fractured septate junction from molluskan ciliated epithelium. The exposed junctional membrane differentiation forms a belt around the cells. Two complementary fracture faces are exposed in the junctional region. Fracture face A (A) contains parallel rows of membrane particles that correspond to the arrangement of the intercellular septa seen in thin sections. Note that the particles within a row maintain a constant spacing without aggregation. Fracture face B (B) contains an arrangement of linear depressions or grooves that complement the particle rows of face A. The particles in the nonjunctional membrane region (X) are randomly arranged. ×60,000.

Desmosomes

Desmosomes are a large class of intercellular contacts that are primarily involved as anchoring sites for filamentous structures and for intercellular adhesion (1,5,7,84–86). Desmosomes are often characterized by a dense plaque along the cytoplasmic surface of the desmosomal membranes and a crystalline modification of the intervening intercellular space (Fig. 12). Filaments are frequently inserted into the dense surface plaque, and the intercellular space (250–350 Å wide) contains a condensation of dense material. A central dense stratum, which is parallel to the membranes, is frequently present in the intercellular space. Tracers, such as lanthanum and ruthenium red, can easily penetrate the intercellular space of the desmosome, thus indicating that the intercellular condensation is partially porous (6,86,87). In some desmosomes, particularly those present in squamous epithelia, the freeze-fractured membranes contain a nonpolygonal arrangement of closely packed particles or granules (80–100 Å in diameter) and short segments of filaments (7,34,88). These fracture face components are present on both the inner and outer membrane fracture faces (A and B). However, it should be noted that in some tissues the desmosomal membrane fracture face components are not readily apparent (11).

Desmosomes have a wide distribution in both vertebrate and invertebrate tissues, and several pleomorphic forms may be present. The desmosomes may exist as bipartite structures or as hemidesmosomes (86). When present as a bipartite structure, the desmosome may be classified as (1) a spot or plaquelike structure, the macula adhaerens; (2) a bandlike structure, the fascia adhaerens; or (3) a beltlike structure, the zonula adhaerens (5). The major differences between these pleomorphic forms lies in their distribution and variations in the modifications of the cytoplasmic surface and the intercellular space.

These structures are clearly not involved as permeability sites for ionic coupling between cells since they can be selectively disrupted without affecting cell-to-cell permeability (36). The desmosomes are resistant to osmotic variations; however, they are extremely sensitive to proteolysis and divalent-cation (calcium and magnesium) concentrations (37,38, 89). Proteolytic treatment or the selective removal of divalent cations results in the disruption of the desmosome by a disintegration of the intercellular condensation. A recent study indicates that desmosomes from different tissues may have varying sensitivities to proteolysis, chelators, and detergents; this suggests that desmosomes may have different chemical compositions (118).

Since desmosomes are abundant in tissues that undergo severe mechanical stress, such as cardiac muscle and stratified squamous epithelium (skin), these structures must have a central role in maintaining cell-to-cell adhesion. Further, the insertion of filaments, such as actin in myocardium and tonofilaments in certain epithelia, into desmosomal plaques also indicates that these structures are involved as important intracellular anchoring sites.

THE ROLE OF GAP JUNCTIONS IN INTERCELLULAR COMMUNICATION

Evidence of Role in Ionic Coupling

WIDESPREAD DISTRIBUTION. There have been a large number of morphological studies that have clearly identified gap junctions in a wide variety of animal tissues (both vertebrate and invertebrate).

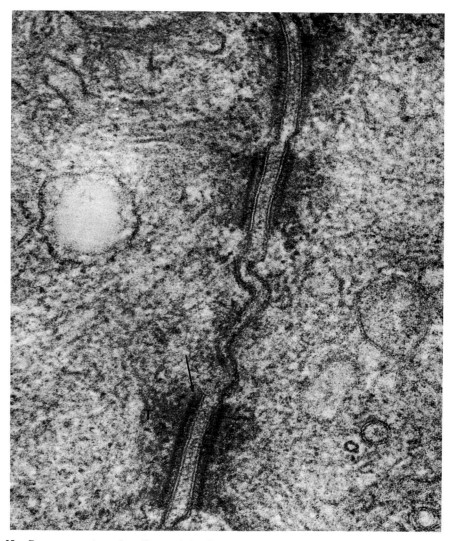

Fig. 12. Desmosomes (macula adhaerens) in the rat intestinal epithelium. The desmosomes occur as symmetrical plaques between two adjacent cells. Their characteristic features include (1) a wide intercellular space ($\sim$ 250–350 Å) containing dense material; (2) two parallel cell membranes; (3) a dense plaque associated with the cytoplasmic surface (arrows); and (4) cytoplasmic tonofilaments (t) that converge on the dense plaque. $\times$156,000.

These studies have included both embryonic and mature tissue, as well as cells in culture. Besides the morphological observations, there have been an equally impressive number of electrophysiological studies, which have demonstrated the widespread occurrence of ionic coupling. In many instances the demonstration of ionic coupling in a tissue has led to the subsequent finding of gap junctions, and vice versa. In general, gap junctions are common plasma membrane specializations between animal cells, and their presence is remarkably coincident with the occurrence of ionic coupling.

MORPHOLOGICAL CORRELATION WITH IONIC COUPLING. A large body of evidence implicating the gap junction as a low-resistance pathway for ionic coupling has been derived from studies that have integrated both morphological and physiological examination of a particular type (tissue) of cellular interaction or a specific cell-to-cell site of interaction.

Convincing studies have demonstrated that, in vertebrate myocardium, myocardial cells are ionically coupled when the gap junctions are intact (36,40). If the gap junctions are "opened" by hypertonic treatment, the cells are uncoupled. However, another cell contact, the desmosome, may be selectively disrupted without altering the ionic coupling between cells (36). Thus the gap junction is an essential feature of myocardial cell coupling.

In the central nervous system of fish, gap junctions have been identified at specific sites of electrotonic (ionic) coupling (2,8,90–93). These studies have demonstrated that the gap junction may serve as the electrical synapse. Extensive studies on the crayfish nervous system have further documented the presence of gap junctions at the site of electrotonic coupling (12,58,60,62,94,95). These studies have also indicated that certain procedures for uncoupling cells may be accompanied by a concomitant alteration in the gap junction (95).

The dipteran salivary gland has provided a unique multicellular system for studying ionic coupling. Many studies have demonstrated that in this gland extensive ionic coupling exists between the epithelial cells (55,61,82) and that the cells are joined by both gap and septate junctions (55,83,96). In this tissue either one or both of the junctions could serve as the low-resistance pathway, and a selective junctional disruption procedure will be necessary in order to clarify the possibilities.

In embryonic tissues gap junctions have been characterized between cells that are ionically coupled (97–102). The most extensive studies have been made on the fish embryo (*Fundulus*) and the chick embryo. Even though ionic coupling data are available on several other embryonic tissues (103–107), there is not sufficient morphological information available at this time to implicate the gap junction in the observed coupling in these other systems.

Recently the gap junction has been thoroughly implicated as a low-resistance pathway from studies on cell cultures. These studies, which have fully characterized the morphological and physiological interactions between a variety of cell types, have clearly resolved that gap junctions are the major cell junction between ionically coupled cells in culture (17–21,108–110). These studies have been extremely valuable since the precise geometry of cellular interaction as well as the homogeneity or heterogeneity of cell junctions within the population can be definitely established. These studies have also demonstrated that gap junctions together with ionic coupling can be present in both normal and transformed cell populations (17,20,111).

Evidence of Role in Ionic and Metabolic Coupling

Gap junctions have recently been implicated in another cell-to-cell interaction that was originally described as "metabolic cooperation between cells" (112). In the phenomenon of metabolic cooperation, or metabolic coupling, a normal or wild-type cell is capable of rendering a metabolic capacity to an enzyme-deficient or mutant cell after the two cell types interact by direct physical contact (112–114). During this metabolic interaction a low-molecular-weight molecule is apparently transmitted from the wild-type cell to the mutant one (115). A complete discussion of this biological phenomenon appears in Chapter 4 of this volume.

In a recent study three cell types were used to examine the possible relationships between two independent phenomena, metabolic and ionic coupling, and the appearance of gap junctions (19). In this system a normal cell type (Don hamster fibroblasts) could incorporate exogenous ^{3}H-hypoxanthine into its nucleic acids, whereas two mutants (DA cells and A9 cells) could not incorporate the label when cultured alone. This insufficiency of the mutant cells is due to the absence of the enzyme inosinic pyrophosphorylase (IPP) in the mutants (19,112).

Cocultivated populations (Don:DA and Don:A9) were examined for metabolic coupling, ionic coupling, and gap junctions. In Don:DA populations metabolic coupling, ionic coupling, and gap junctions were present between the cells (Fig. 13). In this population the gap junctions were similar (thin sections, freeze-fracturing) to those observed between other cells *in vivo* and in culture (Fig. 14). However, in Don:A9 populations metabolic coupling, ionic coupling, and gap junctions were absent between Don and A9 cells. In this cocultivated population both ionic coupling and gap junctions were present between interacting Don cells, but not between Don:A9 and A9:A9 cells. Between the noncommunicating cells a variety of cell contacts were observed (Fig. 15), but none of them was competent for metabolic and ionic coupling. A similar relationship between gap junctions, metabolic coupling, and ionic coupling has recently been reported in a different cell culture system (116,117).

In essence, the observations indicate that under these experimental condi-

tions ionic and metabolic coupling may take place simultaneously if gap junctions are also present. Conversely, metabolic and ionic couplings are not observed if gap junctions are absent. In this case the A9 cells (genetic variants of the mouse L cell line) are apparently noncommunicators or intercellular communication mutants. From this information it may be concluded that a specific structure, the gap junction, is required for the passage of ions and certain small molecules between cells. The qualitative and/or quantitative features of the "channels" within the gap junction still remain an important question, as well as the possibility that gap junctions may completely differ with respect to their "channel" properties.

FUTURE DIRECTIONS FOR STUDY

There is now substantial information to conclude that gap junctions serve as a specialized pathway for communication (ionic and metabolic coupling) between cells. It is also now established that communication may involve at least two types of gap junction (A and B) and that communication may be mediated by ions or, at least in one instance, by a low-molecular-weight metabolic product. The evidence to date does not rule out the possibility that other cellular interactions, such as tight and septate junctions, may also be involved in communication, but it does indicate that communication is present in the absence of these structures.

The ultimate demonstration of gap junctional involvement in cell-to-cell communication will rely on some future developments. In this regard we can anticipate the use of a physiological probe with suitable characteristics that will permit a precise localization (definition) of the cell-to-cell pathways.

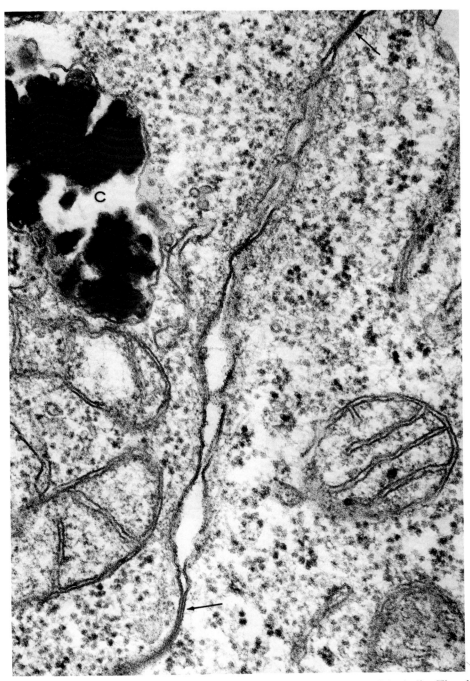

Fig. 13. Interactions between cells that are communicating metabolically and ionically. The electron-dense carmine (C) is used to distinguish the wild or normal cell type. Note the presence of gap junctions (arrows) between these cells. ×84,000. From N. B. Gilula et al., *Nature* **235**, 262 (1972), Fig. 4. Reprinted with permission from *Nature*, Macmillan Journals, Ltd.

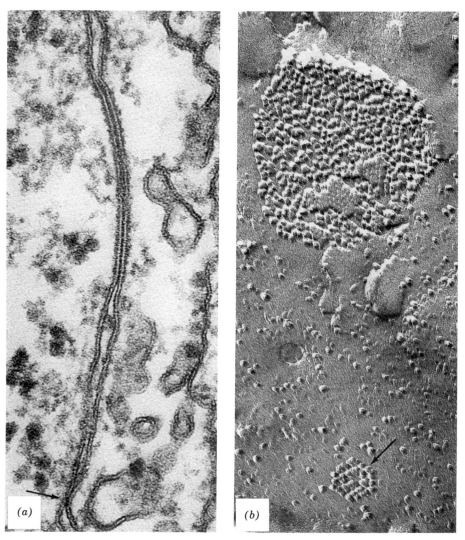

Fig. 14. Gap junctions between communicating cells in culture. (*A*) In thin sections both large and small (arrow) junctional contacts are observed. The large structures can be identified as gap junctions; however, it is generally difficult to characterize the small contacts on the basis of thin-section information alone. ×210,000. (*B*) Freeze-fracture examination facilitates the characterization of both large and small (arrow) gap junctional contacts between communicating cells. ×180,000. From N. B. Gilula et al., *Nature* **235**, 262 (1972), Figs. 5 and 6. Reprinted with permission from *Nature,* Macmillan Journals, Ltd.

Complete chemical and structural characterization of the gap junction will provide important information that can be used to understand the biology of cell-to-cell communication. It will be essential to obtain chemical information that can be used to effectively study the biogenesis and turnover of gap junctions. Through these studies the functional relationship of gap junctions and communication to such phenomena as intercellular synchronization and growth regulation will become clear. In this regard the development of immunological

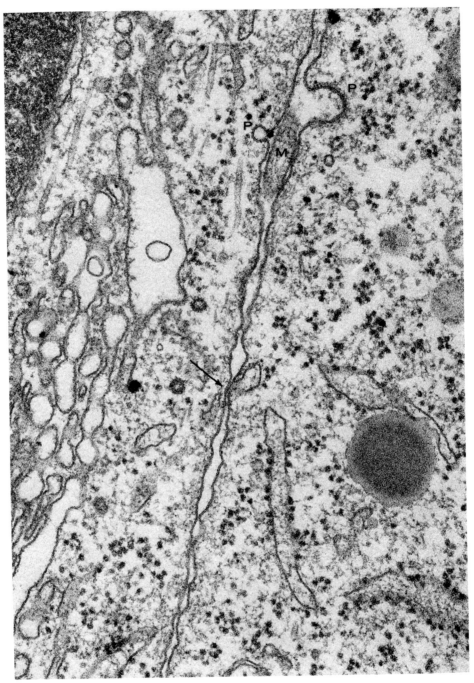

Fig. 15. Interactions between cells that are not communicating metabolically or ionically in culture. Noncommunicating interactions are characterized by the absence of gap junctions and by the presence of interdigitating cell processes or microvilli (M), pinocytosis (P), and close cell approximations that are extensive or focal (arrow). ×80,000.

probes and junctional mutants should be invaluable for dissecting the process of junctional formation between cells.

CONCLUSION

In the past few years a definite relationship has been established between gap junctions and intercellular communication. A major challenge that now exists is to ascertain the relationship of intercellular communication to specific functions (i.e., growth regulation, hormonal stimulation, differentiation, etc.), particularly in nonexcitable cell systems. For this purpose the gap junction provides a unique membrane component that can be characterized and studied in the future as a specific gene product for performing an intriguing function—intercellular communication.

REFERENCES

1. Sjöstrand, F. S., Andersson-Cedergren, E., and Dewey, M. M., *J. Ultrastruct. Res.* **1,** 271 (1958).

2. Robertson, J. D., *J. Cell Biol.* **19,** 201 (1963).

3. Karrer, H. E., *J. Biophys. Biochem. Cytol.* **8,** 135 (1960).

4. Dewey, M. M., and Barr, L., *Science* **137,** 670 (1962).

5. Farquhar, M. G., and Palade, G. E., *J. Cell Biol.* **17,** 375 (1963).

6. Revel, J. P., and Karnovsky, M. J., *J. Cell Biol.* **33,** C7 (1967).

7. McNutt, N. S., and Weinstein, R. S., *Prog. Biophys. Mol. Biol.* **26,** 45 (1973).

8. Brightman, M. W., and Reese, T. S., *J. Cell Biol.* **40,** 648 (1969).

9. Goodenough, D. A., and Revel, J. P., *J. Cell Biol.* **45,** 272 (1970).

10. McNutt, N. S., and Weinstein, R. S., *J. Cell Biol.* **47,** 666 (1970).

11. Friend, D. S., and Gilula, N. B., *J. Cell Biol.* **53,** 758 (1972).

12. Payton, B. W., Bennett, M. V. L., and Pappas, G. D., *Science* **166,** 1641 (1969).

13. Hudspeth, A. J., and Revel, J. P., *J. Cell Biol.* **50,** 92 (1971).

14. Gilula, N. B., and Satir, P., *J. Cell Biol.* **51,** 869 (1971).

15. Peracchia, C., and Robertson, J. D., *J. Cell Biol.* **51,** 223 (1971).

16. Hand, A. R., and Gobel, S., *J. Cell Biol.* **52,** 397 (1972).

17. Johnson, R. G., and Sheridan, J. D. *Science* **174,** 717 (1971).

18. Revel, J. P., Yee, A. G., and Hudspeth, A. J., *Proc. Natl. Acad. Sci. U.S.* **68,** 2924 (1971).

19. Gilula, N. B., Reeves, O. R., and Steinbach, A., *Nature* **235,** 262 (1972).

20. Pinto da Silva, P., and Gilula, N. B., *Exp. Cell Res.* **71,** 393 (1972).

21. Martinez-Palomo, A., *Lab. Invest.* **22,** 605 (1970).

22. Martinez-Palomo, A., Alanis, J., and D. Benitez, *J. Cell Biol.* **47,** 1 (1970).

23. Goodenough, D. A., and Revel, J. P., *J. Cell Biol.* **50,** 81 (1971).

24. Benedetti, E. L., and Emmelot, P., *J. Cell Biol.* **26,** 299 (1965).

25. Benedetti, E. L., and Emmelot, P., *J. Cell Biol.* **38,** 15 (1968).

26. Goodenough, D. A., and Stoeckenius, W˙ʻ *J. Cell Biol.* **54,** 646 (1972).

27. Branton, D., *Proc. Natl. Acad. Sci. U.S.* **55,** 1048 (1966).

28. Pinto da Silva, P., and Branton, D., *J. Cell Biol.* **45,** 598 (1970).

29. Tillack, T. W., and Marchesi, V. T., *J. Cell Biol.* **45,** 649 (1970).

30. Wehrli, E., Muhlethaler, K., and Moor, H., *Exp. Cell Res.* **59,** 336 (1970).

31. Chalcroft, J. P., and S. Bullivant, *J. Cell Biol.* **47,** 49 (1970).

32. Satir, P., and Gilula, N. B., *Ann. Rev. Entomol.* **18,** 143 (1973).

33. Pitelka, D. R., Hamamoto, S. T., Duafala, J. G., and Nemanic, M. K., *J. Cell Biol.* **56,** 797 (1973).

34. McNutt, N. S., Hershberg, R. A., and Weinstein, R. S., *J. Cell Biol.* **51,** 805 (1971).

35. Raviola, E., and Gilula, N. B., *Proc. Natl. Acad. Sci. U.S.* **70,** 1677 (1973).

36. Dreifuss, J. J., Girardier, L., and Forss-mann, W. G., *Pflügers Arch.* **292,** 13 (1966).

37. Muir, A. R., *J. Anat.* **101,** 239 (1967).

38. Berry, M. N., and Friend, D. S., *J. Cell Biol.* **43,** 506 (1969).

39. Goodenough, D. A., and Revel, J. P., *J. Cell Biol.* **50,** 81 (1971).

40. Barr, L., Dewey, M. M., and Berger, W., *J. Gen. Physiol.* **48,** 797 (1965).

41. Barr, L., Berger, W., and Dewey, M. M., *J. Gen. Physiol.* **51,** 347 (1968).

42. Goodenough, D. A., and Gilula, N. B., in *Membranes and Viruses in Immunopathology,* Day, S. B., and Good, R. A., eds., Academic Press, New York, 1972, pp. 155–168.

43. Sommer, J. R., and Spach, M. S., *Am. J. Pathol.* **44,** 491 (1964).

44. Benedetti, E. L., and Delbauffe, in *Cell Membranes. Biological and Pathological Aspects,* Richter, G. W., and Scarpelli, D. G., eds., Williams and Wilkins, Baltimore, 1971, pp. 54–83.

45. Gilula, N. B., unpublished results.

46. Benedetti, E. L., and Emmelot, P., in *Ultrastructure in Biological Systems,* Vol. 4, *The Membranes,* Dalton, A. J., and Haguenau, F., eds., Academic Press, New York, 1968, p. 33.

47. Evans, W. H., and Gurd, J. W., *Biochem. J.* **138,** 691 (1972).

48. Gurd, J. W., and Evans, W. H., *Eur. J. Biochem.* **36,** 273 (1973).

49. Zampighi, G., and Robertson, J. D., *J. Cell Biol.* **56,** 92 (1973).

50. Gilula, N. B., Ph.D. thesis, University of California at Berkeley, 1971.

51. Gilula, N. B., *J. Ultrastruct. Res.* **38,** 215 (1972).

52. Flower, N. E., *J. Cell Sci.* **10,** 683 (1972).

53. Peracchia, C., *J. Cell Biol.* **57,** 66 (1973).

54. Johnson, R. G., Herman, W. S., and Preus, D. M., *J. Ultrastruct. Res.* **43,** 298 (1973).

55. Rose, B., *J. Membrane Biol.* **5,** 1 (1971).

56. Oschman, J. L., and Berridge, M. J., *Tissue and Cell* **2,** 281 (1970).

57. Noirot, C., and Noirot-Timothee, C., *J. Ultrastruct. Res.* **37,** 335 (1971).

58. Peracchia, C., *J. Cell Biol.* **57,** 54 (1973).

59. Flower, N. E., *J. Ultrastruct. Res.* **37,** 259 (1971).

60. Furshpan, E. J., and Potter, D. D., *Nature* **180,** 342 (1957).

61. Loewenstein, W. R., *Ann. N.Y. Acad. Sci.* **137,** 441 (1966).

62. Asada, Y., and Bennett, M. V. L., *J. Cell Biol.* **49,** 159 (1971).

63. Miller, F., *J. Biophys. Biochem. Cytol.* **8,** 689 (1960).

64. Reese, T. S., and Karnovsky, M. J., *J. Cell Biol.* **34,** 207 (1967).

65. Whittembury, G., and Rawlins, F. A., *Pflügers Arch.* **330,** 302 (1971).

66. Machen, T. E., Erlij, D., and Wooding, F. B. P., *J. Cell Biol.* **54,** 302 (1972).

67. Kreutziger, G. O., *26th Proc. EMSA,* p. 234 (1968).

68. Staehelin, L. A., Mukherjee, T. M., and Williams, A. W., *Protoplasma* **67,** 165 (1969).

69. Claude, P., and Goodenough, D. A., *J. Cell Biol.* **58,** 390 (1973).

70. Erlij, D., and Martinez-Palomo, A., *J. Membrane Biol.* **9,** 229 (1972).

71. Wade, J. B., Revel, J. P., and DiScala, V. A., *Am. J. Physiol.* **224,** 407 (1973).

72. DiBona, D. R., and Civan, M. M., *J. Membrane Biol.* **12,** 101 (1973).

73. Wood, R. L., *J. Biophys. Biochem. Cytol.* **6,** 343 (1959).

74. Gilula, N. B., Branton, D., and Satir, P., *Proc. Natl. Acad. Sci. U.S.* **67,** 213 (1970).

75. Danilova, L. V., Rokhlenko, K. D., and Bodryagina, A. V., *Z. Zellforsch.* **100,** 101 (1969).

76. Noirot-Timothee, C., and Noirot, C., *J. Microscopie* **17,** 169 (1973).

77. Friend, D. S., and Gilula, N. B., *J. Cell Biol.* **53,** 148 (1972).

78. Lasansky, A., *J. Cell Biol.* **40,** 577 (1969).

79. Gobel, S., *J. Cell Biol.* **51,** 328 (1971).

80. Satir, P., and Gilula, N. B., *J. Cell Biol.* **47,** 468 (1970).

81. Dan, K., *Int. Rev. Cytol.* **9,** 321 (1960).

82. Loewenstein, W. R., and Kanno, Y., *J. Cell Biol.* **22,** 565 (1964).

83. Bullivant, S., and Loewenstein, W. R., *J. Cell Biol.* **37,** 621 (1968).

84. Odland, G. F., *J. Biophys. Biochem. Cytol.* **4,** 529 (1958).

85. Fawcett, D. W., *Exp. Cell Res. (Suppl.)* **8,** 174 (1961).

86. Kelley, D. E., *J. Cell Biol.* **28,** 51 (1966).

87. Rayns, D. G., Simpson, F. O., and Ledingham, J. M., *J. Cell Biol.* **42,** 322 (1969).

88. Breathnach, A. S., Stolinski, C., and Gross, M., *Micron* **3**, 287 (1972).

89. Overton, J., *J. Exp. Zool.* **168**, 203 (1968).

90. Furakawa, T., and Furshpan, E. J., *J. Neurophysiol.* **26**, 140 (1963).

91. Furshpan, E. J., *Science* **144**, 878 (1964).

92. Bennett, M. V. L., Pappas, G. D., Aljure, E., and Nakajima, Y., *J. Neurophysiol.* **30**, 180 (1967).

93. Bennett, M. V. L., Pappas, G. D., Gimenez, M., and Nakajima, Y., *J. Neurophysiol.* **30**, 236 (1967).

94. Pappas, G. D., Asada, Y., and Bennett, M. V. L., *J. Cell Biol.* **49**, 173 (1971).

95. Robertson, J. D., *Ann. N.Y. Acad. Sci.* **94**, 339 (1961).

96. Loewenstein, W. R., *Fed. Proc.* **32**, 60 (1973).

97. Trelstad, R. L., Hay, E. D., and Revel, J. P., *Dev. Biol.* **16**, 78 (1967).

98. Sheridan, J., *J. Cell Biol.* **37**, 650 (1968).

99. Revel, J. P., Yip, P., and Chang. L. L., *Dev. Biol.* **35**, 302 (1973).

100. Bennett, M. V. L., and Trinkaus, J. P., *J. Cell Biol.* **44**, 592 (1970).

101. Lentz, T. L., and Trinkaus, J. P., *J. Cell Biol.* **48**, 455 (1971).

102. Bennett, M. V. L., *Fed. Proc.* **32**, 65 (1973).

103. Potter, D. D., Furshpan, E. J., and Lennox, E. S., *Proc. Natl. Acad. Sci. U.S.* **55**, 328 (1966).

104. Slack, C., and Palmer, J. F., *Exp. Cell Res.* **55**, 416 (1969).

105. Sheridan, J. D., *Dev. Biol.* **26**, 627 (1971).

106. Tupper, J. T., and Saunders, J. W., *Dev. Biol.* **27**, 546 (1972).

107. Ito, S., and Loewenstein, W. R., *Dev. Biol.* **19**, 228 (1969).

108. O'Lague, P., Dalen, H., Rubin, H., and Tobias, C., *Science* **170**, 464 (1970).

109. Hyde, A., Blondel, B., Matter, A., Cheneval, J. P., Fillous, B., and Girardier, L., *Prog. Brain Res.* **31**, 282 (1969).

110. Goshima, K., *Exp. Cell Res.* **63**, 124 (1970).

111. O'Lague, P., Ph.D. thesis, University of California at Berkeley, 1969.

112. Subak-Sharpe, J. H., Burk, R. R., and Pitts, J. D., *J. Cell Sci.* **4**, 353 (1969).

113. Cox, R. P., Krauss, M. R., Balis, M. E., and Dancis, J., *Proc. Natl. Acad. Sci. U.S.* **67**, 1573 (1970).

114. Fujimoto, W. Y., and Seegmiller, J. E., *Proc. Natl. Acad. Sci. U.S.* **65**, 577 (1970).

115. Pitts, J. D., in *Ciba Foundation Symposium on Growth Control in Cell Cultures*, Churchill, London, 1971, pp. 89–98.

116. Azarnia, R., Michalke, W., and Loewenstein, W. R., *J. Membrane Biol.* **10**, 247 (1972).

117. Larsen, W., Azarnia, R., and Loewenstein, W. R., *J. Cell Biol.* **59**, 186a (1973).

118. Borysenko, J. Z., and Revel, J. P., *Am. J. Anat.* **137**, 403 (1973).

CHAPTER TWO

Electrical Coupling of Cells and Cell Communication

JUDSON D. SHERIDAN

Department of Zoology
University of Minnesota
Minneapolis, Minnesota

Metazoan cells are highly interdependent. This interdependency is seen clearly in the variety of mechanisms for cell communication discussed in this volume. Recently a particularly direct mechanism for communication has been recognized which involves movement of small molecules from cell to cell via specialized membrane junctions. The ultrastructural features of these "low-resistance" junctions have been discussed in detail by N. B. Gilula in Chapter 1. In this chapter I discuss primarily the physiological properties of the junctions as they are currently understood and develop some ideas concerning their function.

GENERAL HISTORICAL BACKGROUND

Junctional communication was first recognized and studied by electrophysiologists interested in the transmission of action potentials between certain nerve cells (1,2), and between cardiac and smooth muscle cells (3,4). These early studies left their imprint on all subsequent work on junctional communication. For example, the terms "electrical coupling" and "low-resistance junctions," originally used to describe relevant features of the junctions between excitable cells, have persisted even though it has been shown quite clearly that similar junctions occur between most nonexcitable cells (5,6). Thus we speak of junctions that "electrically couple" cells, although the junctions between nonexcitable cells presumably do something other than transmit electrical signals. An important clue about possible nonelectrical functions came from the observation that fluorescein injected into one cell moved rapidly to its electrically coupled neighbors (7). This observation suggested that the junctions were permeable not only to small inorganic ions but also to larger molecules. On the basis of the distribution and permeability of the junctions, it was suggested that they might help regulate growth and development by passing small metabolites, nutrients, or special control substances (5,8). This was an exciting prospect and has been the major impetus for research into junctional communication.

Most studies on low-resistance junctions have attempted to answer one or more of the following questions, which will serve as a convenient outline for my discussion:

1. What structure or structures are responsible for the passage of ions and larger tracers?

2. What types of molecules can pass through the junctions?

3. Where do the junctions occur and where are they absent?

4. How are junctions formed?

5. What biological roles do the junctions play?

ULTRASTRUCTURE

Because N. B. Gilula has covered ultrastructure in detail, I shall simply pass briefly over the main conclusions. The only junction firmly implicated in direct transfer of small molecules is the gap junction. Other junctions (e.g., tight junctions and septate junctions) may share some of the physiological properties of gap junctions, but there is no firm evidence that this is so. Therefore I shall assume for the rest of my discussion that junctional communication occurs via gap junctions.

The combined approaches of thin section and *en bloc* stains (9,10), extracellular tracers (10) and negative stains (11), and freeze-fracture (12–15) have given us the following picture of gap junctions. They contain two complementary sets of subunits, one set in each of the junctional membranes. The subunits are frequently packed into regular arrays and interact with their counterparts across a narrowed extracellular space. It is generally assumed, without any direct evidence, that each pair of subunits contains a water-filled channel running from one cytoplasm to the other. These channels are proposed as the routes for small molecules to move from cell to cell.

PERMEABILITY

In my preliminary comments I have said rather glibly that low-resistance junctions are generally permeable to small molecules. In fact much of the work on junctions has been aimed at establishing the truth of this statement, and the evidence is still incomplete. There have been three major approaches, two directly related to junctions and the third more indirect.

The first approach involves the use of electrical methods to follow the movement of small inorganic ions from cell to cell. It is not necessary here to cover the methods in detail, for they have been extensively reviewed elsewhere (see, for example, ref. 2), but only to consider the kinds of information that can be obtained and how it can be interpreted.

The simplest, and least revealing, information is that a given pair of cells are "electrically coupled." This means that a change in the membrane potential of a cell injected with current is accompanied by a similar, but smaller, potential change in another cell. For coupling to occur, some of the injected current, carried by small ions, must somehow flow between the insides of the coupled cells. Provided that the two cells are not surrounded by a common high-resistance barrier, significant electrical coupling requires that the membranes between the two cells have a lower resistance for unit area than the membranes forming the rest of the cells' surfaces. This condition does not demand the existence of continuous intercytoplasmic channels, such as is postulated for the gap junctions, and in fact coupling between early amphibian embryo cells would probably occur even in the absence of gap junctions (16,17). However, the amphibian case is probably unusual, and the presence of coupling can generally be taken to imply the presence of junctions with low electrical resistance, or, in other words, with high permeability to small inorganic ions.

Another kind of information obtained with electrical methods is the degree of coupling, generally given by the ratio of the potential change in one cell divided by the potential change in an adjacent cell injected with current (1,18). This "coupling coefficient" indicates how efficiently potential changes are transferred between the coupled cells: a larger coupling coefficient means greater efficiency of transfer. Because the coupling coefficient increases with increased junctional permeability, *all other factors remaining constant,* it has been widely used to compare the electrical capabilities of differ-

ent junctions. However, *all other factors* seldom remain constant, and as a result changes or differences in the coupling coefficient are not simple functions of junctional resistance (2,19). The coupling coefficient is influenced, for example, by the resistance per unit area of the nonjunctional membranes, by the number of cells coupled to the cells being studied, and by the size and shape of the cells. The effect of these nonjunctional factors can be so severe that an increase in the coupling coefficient can theoretically occur while the junctional permeability actually decreases. Even when supportive electrical data indicate that junctional permeability is changed in the same direction as the change in coupling, the magnitude of the two changes is unlikely to be proportionate.

There is an additional problem. When the coupling coefficient is high, it is not very sensitive to changes in junctional permeability. Doubling junctional permeability (halving the resistance), for example, when the coupling coefficient is 0.95 will increase the coupling only to 0.97, roughly a 2% increase.

The preceding comments suggest that junctional resistance needs to be measured directly. For unambiguous and reliable measurements it is necessary that the coupled cells be adjacent and have only one possible junctional pathway, and that each cell be penetratable by two microelectrodes (1,2). Unfortunately few systems meet these requirements. In their pioneering study on the giant motor synapse of crayfish, Furshpan and Potter (1) found junctional resistances of the order of 5×10^5 ohms, when current was made to flow from the lateral giant fiber to the giant motor fiber ("orthodromic" direction). However, the resistance was many times higher for current flow in the opposite ("antidromic") direction. This "rectification" is not typical of low-resistance junctions, and its source is unknown. Two other systems have yielded comparable measurements of junctional resistance. The crayfish septal synapse, studied extensively by Grundfest, Bennett, and colleagues (20,21), has a resistance in the range 0.3–1.3 $\times$ 10^5 ohms. Reaggregated cells from killifish embryos develop junctions having resistances in the range of 10^6 ohms or more (22).

A simpler approach can be used when the coupled cells are similar in size and the specific resistances of the nonjunctional membranes of the two cells are the same. This situation is approximated by pairs of Novikoff hepatoma cells taken from suspension culture. In studying these cells we have estimated junctional resistances to range from 0.26 $\times$ 10^6 to 181 $\times$ 10^6 ohms (23,24). Presumably this wide range of resistances is due in part to the fact that some of the junctions are in different stages of formation and thus vary in size. (Junctional formation is discussed more extensively in a subsequent section.)

More complex methods have been used to estimate junctional resistances in systems consisting of more than two cells. Some systems (e.g., *Drosophila* salivary gland) behave electrically like finite cables (18,25). These systems have therefore been analyzed by standard cable equations. Other systems (e.g., monolayer cultures of chick embryo cells) have been fitted by more complicated, two-dimensional models (26,27). The equations for these systems are complex and are best solved by computer techniques. Values for junctional resistances in these systems range from 0.5 $\times$ 10^5 to 200 $\times$ 10^5 ohms. These values are, however, only averages and do not reflect the resistance of any particular junction.

A striking feature of the values for junctional resistances is their variability, ranging over more than two orders of magnitude. There are two possible sources of this variability. Either the junctions vary in their specific resistance (resistance for a unit area) or they have equal specific resistance but different areas. The fact that rectification is seen in the giant motor synapses but is absent at most other low-resistance junctions indicates that changes in specific resistance can occur under some conditions. Yet there is no indication that specific resistances vary enough to account for the wide range of junctional resistances observed. For example, even the giant motor synapse shows only about one order of magnitude change in resistance during rectification (1).

There have been many attempts to estimate resistance per unit area. None of these studies has directly compared the area of gap junction between a given pair of cells with the junctional resistance. The closest approach has been made with cardiac muscle, where average junctional resistance values have been compared with estimates of junctional area. These studies yielded a specific resistance value of 4 ohm-cm^2 (28).

There is another way to approach the problem of resistance per unit area. Assumptions can be made regarding the diameter and length of the presumed channels through the subunits of the gap junctions. Then, with estimates of the numbers of subunits per square micron and of the resistivity of the solution in the channels, a theoretical specific resistance can be calculated (29). With channels of 10 μ in diameter one obtains a value of about 0.025 ohm-cm^2, about two orders of magnitude lower than the estimates derived from more direct experimental values. However, this value is only one order of magnitude lower than the recently revised value of Woodbury and Crill (30) for the upper limit of specific resistance required for effective transmission in atrial muscle. Their value is 0.3 ohm-cm^2.

We have tried a third approach, using our values for junctional resistance between pairs of Novikoff hepatoma cells and for junctional areas estimated from electron micrographs (23,24,31). It is most revealing to compare the lowest junctional resistance, 0.5×10^6 ohms, and the largest junctional area, $4 \ \mu^2$. The product of these values gives a calculated specific resistance of 2×10^{-2} ohms-cm^2. In order to have a specific resistance of 3×10^{-1} ohm-cm^2, the junctional areas would have to be 60 μ^2, a value much larger than we have ever seen.

The results to date from the various approaches are not sufficiently reliable to determine whether or not specific resistances of junctions vary, except in the peculiar rectifying junctions. The uniform structure of gap junctions in widely divergent systems is most consistent with a minimal variation in specific resistance, but obviously more direct correlations of junctional resistance and area are needed. This point is an important one in trying to decide the functional significance of variations in junctional area observed in different systems.

Because the electrical methods are esoteric and tend to convince electrophysiologists more than nonelectrophysiologists, it is comforting that other methods for studying junctional permeability to inorganic ions and other small molecules have been devised and have given support to the electrical data. In some ways data obtained with these other methods are more dramatic, but nevertheless they suffer from their own interpretive difficulties.

With all these other methods, the tracer substance is introduced selectively into one cell and then movement of tracer to other cells is monitored, often with direct visual methods. The most direct way to introduce the tracer is to inject it from a micropipette.

The micropipette is filled with an aqueous solution of the tracer that is ejected by applying hydrostatic pressure (7) or electrical potential differences (8,32). The electrical method is more easily controlled and has been the primary method chosen.

The tracers most commonly used are fluorescent molecules whose distribution and movement can be followed directly during the experiment or in sections of fixed material. Fluorescein, used extensively because of its intense fluorescence, high mobility, and ease of injection, has a molecular weight of 330 and a diameter of about 10 Å. It was first shown to move from cell to cell in the insect salivary gland (7) and has since been shown to be transferred between coupled cells in a variety of adult and tissue culture systems from vertebrates and invertebrates (8,32–34).

Fluorescein is not an ideal tracer, however, because it is a weak acid, and its undissociated form, though in low concentration at neutral pH, crosses cell membranes. Thus it has been argued that fluorescein might leak out of the injected cell, accumulate in the extracellular space, and gradually diffuse into the adjacent cell. The failure of fluorescein to move between uncoupled cells argues against this alternative, but obviously a better tracer is desirable. Procion Yellow, a dye of about 550 molecular weight and used originally for marking nerve cells (35), has proved to be such a tracer. In an important study, Payton, Bennett, and Pappas (36) showed that Procion Yellow moved readily through the septal synapses of crayfish giant axons but was effectively excluded by nonjunctional membranes. Other workers have since confirmed these observations, demonstrating transfer in insect salivary gland (37), mammalian cell cultures (33), mammalian salivary gland (38), and fish retina (39).

Other tracers have been used, examples being dansylated amino acids (33) and sucrose (40). Although these tracers have been applied only to a few systems, their variety suggests that the junctions act essentially as sieves, allowing any appropriately sized molecule to pass. If the junctions are molecular sieves, there should be an upper limit on the size of molecules that can pass. This upper limit is unknown, although there are some reasons to believe that it is somewhere around 1000 molecular weight. There has been one report of junctional transfer of small proteins in insect salivary gland (41), but in this experiment there are some questions: was it the protein that moved or just some of the fluorescein isothiocyanate used to label it (42)? A polypeptide of even smaller size ($\sim$ 1800 MW) apparently fails to move (except *after* fixation) through the septal synapse in crayfish axons (43). Certainly, if the gap-junction subunits contain channels, their size would not allow the transfer of molecules larger than 20–30 Å. This would prohibit the movement of all but the smallest proteins.

Additional information about the selectivity of low-resistance junctions for small molecules comes from experiments on "metabolic cooperation" (44). Some of these experiments have already been discussed in Chapter 1 by N. B. Gilula, whose elegant studies strongly implicate gap junctions in the process (45). R. P. Cox and associates discuss metabolic cooperation more fully in Chapter 4 and present evidence of the transfer of small nucleotides, but exclusion of messenger RNA and enzyme. They have shown in previous experiments (46) that another enzyme, glucose-6-phosphate dehydrogenase (G-6-PD) will not transfer in detectable quantities between normal and mutant cells—further indirect evidence

that proteins are excluded by low-resistance junctions.

Recently Kolodny (47) has reported the transfer of various species of RNA and protein between cells in culture (see also Chapter 5 in this volume). However, Pitts and coworkers (48) have evidence suggesting that this apparent transfer of macromolecules depends on nucleotide transfer.

In my discussion of junctional permeability I have implied that the inorganic ions responsible for electrical coupling and the other small tracer molecules move through the same intercytoplasmic channels. Although this conclusion has not been proved, it does have circumstantial support. Without exception, cells that pass tracers are also electrically coupled. Also, disruption of coupling leads to disruption of tracer movement (see, for example, ref. 36). This correlation makes sense. Channels large enough to pass the tracers should certainly have high ionic permeability. However, the converse is not necessarily so. Electrical resistance low enough to couple cells can occur theoretically via channels too small to pass the larger tracers, as long as there are enough channels. The observation that some early embryonic cells that are coupled fail to pass tracers may be a case of this kind (22,49,50; but see ref. 51). Further work on the embryonic systems is needed to confirm that tracer movement is absolutely blocked rather than being merely undetectable due to the large size of the cells and their geometric arrangement or reduced junctional area.

DISTRIBUTION

The list of cells having low-resistance junctions detected by electrophysiological or tracer techniques and/or gap junctions observed with ultrastructural techniques is long and still growing (6,52,53). No multicellular organism has yet been found to lack these junctions. They are found in lower invertebrates (54) and higher vertebrates (55), in embryos (56–60) and in adults (9,61–63), and in tissue culture (8,33,58,64,65) as well as *in situ*. Cells with junctions include those that produce electrical signals (excitable cells) and those that do not (nonexcitable cells). Junctions occur between epithelial cells (see, for example, ref. 61) and between nonepithelial cells (see, for example, ref. 63). Junctions usually connect cells of the same type (i.e., homocellular junctions), but in some cases also connect cells of different types (i.e., heterocellular junctions) (65–67). Furthermore, cells that usually do not come into contact *in situ* can form junctions in culture (54). In short, nearly all cells studied can and do form low-resistance junctions.

There are some cells, however, that form junctions either infrequently or not at all. For example, certain nerve cells form junctions, but most apparently do not, at least in the mature organism. Skeletal muscle cells completely lack junctions, and if blood cells form junctions at all, they do so only under unusual circumstances (68).

FORMATION OF JUNCTIONS

We still know very little about the way low-resistance junctions form. Most of our information is very descriptive and not refined at the molecular level. For example, we know that cells coming into contact *in vitro* can form low-resistance junctions within a relatively brief, but variable, time (51,60,69,70). Some recent experiments have begun to define the time course of formation and in one case provide evidence for the steps involved.

DeHaan and Hirakow (71) used cardiac myoblasts to study formation. Anal-

ysis in this system is facilitated by the presence of spontaneous cell contractions that can be observed directly and depend on endogenous electrical signals (action potentials). When two cells producing action potentials at different frequencies and thus beating at different rates come into contact, their electrical activities and contractions become synchronized after some variable time. The appearance of synchrony signals the development of a low-resistance junction between the two cells. Electron microscopy of the interface between recently synchronized pairs of cells reveals regions of close contact, possibly gap junctions, Junctions detected by appearance of synchrony and by electron microscopy can occur in as short a time as a few minutes after contact is observed. Yet, with some pairs of cells, junctions take more than an hour to form after initial contact. Sources of this variability are not known, and these studies have provided little information about the membrane events occurring before or during formation.

In a more recent series of experiments Yee (72) has used freeze-fracture techniques to study the junctional changes occurring after partial hepatectomy. She reports that shortly after hepatectomy the typically extensive hepatocellular gap junctions are greatly reduced in number and size. These changes appear to coincide with the phase of rapid mitotic activity. Following the rapid growth phase, small gap junctions appear, often in association with elements of the zonulae occludentes, which are also developing at this time. Over the next 4–9 hours, the gap junctions gradually become larger and more numerous, finally reaching a size and distribution similar to that in the intact liver.

We have recently begun a study of junction formation between reaggregating Novikoff hepatoma cells (73). Using a combination of ultrastructural and physiological approaches we have obtained evidence of a stepwise process of membrane changes resulting in fully developed gap junctions. We have postulated the following steps:

1. In localized regions of apposition, flattened areas of membrane containing 100 Å particles but nearly devoid of other particles occur in both membranes. The particles may interact across the extracellular space, which at this stage is on the order of 100–200 Å.

2. The interaction of the particles leads to narrowing of the extracellular space to a width of about 20–40 Å.

3. The particles then begin to aggregate into small, ordered arrays (small "gap" junctions).

4. Finally, the small aggregates fuse to form larger aggregates.

Somewhere between steps 3 and 4 the pairs of apposed, interacting particles form intercytoplasmic channels.

Our parallel physiological studies indicate that the percentage of coupled cells and the efficiency of coupling (coupling coefficient) increase with time. Furthermore, we have shown that most of the coupled cells can exchange fluorescein. The lack of exchange between some of the coupled cells is likely due to the greater sensitivity of the electrical than the tracer methods. In preliminary studies we have found that the development of electrical coupling is not prevented by complete inhibition of protein synthesis or of energy metabolism (74).

BIOLOGICAL ROLE(S) OF LOW-RESISTANCE JUNCTIONS

Our understanding of the biological roles played by low-resistance junctions has lagged far behind our understanding of their ultrastructure, permeability, and distribution. It is clear that they transmit electrical signals between certain excitable cells, such as cardiac and smooth muscle cells, and some nerve cells. Yet most cells connected by low-resistance junctions do not produce electrical signals, and in these cases the junctions must have some nonelectrical function.

A number of different approaches have been used in searching for these nonelectrical functions. One popular approach has been to look for junctional alterations in systems displaying abnormal cellular interactions (e.g., cancer). This approach has yielded interesting but mixed results with as yet no unequivocal evidence of a biological role of junctions (see refs. 8,24,75, and 76 for reviews).

A second approach has been to study embryos for changes in junctional distribution that might be correlated with critical phases of development (56–60). Aside from demonstrating the interesting, but not unexpected, loss of junctions between cells taking different pathways of differentiation (see, for example, ref. 56), this approach has not revealed a clear developmental role for junctions.

A third approach has been to look for junctional changes induced by hormone treatment. Some hormones elicit changes in junctions; for example, estrogens lead to increased size and frequency of gap junctions in ovaries of hypophysectomized rats (77,78) and cause apparent junctional proliferation in intermolt horseshoe crabs (Quick and Johnson, personal communication). However, the biological meaning of such changes is not yet known.

R. P. Cox and associates discuss a fourth approach in Chapter 4. Their work, and that of Pitts and colleagues, on metabolic cooperation between mutant and normal cells has provided some of the clearest evidence of a biological role of low-resistance junctions.

I have approached the problem in a slightly different manner that I feel helps to tie together the other approaches and suggests further directions for investigation. I have discussed these ideas in depth elsewhere (79), and here I wish to summarize the arguments by discussing some general conceptual principles and some specific examples (see refs. 8, 42, and 80 for discussions of similar ideas).

I have discussed evidence that low-resistance junctions allow passive diffusion of small molecules (<1000 MW) directly from cell to cell. However, in order for net diffusion to take place, there must first be differences in concentrations of small molecules in adjacent cells. Such differences can theoretically be traced back to differences in the genomes, in the parts of the genome expressed, in the quantities of gene product (e.g., proteins), or in the environments of the cells. Whatever the source of the difference in concentration of small molecules, diffusion via the junctions will have the same qualitative effect—to decrease the differences. That is, the cell with the higher concentration will lose molecules to the cell with the lower concentration. The quantitative effect, however, will vary depending on what might generally be termed the relative "capacities" of the cell or cells acting as "sources" and the cell or cells acting as "sinks." For example, if the capacities are equal, transfer will result in complete abolition of the difference, and the concentrations will equilibrate halfway between the initial concentrations of the source and sink. Usually, however, the capacities will be unequal, and the equilibrium concentration will be closer to the initial concentration of the cell(s) with the larger capacity. This argument is oversimplified, and to be complete, dynamic changes in production and degradation of molecules in both sources and sinks should be included, but the basic result would be similar. Depending on the relative capacities (with or without dynamic features), the junctions will serve in some cases to deplete the concentration in the source or to increase the concentration in the sink. For example, if a change in the concentration of small molecules occurs in a single cell, the change will be damped by junctional transfer to adjacent cells. If, on the other hand, the change occurs in a large number of cells, junctional transfer will pass that change on to adjacent cells.

To make these general concepts more concrete, I would like to consider three molecules that are likely to be transferred via junctions and two specific systems in which this transfer might have important functional consequences. The molecules are cyclic adenosine 3′,5′-monophosphate (AMP), cyclic guanosine 5′-monophosphate (GMP), and Ca^{2+}, which have all been implicated in the control of various processes carried out by cells having "low-resistance" junctions and which all fall within the size range of molecules permeating the junctions (51, 63,81). The two systems I will discuss are vascular smooth muscle and bone.

The contraction of vascular smooth-muscle cells is under sympathetic nervous control (see ref. 82 for review). The nerves release norepinephrine, which then combines with α-receptors to induce contraction or β-receptors to induce relaxation. In simple terms, the contraction is correlated with, and perhaps dependent on, an increase in intracellular Ca^{2+} and cyclic GMP (83). The change in intracellular Ca^{2+} is presumably due in part to a change in membrane potential; that is, depolarization of the membrane produces an increase in Ca^{2+} permeability and Ca^{2+} enters the cell. However, some increase in Ca^{2+} permeability may result from the direct effect of norepinephrine on α-receptors. The cause of the change in cyclic GMP level is unknown but may be a direct effect of the norepinephrine on the α-receptors, an indirect effect of a change in membrane potential, or an indirect effect of the increase in intracellular Ca^{2+}. Relaxation is correlated with an increase in cyclic AMP levels caused presumably by the direct action of norepinephrine on the β-receptors (84).

The main argument for movement of cyclic AMP (and perhaps cyclic GMP and Ca^{2+}) via junctions connecting the smooth-muscle cells comes from a consideration of the innervation of the blood vessels. All of the nerve endings occur in the adventitia, and none penetrate the smooth-muscle layer (media) (82). Therefore, in order to produce contraction and/or relaxation of all muscle cells, the effect of the nerves must be distributed from the outer smooth-muscle cells to the inner ones. Since contraction appears to be controlled primarily by changes in membrane potential, electrotonic transmission of synaptic potentials and action potentials via junctions should provide one means of distribution. To the extent that norepinephrine causes direct increase in Ca^{2+} permeability and increase in cyclic GMP production, junctional transfer of these two molecules might also be important. Since increased production of cyclic AMP is likely to result from a direct effect of norepinephrine on β-receptors, junctional transfer of cyclic AMP appears to be necessary for the β-effects to be distributed.

A second system in which junctional transfer of cyclic nucleotides and Ca^{2+} seems likely is bone. Osteocytes, and their long processes, form an extensive cellular network within the lacunae and interlacunal canaliculi of bone. Adjacent osteocytes are connected to each other by elaborate junctions that have the appearance of gap junctions (85) (*en bloc* staining has not been used). Since the osteocytes and their processes appear to be tightly adherent to the bone matrix, there seems to be little extracellular space for the transport of nutrients and other molecules between the deeper osteocytes and the endosteal or periosteal blood supply. It has therefore been suggested (85) that the transport of these molecules occurs via cell processes and the junctions connecting them.

An important function of the osteocytes is the laying down and resorption of Ca^{2+}. The resorptive function is under the control of parathyroid hormone (PTH), and although the precise mechanism of the control is not yet known, there is evidence that part of the PTH effect is mediated by cyclic AMP (86). If the deeper osteocytes obtain nutrients via a junctional transport system, they might also receive cyclic AMP from the peripheral osteocytes directly affected by PTH. Furthermore, the Ca^{2+}, removed from the bone by the deeper osteocytes, might use the same junctional transport system to reach the blood.

CONCLUSION

We are still probing for clues to the biological significance of specialized, low-resistance junctions. Their early appearance in phylogeny and presence with minimal alterations (at least at the ultrastructural level) throughout the metazoa implies fundamental importance. The importance should be related to the tendencies of adjacent cells to have different concentrations of small molecules such as ions and metabolites. In their most primitive role the junctions might serve to smooth out these differences, thus increasing the homogeneity of the intracellular compartments. More specialized roles (e.g., in the transmission of action potentials between cardiac muscle cells or in the transfer of growth-regulating molecules such as cyclic nucleotides) may then represent adaptations of the primitive homeostatic function.

ACKNOWLEDGEMENTS

I thank Dr. R. G. Johnson, Dr. M. Epstein, and Ms. M. G. Hammer for fruitful discussions and Ms. G. Busack for preparation of the manuscript. Some of the studies reported were supported in part by USPHS grant CA11114 and Research Career Award K4-CA-70, 388.

REFERENCES

1. Furshpan, E. J., and Potter, D. D., *J. Physiol.* **143,** 289 (1958).
2. Bennett, M. V. L., *Ann. N.Y. Acad. Sci.* **137,** 509 (1966).
3. Woodbury, J. W., and Crill, W. E., in *Nervous Inhibition,* Florey, E., ed., Pergamon Press, New York, 1961.
4. Nagai, T., and Prosser, C. C., *Am. J. Physiol.,* **204,** 910 (1963).
5. Loewenstein, W. R., *Ann. N.Y. Acad. Sci.* **137,** 441 (1966).
6. McNutt, S., and Weinstein, R., *Prog. Biophys. Mol. Biol.* **26,** 45 (1973).
7. Kanno, Y., and Loewenstein, W. R., *Nature* **212,** 629 (1966).
8. Furshpan, E. J., and Potter, D. D., in *Current Topics in Developmental Biology,* Vol. 3, Moscona, A. A., and Monroy, A., eds., Academic Press, New York, 1968, p. 95.
9. Robertson, J. D., *J. Cell Biol.* **19,** 201 (1963).
10. Revel, J. P., and Karnovsky, M. J., *J. Cell Biol.* **33,** C7 (1967).
11. Benedetti, E. L., and Emmelot, P., *J. Cell Biol.* **26,** 299 (1968).
12. Kreutziger, G. O., *Proc. Electron Microscope Soc. Am.* **26,** 234 (1968).
13. Goodenough, D. A., and Revel, J. P., *J. Cell Biol.* **45,** 272 (1970).
14. Chalcroft, J. P., and Bullivant, S., *J. Cell Biol.* **47,** 49 (1970).
15. McNutt, N. S., and Weinstein, R. S., *J. Cell Biol.* **47,** 666 (1970).
16. Woodward, D. J., *J. Gen. Physiol.* **52,** 509 (1968).
17. Bluemink, J. G., and de Laat, S. W., *J. Cell Biol.,* **59,** 89 (1973).
18. Loewenstein, W. R., and Kanno, Y., *J. Cell Biol.* **22,** 565 (1964).
19. Sheridan, J. D., *Am. Zoologist,* **13,** 1119 (1973).
20. Kao, C. Y., and Grundfest, H., *Fed. Proc.* **15,** 104 (1956).
21. Asada, Y., and Bennett, M. V. L., *J. Cell Biol.* **49,** 159 (1971).
22. Bennett, M. V. L., Spira, M. E., and Pappas, G. D., *Dev. Biol.* **29,** 419 (1972).
23. Sheridan, J. D., *J. Cell Biol.* **55,** 236a (1972).
24. Sheridan, J., and Johnson, R., in *Molecular Pathology,* Good, R., and Day, S., eds., Thomas, in press.
25. Loewenstein, W. R., Nakas, M., and Socolar, S. J., *J. Gen. Physiol.* **50,** 1865 (1967).
26. Van Heukelom, J. S., van der Gon, J. J. D., and Prop, F. J. A., *J. Membrane Biol.* **7,** 88 (1972).
27. Jongsma, H. J., and van Rijn, H. E., *J. Membrane Biol.* **9,** 341 (1972).
28. Spira, A. W., *J. Ultrastruct. Res.* **34,** 409 (1971).
29. J. D. Sheridan, unpublished calculations.
30. Woodbury, J. W., and Crill, W. E., *Biophys. J.* **10,** 1076 (1970).
31. Johnson, R. G., *J. Cell Biol.* **55,** 1269 (1972).
32. Pappas, G. D., and Bennett, M. V. L., *Ann. N.Y. Acad. Sci.* **137,** 495 (1966).
33. Johnson, R. G., and Sheridan, J. D., *Science* **174,** 717 (1971).
34. Azarnia, R., and Loewenstein, W. R., *J. Membrane Biol.* **6,** 368 (1971).
35. Stretton, A. O. W., and Kravitz, E. A., *Science* **162,** 132 (1968).
36. Payton, B. W., Bennett, M. V. L., and Pappas, G. D., *Science* **166,** 1641 (1969).
37. Rose, B., *J. Membrane Biol.* **5,** 1 (1971).
38. Hammer, M. G., and Sheridan, J. D., *Abstr. ASCB Meeting,* 1971, p. 116.
39. Kaneko, A., *J. Physiol.* (*London*) **213,** 95 (1971).
40. Bennett, M. V. L., and Dunham, P. B., *Biophys. J.* **10,** 117a (1970).
41. Kanno, Y., and Loewenstein, W. R., *Nature* **212,** 629 (1966).
42. Loewenstein, W. R., *Dev. Biol. Suppl.* **2,** 151 (1968).
43. Bennett, M. V. L., Feder, N., Reese, T. S., and Stewart, W., *J. Gen. Physiol.* in press, 1973.

44. Subak-Sharpe, H., Burk, R., and Pitts, J., *Heredity (London)* **21,** 342 (1966).

45. Gilula, N. B., Reeves, O. R., and Steinbach, A., *Nature* **235,** 262 (1972).

46. Cox, R. P., Krauss, M. R., Balis, M. E., and Dancis, J., *Exp. Cell Res.* **74,** 251 (1972).

47. Kolodny, G. M., *Exp. Cell Res.* **65,** 313 (1971).

48. Pitts, J., personal communication.

49. Slack, C., and Palmer, J. P., *Exp. Cell Res.* **55,** 416 (1969).

50. Tupper, J. T., and Saunders, J. W., Jr., *Dev. Biol.* **27,** 546 (1972).

51. Sheridan, J. D., *Dev. Biol.* **26,** 627 (1971).

52. Satir, P., and Gilula, N. B., *Ann. Rev. Entomol.* **18,** 143 (1973).

53. Friend, D., and Gilula, N. B., *J. Cell Biol.* **53,** 758 (1972).

54. Hand, A. R., and Gobel, S., *J. Cell Biol.* **52,** 397 (1972).

55. Brightman, M. W., and Reese, T. S., *J. Cell Biol.* **40,** 648 (1969).

56. Potter, D. D., Furshpan, E. J., and Lennox, E. S., *Proc. Natl. Acad. Sci. U.S.* **55,** 328 (1966).

57. Sheridan, J. D., *J. Cell Biol.* **37,** 650 (1968).

58. Ito, S., and Hori, N., *J. Gen. Physiol.* **49,** 1019 (1966).

59. Tupper, J., Saunders, J. W., and Edwards, C., *J. Cell Biol.* **46,** 187 (1970).

60. Bennett, M. V. L., and Trinkaus, J. P., *J. Cell Biol.* **44,** 592 (1970).

61. Loewenstein, W. R., Socolar, S. J., Higashino, S., Kanno, Y., and Davidson, N., *Science* **149,** 295 (1965).

62. Penn, R. D., *J. Cell Biol.* **29,** 171 (1966).

63. Sheridan, J. D., *J. Cell Biol.* **50,** 795 (1971).

64. Borek, C., Higashino, S., and Loewenstein, W. R., *J. Membrane Biol.* **1,** 274 (1969).

65. Michalke, W., and Loewenstein, W. R., *Nature* **232,** 121 (1971).

66. Garant, P. R., *J. Ultrastruct. Res.* **40,** 333 (1972).

67. Johnson, R., Herman, W., and Preus, D. *J. Ultrastruct. Res.* **43,** 298 (1973).

68. Hülser, D. F., and Peters, J. H., *Exp. Cell Res.* **74,** 319 (1972).

69. Loewenstein, W. R., *Dev. Biol.* **15,** 503 (1967).

70. Ito, S., and Loewenstein, W. R., *Dev. Biol.* **19,** 228 (1969).

71. DeHaan, R. L., and Hirakow, R., *Exp. Cell Res.* **70,** 214 (1972).

72. Yee, A., *J. Cell Biol.* **55,** 294a (1972).

73. Johnson, R., Hammer, M., Sheridan, J. D., and Revel, manuscript in preparation.

74. Epstein, unpublished observations.

75. Loewenstein, W. R., *Fed. Proc.* **32,** 60 (1973).

76. Martinez-Palomo, A., *Pathobiol. Ann.,* 1971, p. 271.

77. Merk, F. B., Botticelli, C. R., and Albright, J. T., *Endocrinology* **90,** 992 (1972).

78. Merk, F. B., and McNutt, N. S., *J. Cell Biol.* **55,** 511 (1972).

79. Sheridan, J. D., *Proc. Int. Congr. Dev. Biol.* in press.

80. Socolar, S. J., *Exp. Eye Res.* **15,** 693 (1973).

81. Crick, F., *Nature* **225,** 420 (1970).

82. Bennett, M. R., *Autonomic Neuromuscular Transmission,* Cambridge University Press, London, 1972.

83. Goldberg, N. D., Haddox, M. K., Dunham, E., Lopez, E., and Hadden, J. W., in *The Cold Spring Harbor Symposium on the Regulation of Proliferation in Animal Cells,* Clardson, B., and Baserga, R., eds., Academic Press, New York, 1973.

84. Andersson, R., *Acta Physiol. Scand.* **87,** 84 (1973).

85. Holtrop, M. E., and Weinger, J. M., in *Calcium, Parathyroid Hormone and the Calcitonins,* Talmage, R. V., and Munson, P. L., eds., Excerpta Medica, Amsterdam, 1972.

86. Borle, A. B., in *Calcium, Parathyroid Hormone and the Calcitonins,* Talmage, R. V., and Munson, P. L., eds., Excerpta Medica, Amsterdam, 1972.

CHAPTER THREE

Some Aspects of Neuromuscular Junction Formation

GERALD D. FISCHBACH

Department of Pharmacology
Harvard Medical School
Boston, Massachusetts

The transmitting junction between a motoneuron and a muscle cell differs from an interneuronal synapse in several ways: nerve and muscle cells are derived from different germ layers, nerve and muscle membranes are separated by a 500 Å cleft that is filled with a perhaps unique amorphous extracellular matrix, and the nerve terminal is covered by a type of glial cell that is not found in the central nervous system. Nevertheless, since virtually every principle governing the function of chemically mediated interneuronal synapses was either first or most clearly defined at the peripheral neuromuscular junction [see reviews by Katz (1) and Eccles (2)], it seems likely that description of neuromuscular junction formation will apply to central connections as well. Indeed, many questions relevant to the establishment of nerve–muscle contacts are simply a subset of those relating to the recognition and specific interaction of other types of cell.

With this in mind, the impossibility of a broad, yet finite, review of neuromuscular junction formation should be quite obvious. It is useful, however, to review with a rather speculative bent some recent experiments in the light of relevant questions raised in the early embryologic literature. Discussion will be based on vertebrate motoneuron–skeletal muscle junctions where each fiber is innervated by a single neuron at one site termed the motor endplate and where transmission is mediated by acetylcholine. Data are sparse, however, and other systems will be cited where appropriate. Attention will be focused on three types of study: regeneration of adult motor nerves; embryonic neuromuscular junctions; and nerve–muscle contacts that form in tissue culture.

IS THERE A LATENT PERIOD BETWEEN NERVE–MUSCLE CONTACT AND SYNAPSE FORMATION?

In most studies of regenerating adult motor nerves, neuromuscular transmission is restored several days to months after the injury, depending on the species, the site of injury, and on whether the nerve was cut or crushed. Most of the delay can be accounted for by the sorting out of severed axons at the site of injury and their slow rate of growth back to the denervated muscle (3). Fex and associates (4,5) took advantage of the fact that a foreign nerve will innervate a muscle if and only if the original nerve is cut* and circumvented this long delay in rats by implanting the proximal stump of the severed peroneal nerve into the gastrocnemius muscle for varying periods before cutting the gastrocnemius nerve. Synaptic transmission, detected by

---

* The inability of most muscles to be innervated by more than one neuron simultaneously has been known for more than 50 years (6,7). Harrison (6) proposed an analogy between a singly innervated mammalian muscle and a fertilized egg, and a more profound explanation has not been offered since then. Polyinnervation of skeletal muscle has been described in invertebrates (8) and can be induced by certain experimental manipulations in higher vertebrates. Polyinnervation may also occur during the normal development of vertebrate neuromuscular junctions (see p. 55).

measuring twitch tension, was established 2 days after sectioning the native nerve when the peroneal nerve was preimplanted for 2 weeks whereas transmission was not established for 12 days when the foreign nerve was implanted at the time of gastrocnemius nerve section. This minimum latency is probably an overestimate. As discussed below, subthreshold activity may be present before twitches can be evoked. Furthermore, since the foreign nerve was implanted several millimeters from the original endplate (and synapses formed at that site), some time must have been required for the appearance of receptors beneath the new terminals.

Studies of young embryos also suggest that synapses form shortly after exploring motor fibers contact muscle cells. Cells destined for the lateral motor columns are the first to migrate away from the germinal epithelium lining the ventricular surface of the primitive neural tube (9–11). Lyser (12) found by electron microscopy that some cells in the most cephalic† segment of the spinal cord of 2-day chick embryos had formed axons, and primitive motor root bundles exciting from the cord were identified in 3-day embryos. Silver-impregnated axons have been identified in adjacent myotomes as early as 72 hours after incubation (13–15).

Hamburger (16) has observed head and neck movements in $3\frac{1}{2}$-day chick embryos that he considered neurogenic because they were coarse and episodic. Embryonic muscle fibers are often joined by "close" junctions (17), so it is conceivable that the movements resulted from spontaneous action potentials arising in single uninnervated fibers. However, single-unit discharge and complex bursts of electrical activity have been recorded in the brachial regions of 4-day embryonic

---

† There is a marked rostrocaudal gradient in maturation of the chick neural tube (9).

chick cords that coincide with the movements (18,19), and this, together with the correspondence between onset of movements and ingrowth of motor axons, makes Hamburger's explanation more likely.

Thus data from both regeneration in the adult and from embryos indicate that critical events may occur very soon after nerve–muscle contact. Further analysis would be greatly simplified in a situation in which the advancing motor growth cone could be directly observed over a period of several hours and tested with electrophysiologic techniques sufficient not only to demonstrate the existence of a synapse but also to clearly establish the absence of synaptic transmission. Recently several reports of functional neuromuscular junction formation in different types of culture system have appeared (20–28), and some of these *in vitro* systems permit the requisite direct observation, improved time resolution, and intracellular recording. Although the earliest events have not yet been described in culture, several interesting observations have been made. First, as expected, synaptic transmission can be detected soon after nerve–muscle contact. Neurons dissociated from 7-day chick spinal cords and added to previously plated muscle cells settle and extend processes that contact nearby myotubes in the first 24–48 hours. We have detected synaptic interaction between nerve and muscle cells as early as 45 hours after seeding with neurons (29). The time required for the neurons to settle, recover from the dissociation procedure, and extend processes to nearby muscle cells accounts for most of this interval. Less than 10% of the neurons in the initial suspension of spinal cord cells that survive longer than 2 weeks in culture form functional connections with the muscle cells (25,26), and this sampling problem has hampered further analysis of the earliest stages of synapse formation.

Cohen (27) found that neurons extending from intact slices of frog spinal cord tissue form functional contacts on dissociated muscle cells within 24 hours after addition of myoblasts to the established explants. Indirect evidence has been presented by Chamley, Campbell, and Burnstock (30) that exploring nerve fibers from sympathetic ganglia explants form synapses on smooth-muscle cells within 1 hour after contact. The smooth-muscle cells in these cultures "beat" spontaneously at a low rate, and the rate doubled as individual cells were palpated by exploring growth cones over a 50 minute period.

In the absence of an unambiguous morphological or biochemical sign the only clear criterion for nerve–muscle junction formation is a functional one—chemically mediated synaptic potentials. In order to demonstrate chemical transmission, in most circumstances, the motoneuron must be electrically excitable; it must synthesize, store, transport, and release transmitter molecules; and the muscle membrane must be able to respond to the transmitter. Thus function may, in fact, be a relatively late stage in nerve–muscle junction formation.

Recent studies of the ultrastructure of growth cones of sensory and sympathetic ganglion cells are consistent with the perhaps obvious notion that an irreducible latent period, or a stage of "recognition," precedes the onset of chemical transmission. The thin, motile pseudopodia that extend 5–10 μ from the flat base of the growth cone are packed with a lattice of 40–60 Å filaments (31–34). Apparently the only other structure present in these microspikes are elongated vesicles, which, at one time, may have been connected to the smooth endoplasmic reticulum in the axon. Clusters of small (450–500 Å) synaptic vesicles have not been found in the pseudopodia or in the flat base. If transmitter is released in the form of discrete quanta at the earliest synapses (see p. 49) and if these quanta are equated with synaptic vesicles, a certain delay for formation of these vesicles might be expected. Bray (35) and others have raised the possibility that the membrane bounding the elongated vesicles is added to the surface membrane at the very tips of the motile pseudopodia (see also ref. 36). It may be that, after the tip recognizes a muscle membrane and stops "growing," the elongated vesicles form small, round synaptic vesicles (cf. ref. 37).

An intriguing observation in this regard is the fact that in the experiments of Fex and associates (4,5) new synapses were formed in the immediate vicinity of the implanted nerve terminals even when the foreign nerve was implanted 12 weeks prior to section of the original nerve. Some interaction between nerve and muscle may have played a role in preventing migration of the nontransmitting terminals. The ultrastructure of these dormant contacts is unknown. As a limiting statement it seems clear that function is not a prerequisite for recognition. Neuromuscular junctions form *in vitro* in the presence of high concentrations of curare (27,38).

Is there an electrophysiologic correlate of the recognition event? One speculation based on simultaneous intracellular stimulation and recording in chick neuron and muscle cell cultures can be offered. In a few nerve–muscle pairs, in relatively mature cultures, injection of current into one cell (neuron or muscle) produced an electrotonic potential in the other (26). In all save one of these electrically coupled pairs, adequate stimulation of the neuron produced a typical endplate potential (epp) in the muscle, and it is tempting to speculate that the two phenomena are related. Perhaps the reason that electrical coupling is rare in older cultures is that such contacts represent an early and transient stage in the formation of a chemical synapse.

Current generated in adult motor nerve terminals during an action potential or graded electrotonic polarization does not polarize the postsynaptic membrane (39). In the adult, presynaptic and postsynaptic membranes are separated by a 500 Å cleft that is filled with a thick amorphous external lamina. Muscle fibers that develop in fibroblast-free cultures (treated with cytosine arabinoside) are not completely enclosed in an external lamina, so nerve fibers may form a more intimate relation with the muscle membrane. Kelly and Zacks (40) found "close" junctions marked by a narrow gap and increased membrane density between fine nerve fibers and muscle cells in 16-day embryonic rat intercostal muscles. The muscles were not covered by a basement membrane at this stage. Interestingly, these contacts were located near the middle of the muscle, which is the location of endplates in mature fibers. James and Tresman (41) found that nerve processes from spinal cord explants deeply indented nearby muscle fibers. True gap junctions, which have been correlated with electrical coupling in other tissues, were not described in either of these studies.

Many types of embryonic or transformed cells form low-resistance junctions on contact *in vitro,* and coupling is quite common between cells of the same or even different germ layers in intact embryos [see Chapter 2 as well as reviews by Furshpan and Potter (42) and Bennett (43)]. It seems unlikely, however, that the coupling between nerve and muscle simply reflects a general property of relatively immature cell membranes: the great majority of "touching" nerve–muscle pairs were not electrically coupled. A strikingly high incidence (20%) of electrical coupling between muscle cells derived from dystrophic embryos and spinal cord neurons dissociated from normal or dystrophic embryos has been reported (44). It may be, as suggested by the authors, that the membranes of dystrophic muscle fibers are relatively immature, and synaptogenesis may be "stuck" in an early stage. What is required is to show that electrical coupling necessarily precedes the onset of chemical transmission at the earliest contacts and that coupling is invariably followed by the formation of a chemical synapse.

Neurons dissociated from 7-day chick cords are quite small (10–25 μ) immediately after plating and are difficult to penetrate with microelectrodes. Preliminary experiments in which migration into nerve cells of the fluorescent dye Procion Yellow following injection into muscle fibers have yielded negative results. However, further experiments are warranted because there is some indication that junctions between electrically coupled embryonic cells may be less permeable to low-molecular-weight dyes than junctions between adult cells (see ref. 43 for review) and because it is quite possible that none of the neurons examined was destined to form chemical junctions.

Electrical coupling and gap junctions have been detected between myogenic cells *in vitro* prior to fusion of the membranes (45). Although the authors did not investigate whether coupling was a necessary or sufficient condition for fusion, this process may signal recognition of one muscle cell by another.

TRANSMITTER RELEASE AT EARLY CONTACTS

Spontaneous and Stimulus-Evoked Potentials

At the adult neuromuscular junction, transmitter is released in the form of multimolecular packets, or quanta. Each stimulus-evoked endplate potential (epp) is composed of an integral number of quanta, and, in the absence of nerve impulses, individual quanta "leak" from motor nerve terminals and generate miniature endplate potentials (mepps) (1,46). Evidence of quantal release of transmitter at newly formed junctions has been obtained by intracellular recording from adult muscle soon after reinnervation by regenerating motor nerves (47–54) from embryonic or neonatal muscle (55) and from muscle grown in culture and innervated by slices of spinal cord tissue (22–24,27) or by dissociated spinal cord cells (25, 26).

Mechanisms underlying spontaneous and stimulus-evoked transmitter release are thought to be identical. The synchronous release of transmitter effected by the influx of Ca^{2+} during the presynaptic action potential is probably due to a transient increase in frequency of the same critical events that govern the ongoing leak of quanta. Nevertheless, differences in the time of appearance of mepps and epps, and a discrepancy in size between mepps and the smallest unit evoked by a stimulus during the course of innervation have been reported.

Mepps reappear and increase in frequency before epps can be evoked soon after reinnervation of adult amphibian, avian, and mammalian endplates (47–49, 52,53). In the frog, the situation is complicated by the fact that, following a silent period beginning shortly after denervation and lasting for a few days, mepps reappear before the nerve has reestablished contact with the muscle fiber (56). Quanta are released, apparently, from Schwann cells that displace the degenerated nerve terminals and lie immediately adjacent to the postsynaptic folds. However, mepps derived from the newly arrived nerve terminal can be distinguished from the glial mepps. They are larger and more frequent, on the average, and their frequency is increased by raising extracellular K+, La³+, or osmolality—or by repetitive stimulation of the motor nerve (47,48, 56). Schwann cells do not remain at denervated mammalian endplates, and, in the absence of reinnervation, mepps do not reappear (57).

This relative delay in the appearance of evoked epps may simply be due to the failure of the presynaptic action potential to be conducted into the finest regenerated nerve terminals rather than to an uncoupling of transmitter secretion from membrane depolarization. The fact that at amphibian junctions epps reappear in an all (full size)-or-none fashion is consistent with this interpretation (47). Dennis and Miledi (49) found that epps could be evoked soon after mepps had reappeared by local depolarization of "nontransmitting" terminals or by stimulating the motor nerve with a pair of pulses. Extracellular recordings showed that the conditioning stimulus had allowed the spike set up by the second shock to invade the otherwise silent nerve terminals. The situation may be different at mammalian junctions. The first epps recorded in fibers in the rabbit diaphragm that are relatively far from the site of entry of the phrenic nerve into the muscle (presumably the most recently innervated) were invariably subthreshold, whereas epps in more proximal fibers were larger and nearly always triggered spikes and twitches (53). Incomplete invasion of the nerve terminals by the presynaptic action potential may account for these subthreshold responses as well as the delay in appearance of epps, but further experiments are needed.

The need for considering alternate hypotheses is underlined by the finding that, immediately after the reappearance of the first epps at amphibian junctions, the modal amplitude of mepps was significantly less than that of the first peak in the epp histogram (48,50). This size discrepancy decreased with time and was no longer apparent 12 days after the reappearance of epps. A similar discrepancy has also been observed at intact endplates poisoned with botulinum toxin (58). The small mepps observed early in reinnervation was probably not derived from Schwann cells (although they did not increase in frequency as much as the larger mepps following tetanic stimulation of the nerve or increase in extracellular La³+). They were probably not due to electrotonic decrement of mepps arising at some distance from the point of evoked transmitter release: both spontaneous and evoked release were detected at the same site by extracellular recording (although it is not clear whether the extracellularly recorded mepps accounted for most of the mepps detected with an intracellular electrode). Several explanations were considered by the authors. Within the context of the vesicle hypothesis, spontaneously released vesicles may be only partially filled or partially emptied, or they may be released at some distance from the postsynaptic receptors, or a normal sized mepp may correspond to the release of more than one vesicle.

Initial studies of quantal release at junctions that form *in vitro* (26) indicate that the size of mepps is the same as the basic unit of evoked transmitter release. Small differences would not have been detected in these experiments, however, and more detailed studies of precisely located synapses are required.

Precocious mepps have not been detected at new junctions that form *in vitro*. Quite the opposite sequence was suggested in chick dissociated cell cultures where in many innervated fibers no mepps were detected (even though the oscilloscope was closely monitored for 1–2 minutes) until one or a few epps were elicited by stimulation of a nearby neuron or nerve process (26). Mepps usually appeared on the falling phase of epps, and a single epp was often followed by a prolonged shower of mepps, so it seems unlikely that mepps were missed simply because they were released at a distant site.

A similar but smaller increase in mepp frequency has been noted at adult neuromuscular junctions, and the time course parallels that of facilitation of a test epp evoked at short intervals after a single conditioning stimulus (see ref. 59). Bennett, McLachlan, and Taylor (53) found that epp facilitation determined by the two-shock technique was most marked at early stages of reinnervation of adult endplates. In terms of the residual Ca^{2+} hypothesis for facilitation of transmitter release (60), it may be that Ca^{2+} entering during the presynaptic spike is less rapidly removed from new than from mature nerve terminals. Interestingly, mitochondria, which may play a major role in binding Ca^{2+} in adult terminals, are apparently reduced in number in embryonic nerve terminals in the central nervous system (31) and are rare in growth cones of cultured sympathetic neurons (32,33).

Quantum Content

At newly formed junctions in adult muscles, in embryos, and in culture, estimates of the level of quantal release based either on mepp frequency or statistical analysis of series of evoked epps are low compared to values recorded at adult neuromuscular junctions (24, 26, 27,51–53,55,61). In cell cultures, for example, estimates were about 1–10% of values measured at adult neuromuscular junctions (see ref. 62). Transmitter output increased slowly at neonatal endplates in the rat diaphragm: mepp fre-

quency did not reach adult values until 2 weeks after birth (55). The increase is apparently more rapid at contacts that form in spinal cord explant–muscle cultures. The mean quantum content increased about 50-fold (—1–50) within 7 days in rat cultures (24) and Cohen (27) found a marked decrease in epp amplitude variation (indicative of an increase in quantum content) after only 24 hours in amphibian cultures. No obvious increase in quantum content was noted over a 2–3 week period in dissociated muscle–spinal cord cell cultures (26).

The quantum hypothesis is based on the assumption that transmitter release at each site in response to successive stimuli represents a series of Bernouilli trials—that is, on the assumption that individual quanta are released with a fixed probability p from a constant (replenished) store N, that successive stimuli represent independent trials, and that the probability of release of a given quantum is independent of the release of others [see reviews by Katz (1), Martin (62), Ginzborg (63), and Kuno (64)]. Thus the number of epps composed of different numbers of quanta should be described by the binomial probability law. In the original formulation of the quantum hypothesis the level of transmitter release was reduced by lowering the Ca^{2+} concentration and/or raising the Mg^{2+} concentration in the bathing medium and the observed distribution of epp amplitudes was accurately predicted by the Poisson probability law, which can be derived from the binomial density function by assuming that $p \rightarrow 0$ as $N \rightarrow \infty$ (65).

At adult neuromuscular junctions, transmitter is not released from all points of the motor nerve terminals that lie in close apposition to the subsynaptic membrane. Del Castillo and Katz (66) found by extracellular recording of synaptic currents that transmitter is released from discrete "active sites" along the relatively extended frog motor nerve terminals. Similar estimates have been obtained at the more compact mammalian endplates. When epps are recorded with intracellular microelectrodes, all of the active sites are "monitored" and N refers to the total store of quanta available for release and p to the mean probability.

The low mean quantum content at new junctions might reflect a low store of the transmitter available for release (N). In the rat, mepp frequency and epp quantum content are correlated with endplate size and presumably with number of active sites within the nerve terminal arbor (67). The morphology of regenerated adult or embryonic motor nerve terminals or of nerve endings on muscle fibers in young cultures is relatively "simple." Silver stains and the scanning electron microscope indicate that only one or a few varicosities or terminal boutons characterize presumptive synapses (see ref. 68, Chapter XI; and ref. 69). Thus perhaps the simplest explanation for the low quantal output of newly arrived motor terminals is that they contain relatively few active sites.

It is not yet clear, however, whether varicosities or terminal swellings at new contacts are, in fact, sites of transmitter release or whether they are the only sites of transmitter release and other explanations for a low N should be borne in mind. There may be as many release sites along a single 300 μ process as there are in the terminal arbor of a mature endplate. Dennis and Miledi (50) found by extracellular recording from different points along single nerve endings at newly reinnervated endplates in frog muscle that the frequency of mepps and the probability of evoked transmitter release were greater from the proximal terminal than at the distal end. This gradient warns against quick

acceptance of the "simple morphology" explanation. Also, nerve endings in reinnervated chick anterior latissimus dorsi (ALD) muscles are probably not very different from the simple *en-grappe* boutons in control muscles, yet the mepp frequency soon after reinnervation is less than 10% of controls (52).

One intriguing possibility is that N is limited not by the number of active sites but by the rate of acetylcholine synthesis—that is, by the activity of the enzyme choline acetylase or availability of the substrate. Even at adult junctions where N is only a small fraction of the total available store of preformed (and packaged) transmitter (70), the synthesis of new acetylcholine plays a major role in maintenance of transmitter output during periods of repetitive stimulation. Experiments in which the release of labeled acetylcholine was measured are consistent with the conclusion that about one-half of the acetylcholine released during stimulation at 10–20 impulses per second is newly synthesized from choline (71,72). The recent demonstration that the activity of choline acetylase in cocultures of mouse muscle and spinal cord cells increases dramatically over a period of 3 weeks to a level that was 10 times greater than that found in cultures of spinal cord cells alone (73) is consistent with the notion that the activity of choline acetylase is limiting at early contacts. The increase in activity was paralleled by an increase in the incidence of nerve–muscle synaptic interactions, and Giller et al. (73) suggested that the two phenomena might be related. The specific activity of choline acetylase measured in axial and limb muscles of 3–4 day chick embryos, and presumably restricted to exploring motor nerve fibers, is low and increases to a peak by 12 days (74,75). This increase is quite remarkable considering the rapid growth of muscle and other nonneural tissue during this period. It is not clear whether this increase in enzyme activity reflects an increase in the number of choline acetylase molecules, or, if it does, whether the increase is simply related to the increase in the number and volume of functional endings.

Obviously several factors might regulate the level of choline acetylase on newly formed synapses. The ability of inactive or "disused" neuromuscular junctions to maintain quantal output at high stimulus frequencies is decreased (76), and the activity of choline acetylase per endplate is decreased by 50% (77), so some aspect of impulse activity and/or transmitter release may be a regulatory factor.

The low quantal output may be due to a decrease in fractional release (p) at new contacts. No direct estimates of p have been reported, but the prominent delayed release of mepps and facilitation of a test epp after a conditioning stimulus (see p. 48) may reflect a decrease in p because facilitation is augmented at adult junctions when p is decreased by decreasing the Ca^{2+}/Mg^{2+} ratio in the bathing medium (78). At the squid giant synapse, the depolarization attained by the presynaptic action potential falls on the steep part depolarization-transmitter output relation (79), so considering the description of faulty spike electrogenesis in regenerated amphibian nerve terminals (49), a decrease in fractional release at early contacts is not unlikely.

The esimates of quantum output at junctions in cell culture were based on the Poisson distribution but would not be greatly altered if the binomial statistics were assumed. However, in addition to the possibility greater accuracy, the binomial distribution offers the advantage that both N and p can be estimated directly (63,80).

The preceding discussion of quantal transmitter release at young contacts by no means excludes the possibility of nonquanta release at the same or at earlier contacts. In addition to more critical examination of the statistics of successive epp amplitudes, nonquantal release might be detected by the analysis of postsynaptic membrane noise recently described by Katz and Miledi (81). They found a significant increase in membrane potential fluctuations or noise when adult muscle fibers were depolarized by graded doses of acetylcholine: a 10 mV depolarization was associated with an increase in the rms value of noise of about 50 μV. Although the depolarization due to a single acetylcholine–receptor interaction was below the resolution of the recording system, the size of this elementary event was estimated on the basis of Campbell's theorem and together with a knowledge of the membrane time constant was used to calculate that a depolarization of 10 mV would be caused by summation of random elementary events occurring at a mean rate of 2–5 $\times$ 10^6/per second. Interestingly, this rate is comparable to estimates of nonquantal spontaneous release of acetylcholine (presumably from preterminal axons) at individual adult endplates [$-$ 2 $\times$ 10^6 molecules per endplate per second; see review by Potter (82)]. If this release occurs under the "normal" conditions of tissue culture, and if it is restricted to nerve endings, it may be detected by recording membrane potential and noise of single muscle fibers as they are palpated by exploring motor nerve growth cones.

SPECIFICITY

Questions of specificity in neuromuscular junction formation can be asked on several levels. A few very general points are worth mention. First, it seems clear that not every neuron can form a functional contact on a muscle cell. Despite early arguments to the contrary, sensory fibers cannot be made to innervate previously denervated adult skeletal muscle (83) and embryonic sensory ganglion cells grown in tissue culture do not transmit impulses to muscle fibers which they contact (26). Although some neurons dissociated from embryonic chick or rat spinal cords form chemical synapses *in vitro* on previously plated muscle cells, the great majority do not (25,26, 44). None of the many different cell lines derived from the original C1300 neuroblastoma tumor establish functional contacts with muscle fibers derived from primary or transformed myoblasts (84,85). Muscle cells do not, therefore, induce mechanisms for the synthesis, storage, and release of acetylcholine in contacting or closely apposed neurons, and noncholinergic neurons do not induce appropriate receptors in muscle cells.

The ability to innervate skeletal muscle cells is not limited to cells in the motor columns of the spinal cord or brain stem, however. Some fibers in the vagus nerve of the frog (presumably cholinergic axons that previously innervated cholinoceptive postganglionic neurons) can be made to innervate a skeletal muscle transplanted to the thorax (51,86–88). It is conceivable, therefore, that any neuron capable of synthesizing acetylcholine is also able to "recognize" and transmit impulses to a skeletal muscle cell. An argument that something more is required might be made from the fact that neuroblastoma clones that contain high amounts of the enzyme choline acetylase fail to innervate muscle fibers.

It is also clear that motoneurons can innervate cells other than skeletal muscle fibers. Collateral branches of motor axons feed back on small interneurons (Renshaw cells) within the spinal cord, and severed motor axons can be made

to innervate smooth muscle in the nictitating membrane (89) or superior cervical ganglion cells (88). Thus it is possible that motor axons will transmit impulses to any cell that will listen— that contains acetylcholine receptors.* The relations between receptors and innervation will be discussed in more detail.

Between Muscles

The fact of coordinated movement implies that the motoneuron–muscle wiring diagram is extremely precise. The debate as to whether this precision is established by specific chemical factors that mediate recognition between motor nerves and individual muscles or muscle groups, or by some more general means, such as timed outgrowth of motor axons or selective degeneration of connections that initially form at random, dates from the work of P. Weiss and R. Sperry in the 1940s. Reviews by Sperry (94), Mark (95), and Gaze (96) should be consulted. Until recently, the weight of evidence, based on studies of regeneration of motor nerves in adult animals, was against the existence of muscle-specific chemoaffinities. Motor nerves, in every species investigated, can be forced to innervate "foreign muscles" that have been disconnected from their original innervation. Thus, if specific recognition factors exist, they do not permit absolute distinctions between different muscles.

On the other hand, studies in several species have shown that motoneurons exhibit some measure of selectivity when

---

* Several cell types other than nerve or muscle possess acetylcholine receptors: fibroblasts (90), lymphocytes (91), fat cells (92), and hematopoietic stem cells (93). Cholinergic neurons obviously do not form lasting relations with any of these cells, and the role of these receptors remains unclear.

"offered a choice" between different muscles. When nerves to antagonistic pectoral fin (97) or extraocular muscles (98–100) of teleost fish were cut and allowed to regrow into the entire muscle group, normally coordinated movements were restored. In the case of the fin muscles it was possible to divert the original nerve in such a manner that normal function never returned (97). Thus, as in Sperry's experiments with mammals, there is no evidence in fish that recovery is due to a reorganization of central connections, and the most likely explanation for the return of normal function is that peripheral axons recognize and innervate appropriate muscles within the denervated group. Selective reinnervation was evident even when the wrong nerve was given a head start by resecting a segment of the original nerve. Marrotte and Mark (99) found that the superior oblique nerve took control of the superior oblique muscle from the inferior oblique nerve as late as 3 weeks after the latter had established synapses (see ref. 101 for analogous experiments on mammalian sympathetic ganglia).

The mechanism underlying the sorting out of connections in fish remains obscure. Apparently the once functional foreign nerve does not degenerate after takeover by the original nerve: non-transmitting axons continue to conduct impulses (102), and the expected degenerating foreign nerve terminals could not be demonstrated by electron microscopy in the recaptured muscle (100). An analogous takeover of muscle fibers by appropriate nerves was recently demonstrated in young salamanders (103). In this study the ventral root that contained fibers of the nonfunctional foreign nerve could be identified, and 2–3 days after it was cut degenerating terminals were readily identified in the territory that was recaptured by the

appropriate nerve.* Several alternatives might account for the inability of inappropriate nerve terminals to transmit impulses: action potentials, though conducted in the axons, might not invade the distal nerve terminals; the secretion of transmitter might be uncoupled from depolarization; there may be no receptors beneath the nerve terminals. It would seem that these and several other possibilities might be tested with standard intracellular microelectrode techniques.

The most striking examples of selective reinnervation in higher vertebrates involve competitive regeneration of motor nerves to fast (twitch) muscles and slow (tonic) muscles. In chicks the singly innervated, fast PLD muscle is somewhat resistant to innervation by the nerve to the neighboring multiply innervated, tonic ALD muscle. Feng, Wu, and Yang (104) found that several weeks after severing the common nerve trunk that contains the nerves to both muscles, stimulation of the ALD nerve (which can be separated from the PLD nerve within the main trunk) produced a normal electromyographic response in the ALD, but none at all in the PLD. The ALD nerve failed to innervate the PLD muscle even after the central stump of the mixed nerve was sutured to the distal stump of the PLD nerve. As in the experiments with fish, impulses conducted in the nontransmitting axons were recorded in the regenerated distal nerve stump. Conversely, the ALD muscle resists innervation by the PLD nerve. Bennett, Pettigrew, and Taylor (52) were able to hyperinnervate ALD mus-

---

* The survival of nonfunctional nerve terminals is probably not unique to fish or salamanders. As already discussed, foreign nerves implanted into an innervated mammalian muscle survived in some cases for as long as 12 weeks without establishing functional contacts on the muscle (4).

cles with both ALD and PLD nerves but found that the latter innervated only about 15% of the fibers (judged by visual observation of contraction) whereas nearly 100% were innervated by the ALD nerve.

In analogous experiments in toads involving the fast twitch sartorius and the mixed anteror semitendinosus muscles, Hoh (105) found that small-diameter motor nerve fibers that normally innervate slow muscle fibers in the semitendinosus reinnervated the same fibers rather than fast fibers in the same muscle or in the neighboring sartorius. Slow nerve fibers, distinguished by their higher threshold to electrical stimulation, branched at the site of injury and grew into the sartorius muscle, but in competition with fast nerve (lower threshold) axons failed to establish transmitting junctions. Even when slow motor axons were forced to innervate fast twitch fibers by inserting the nerve directly into a denervated fast muscle, a certain "incompatability" between nerve and muscle was detected (106). Tension was not maintained during repetitive stimulation of the inappropriate slow nerve, and connections between the slow nerve and fast muscle were apparently not maintained after several months. In the two cases in which the slow nerve–fast muscle contacts were not maintained, stimulation of the original fast nerve (which had maintained connection with the slow muscle) evoked a twitch in the fast muscle. Apparently the fast muscle had "rejected" slow axons and had "attracted" axons in the fast nerve.

Additional, indirect, evidence of selective reinnervation of frog slow muscle fibers was presented by Miledi, Stefani, and Steinbach (107). Normally innervated tonic muscle fibers in the frog do not generate action potentials after indirect or direct stimulation, and this passive behavior is somehow determined by the slow motoneurons. After denervation, depolarizing pulses elicit action potentials, and this newly begotten spike electrogenesis persists even after the fibers are innervated by axons that ordinarily innervate fast twitch muscles. Several weeks after cross-innervation the slow fibers become passive once again; considering the regulatory role of the slow motor axons, this reversion may indicate reinnervation of the muscle by the original axons.

The case for selective reinnervation of different fast twitch muscles in higher vertebrates is weak. Normal function is usually not restored after injury to mixed nerves supplying antagonistic muscles [see review by Sperry (94)]. Return of normal coordination may demand a high degree of fidelity, however, and its absence does not imply that neuromuscular connections are reformed entirely at random. In every cross-innervation study in which nerves to singly innervated muscles are switched, the degree of innervation achieved by the foreign nerve is never as great as that of normally innervated muscles or muscles reinnervated by their own nerves (see, for example, refs. 5 and 108).

In a remarkable early study on rabbits Elsberg (7) reported that "The normal nerve will regain its motor connections with the muscle fibers and will in some way prevent a foreign nerve which has been implanted at the same time from making any effective neuromuscular connections." Weiss and Hoag (109) objected to this conclusion and pointed out that in some of Elsberg's experiments the original nerve was guided by the persistent "Schwann tubes" in the distal stump and was thus at an advantage. But Elsberg had obtained the same result when the severed original nerve was implanted in "another part" of the muscle. Weiss and Hoag (109) allowed a native and foreign nerve to compete by inserting their proximal stumps into the two arms of a Y-shaped arterial sleeve. The stem was directed into one of the denervated muscles. No consistent preference for the native nerve could be demonstrated, but in every case either the native or the foreign nerve dominated, and this might be interpreted as a type of competitive interaction. Bernstein and Guth (110) assayed for selective reinnervation of rat plantarus and soleus muscles by determining if the unequal contributions of different lumbar ventral roots to the muscles were restored after cutting or crushing the sciatic or tibial nerve. They were not. As the authors point out, the conclusion is not clear, however, because specific muscle affinities may not be determined solely by location within the lateral motor column.

Thus it seems that question of specific affinities between individual muscles and motoneurons requires further study. Mark (95) has raised the interesting possibility that polyneuronal innervation may be a requirement for the demonstration of specificity, that is, that a muscle must be able to simultaneously sample more than one motoneuron in order to decide between them. Perhaps the process of selection procedes in several stages, and, as a first step, any motor nerve terminal may occupy an exposed or empty postsynaptic site on the muscle membrane. If the terminal cannot be displaced and if no other sites are available, then the first to arrive will be "served." This model may account for some of the confusing results following attempts at selective reinnervation of adult mammalian muscle. After cutting mixed nerves in the rat, Bernstein and Guth (110) found only one endplate and only one nerve terminal per endplate on each fiber. There is physiologic evidence for multi-

ple innervation (at more than one site) or polyinnervation (by more than one nerve) during reinnervation of denervated adult mammalian muscle (see discussion of hyperinnervation in the next section).

There is evidence, on the other hand, that muscles which in the adult are singly innervated are, in the embryo or shortly after birth, both multiply and polyinnervated. Redfern (111) recorded epps of different size and latency in the same muscle fiber (in neonatal rat diaphragm) that were evoked at different stimulus strengths and, therefore, were most likely due to stimulation of axons with different thresholds. Examples of multiple innervation have been reported in spinal cord–skeletal muscle cultures, but it is not yet clear whether the same muscle fiber can be innervated by more than one neuron (24,26). Chamley, Campbell, and Burnstock (30) found that some smooth muscle cells maintained in culture receive "stable contacts" from more than one axon that migrates from a nearby sympathetic ganglion explant. In this system polyinnervation was achieved only when the axons arrived at the muscle cell at the same time or within about 1 hour of each other. A similar critical interval may govern polyinnervation of skeletal muscle. In any case these considerations and Mark's model reemphasize the caution needed in accepting regeneration in adults as an exact model for initial development of nerve–muscle junctions.

Topographic

There is a striking tendency for adult motor axons to reinnervate muscles at sites that were previously innervated, that is, the old endplates (3,47,53,61, 112–115). An affinity for old postsynaptic sites is evident even when a normally multiply innervated muscle is reinnervated by an axon that ordinarily forms a single endplate or when a singly innervated muscle is reinnervated by an axon that ordinarily forms several, distributed, synaptic contacts contacts (52, 116,117). Likewise, when denervated skeletal muscles are reinnervated by cholinergic preganglionic fibers, synapses are apparently restricted to the persistent junctional folds that mark the old endplate (51,54,86). This remarkable topographic specificity is not simply explained by nerves growing down residual Schwann tubes because the same correspondence was found after the intramuscular nerves were excised (47) or when the nerve was cut and implated on the side of muscle opposite the normal point of nerve entry (53).

This affinity for former endplates can be overcome. Foreign nerves will "hyperinnervate" already innervated muscles at extra-endplate sites if the muscle fibers are injured (118), poisoned with botulinum toxin (119), or disused (120). Moreover, a few ectopic endplates have been found following normal (undirected) regeneration of motor nerves (3,112), and new endplates are the rule after a nerve is inserted or preimplanted into, or placed on the surface of denervated muscles at some distance from the original endplates (5,113,119,121,122). The extent to which the formation of new sites after direct insertion of regenerating nerves is due to local injury to muscle membranes (despite the use of thrombin clots) is not clear. In any case, these exceptions are either rare or dependent on unusual circumstances and do not invalidate the general rule that undirected motor growth cones will seek out and synapse on previously differentiated, unoccupied postsynaptic sites.

POSTSYNAPTIC FACTORS IN JUNCTION FORMATION

Several features distinguish the denervated postsynaptic membrane in most of the studies cited, and any or all of them may be related to the "attraction" of motor nerve growth cones. Primary and secondary postsynaptic folds persist, and the membrane along the folds appears relatively dense long after the nerve terminals have disintegrated (56,57). In the frog, but not in the rat, Schwann cells that once covered the intact nerve terminal lie immediately adjacent to the "empty" junctional folds. Endplate acetylcholinesterase, detected histochemically, is reduced after denervation, but some reaction product can be detected after several weeks in mammals (52,53, 123) and several months in amphibia (51,124). A particularly intriguing characteristic of the denervated endplate is that it remains more sensitive to acetylcholine (125–127) and contains more α-bungarotoxin-binding sites (128,129) than other regions of the muscle membrane. Even though the extrajunctional membrane develops significant sensitivity to acetylcholine within a few days after denervation, it remains 5–100 times less sensitive than the denervated endplate.

Acetylcholine Receptors

The appearance of "unoccupied" acetylcholine receptors is a common factor in all of the examples cited in which a motor nerve innervated a denervated muscle, a disused or injured muscle, or a muscle poisoned with botulinum toxin. But since receptors are required for the demonstration of synaptic transmission, this correspondence obviously cannot be considered evidence of a role of acetylcholine receptors in synapse formation.

The possibility that acetylcholine receptors are causally related to, or at least mark the spot of, future nerve–muscle junction formation has been raised by recent observations on the appearance and distribution of receptors on muscle fibers that form in tissue culture. Receptors appear early in myogenesis under normal growth conditions. The sensitivity of muscle fibers and the amount of ^{125}I-α-bungarotoxin binding increase sharply shortly after the mononucleated myogenic cells begin to fuse (130,131). As expected from Holtzer's notion (see ref. 132 for review) of the key role of a unique mitosis that governs the appearance of cell specific "luxury" molecules in skeletal muscle, acetylcholine receptors can be detected in mononucleated precurser cells. Fambrough and Rash (133) measured depolarizing* potentials following iontophoresis of acetylcholine in a few "elongated" mononucleated myogenic cells grown under normal conditions, and the increase in α-bungarotoxin binding, which normally lags a few hours behind fusion, is unaffected when fusion is prevented by reducing the Ca^{2+} concentration in the growth medium (134). Thus the appearance of acetylcholine receptors is probably not rate limiting in neuromuscular junction formation. Functional nerve–muscle contacts have, in fact, been found on mononucleated muscle cells in frog (27) and insect (28) cultures and on very young (three or four nuclei) myotubes in rat cultures (24). Although it seems likely that the early appearance of receptors *in vitro* can be extrapolated to embryonic muscles, a recent study (75) found that the "specific activity" or receptors in young embryonic chick muscle

was low and did not increase until after motor nerves (detected by an increase in the activity of the enzyme choline acetylase) had arrived. It is difficult to evaluate the receptor specific activity in view of the rapid changes in muscle mass and surface area, and more studies are required.

In addition to the early appearance of receptors, a particularly intriguing finding is that the distribution of acetylcholine sensitivity over uninnervated, mature muscle fibers in primary cultures is not uniform. When small patches of chick muscle membranes were tested with brief (1–2 msec) iontophoretically applied pulses of acetylcholine, one or more relative peaks of sensitivity, 3–10 times greater than values measured at nearby points, where detected on nearly every fiber (136,137). The peaks, or "hot spots," were extremely sharp and were lost after movement of the acetylcholine electrode by less than 10 μ. The relative nature of the peaks should be stressed. The mean "background sensitivity" of different fibers ranged between 10–1000 mV/nC, and peaks ranged between 50–7000 mV/Nc.†

The dense patches of grains found in autoradiographs of chick, mouse, and rat muscle cultures exposed to ^{125}I-α-bungarotoxin seem to be quite consistent with the electrophysiologically defined hot spots (130,137,138). Preliminary experiments in which the same fibers tested by acetylcholine iontophoresis were relocated in autoradiographs indicate that patches and hot spots do, in fact, correspond (139,140).

Taken together with the fact that regenerating nerves in adult animals

---

* Hyperpolarizing patentials have been found in transformed mononuclated muscle precursor cells (135).

† Active muscle fibers are less sensitive to acetylcholine than inactive ones (141–144). It is possible, therefore, that much of this variation is due to differences in the degree of spontaneous twitching of different fibers.

synapse at sites on denervated muscle fibers that are most sensitive to acetylcholine, the existence of hot spots on newly formed muscle fibers that have never been innervated raises the possibility that motor axons search along the muscle fiber for patches of membrane that contain a relatively high density of acetylcholine receptors. Interestingly, the density of α-bungarotoxin-binding sites estimated by autoradiography within a hot spot on cultured muscle fibers is approximately the same (allowing for the increased area of membrane associated with postjunctional folds) as that within an adult mammalian endplate ($10^4/\mu^{-2}$) (128,138,145).

The suggestion that hot spots on uninnervated fibers mark sites of future synapse formation is supported by the finding that hot spots detected on innervated fibers in combined muscle–spinal cord cell cultures were often located in the vicinity of fine nerve endings (137). The term "vicinity" is used advisedly because in many instances it was impossible with the phase-contrast illumination employed in these experiments to unambiguously identify nerve terminals, especially when they lay over thick, refractile muscle fibers. Moreover, fine neurites often course along muscle fibers for several hundred microns, and since synaptic potentials were recorded with intracellular microelectrodes (after the stimulation of a nerve cell body or relatively large nerve process), the exact points of transmitter were not identified. Thus it is not yet clear that hot spots are precisely located at sites of transmitter release or whether all sites of release are associated with hot spots. After 1–2 weeks in culture, most fibers contain several physiologically or autoradiographically defined hot spots, and this multiplicity is consistent with the fact that the fibers are often innervated at several different sites (see p. 55).

Similar relative peaks of sensitivity have been identified in combined cultures of muscle fibers derived from a transformed line of myoblasts and a continuous line of neuroblastoma cells (84,135). Muscle cells in this system are flat and less refractile than fibers that form in primary cultures, and the hot spots were located unambiguously at points of contact with nerve fibers. It is not clear even in this case, however, if they were restricted to the vicinity of the nerve terminal or if other sites beneath the preterminal neurite were also more sensitive.

In the chick cultures hot spots were often located in the immediate vicinity of hypolemmal, bulging muscle nuclei (137); this adds to the suspicion that they mark future synaptic sites. Accumulations of subsarcolemmal nuclei are found at adult mammalian, avian, and amphibian endplates—and silver stains of regenerated adult axons, embryonic axons, or axons that issue from spinal cord explants *in vitro* indicate that they terminate on or near muscle nuclei (13,112,146). Tello noted (cited in ref. 68, p. 269) that in those exceptional cases where the terminal nerve sprouts missed the "granular cytoplasm" of the original endplate they ended on nearby "multiplied nuclei of the degenerated muscle fiber," and he speculated that "these tendencies show that not only the legitimate plate but all nuclear accumulations of the contractile fiber have a neurotropic influence." We have found many instances in which silver-impregnated terminals of isolated chick spinal cord neurons "tag" or "embrace" nuclei in nearby muscle cells (147), but this relationship has not yet been systematically investigated.

There is hardly any need to point out the tenuous nature of the suggestion that a critical density of acetylcholine receptors are required for neuromuscular junction formation. First of all, it is not clear that hot spots reflect a true increase in receptor density. Several specializations of the surface membrane of cultured muscle cells have been described, including patches of microvilli (69,148), coarse membrane folds, and an elaborate subscarolemmal tubulovesicular network that communicates with the extracellular space (149,150), and each may expose more receptors to iontophoretically applied acetylcholine or to α-bungarotoxin. High-resolution scanning electron microscopy or the use of ferritin–α-bungarotoxin conjugates in conjunction with acetylcholine iontophoresis may decide between these alternatives (148,151). Second, a hot spot may contain the same number but qualitatively different acetylcholine receptors—either in terms of ligand binding or in the transducer mechanism (ion channels) to which the binding site is coupled. Interestingly, both pharmacologic (152) and physiologic (153) differences between endplate and nonendplate, or extrajunctional, receptors have been described.

Third, although relative peaks of acetylcholine sensitivity exist on cultured muscle fiber prior to innervation, it should be quite obvious that the precise relation between hot spots and ingrowing neurites is by no means clear. It is quite possible, as suggested by Harris et al. (135) in discussing the muscle–neuroblastoma cultures, that hot spots are somehow caused or "induced" by the nerve fibers. Although some variation in acetylcholine sensitivity along individual uninnervated fibers derived from the L-6 myoblast line was noted, (84) the relative peaks (up to five times the background levels in one series) were not as striking as those located near neuroblastoma nerve endings. Steinbach and associates (84) assayed relatively short segments of L-6 myotubes (100–200 μ), so it is possible that more striking peaks, unrelated to

neuroblastoma cells, were missed. Alternatively, it may be that L-6 fibers are less "mature" than those that form in primary cultures. The number of ^{125}I-α-bungarotoxin grain clusters per fiber increases with time in primary cultures (138).

Landmesser (86) found that, after the reinnervation of skeletal muscle by preganglionic axons in the vague nerve, synaptic potentials are 10 times more sensitive to the ganglionic blocking drug hexamethonium than are controls evoked at normal motor endplates. This demonstration that the pharmacologic properties of the receptor may be influenced by the type of innervation adds some weight to the induction hypothesis.*

Finally, the correspondence between hot spots and muscle nuclei and functional nerve terminals is not invariant and must be better defined. The membrane over obvious "bulging" nuclei was more sensitive than that over nearby

---

* Even though impulses were not transmitted between neuroblastoma cells and muscle fibers, Harris et al (135) proposed that the appearance of hot spots represents an early stage in synapse formation. This argument is based on Miledi's (47) observation that, in a few reinnervated, adult amphibian muscle fibers the acetylcholine sensitivity became restricted to the old endplate prior to the reappearance of stimulus-evoked synaptic transmission. However, the two situations may not be analogous. In the adult the peak of sensitivity at the endplate is accentuated because of a decrease in extrajunctional sensitivity rather than the appearance of new receptors at the site of innervation, and it is not clear that this is the case in the neuroblastoma–muscle cultures. Low background values were recorded in a subsequent study (84) by measurements made, in this case, after chronic exposure to cobra α-neurotoxin, and reversal may not have been complete. In primary chick spinal cord muscle cell cultures the mean background or "extrajunctional" sensitivity of innervated fibers was not less than that of uninnervated fibers in control cultures (137). Thus, even if hot spots are caused by nerve fibers, it does not follow that this interaction is a stage in synapse formation.

nonnuclear regions in 37 of 48 paired comparisons, and in a few instances definite hot spots have been detected over apparently nucleus-free segments. Although hot spots in some cases have been relocated in the same position after 1–2 hours, it is not clear whether they are "stable" over longer periods of time. Some nuclei migrate for several hundred microns within a muscle fiber and it would be of interest to determine whether hot spots migrate as well. Are all sites of transmitter release associated with hot spots and vice versa? What happens to hot spots on uninnervated fibers or to those in innervated fibers that are not contacted by neurites? Do hot spots occur in embryonic muscle fibers or are they merely an interesting tissue culture phenomenon?

Whatever the outcome, whether motor nerve growth cones seek out hot spots or cause them, the question of whether acetylcholine receptors serve as recognition molecules will remain unanswered. Patches of membrane that are more sensitive to acetylcholine may differ from the remainder of the membrane in other ways. It is interesting in this regard that Hirano (154) observed small segments of embryonic chick muscle membrane that appeared relatively dense even though the nearest ingrowing neurite was 2000–3000 Å away. Similar densities have been observed in embryonic cerebellar cells prior to innervation (155), and on the basis of their size and proximity to nearby coated vesicles it was suggested that they resulted from fusion of the vesicles with the surface membrane. The relation between hot spots and some muscle nuclei is interesting in this regard because, as in other cells, Golgi complexes are prominent in the perinuclear cytoplasm (148). Perhaps new receptors and other molecules are processed in the Golgi complex and added to the surface membrane via vesicles derived from Golgi cisternae (156,157). It may be that some of the membrane bounding the elaborate tubulovesicular network in cultured fibers (150) is not invaginating from the surface but is headed in the opposite direction.

New neuromuscular junctions form in muscle–spinal cord explant cultures in the presence of high concentrations of curare (27,38), so it seems unlikely that the acetylcholine binding site is involved in recognition. Similarly the correspondence between neuroblastoma nerve endings and the relative peak of acetylcholine on L-6 muscle cells is not affected when the cells are grown in the presence of cobra toxin, which, like bungarotoxin, binds specifically but less tightly to acetylcholine receptors (84,131,135). As pointed out by Cohen (27), these experiments do not rule out the possibility that other sites on the receptor molecule are required for synapse formation. An antibody to partially purified acetylcholine receptors has recently been prepared (158), and it would be of interest to determine whether this larger molecule, which may cover more of the receptor, prevents the formation of functional contacts.

Acetylcholinesterase

Acetylcholinesterase is present in the muscles of young embryos and in muscle fibers that form *in vitro*. The activity of the enzyme increases dramatically shortly after the fusion of myoblasts (159–162). Foci of intense acetylcholinesterase activity identified by light-microscopy histochemical techniques have been identified in embryonic muscle and in muscle fibers cultured in the presence of spinal cord (146,159). Although it is not clear in either instance that the foci preceded innervation, their resemblance to endplate primordia prompted the suggestion that they may determine the site of future synapse formation. Similar spots appear after the denervation of young (but not old) mouse muscle (163) and in muscle cells grown *in vitro* in the absence of neurons (164). It is interesting, in light of the previous discussion of muscle nuclei, that most spots of intense acetylcholinesterase activity in embryonic muscles were located near "sole plate" nuclei (159) and reaction product identified by electron microscopy is especially prominent in the nuclear envelope of embryonic muscle (165). In the denervated muscles of young mice the active foci seemed to be located over "bulges" in the fiber (163).

Several studies have shown, however, that acetylcholinesterase is a relatively late-appearing component of newly formed junctions. Histochemical evidence for the localization of the enzyme does not appear for several days to several weeks after axons arrive at adult denervated endplates, at embryonic contacts or at junctions that form *in vitro* (24,26,55,86,166–169). The results of electrophysiologic experiments are consistent with these negative findings. Spontaneous or stimulus-evoked synaptic potentials are prolonged soon after reinnervation of adult endplates (47, 53) in embryonic muscle (55) and in muscle fibers innervated *in vitro* by dissociated spinal cord cells (26) or by neurons in spinal cord explants (21,22, 24). In every instance where they were tested, drugs known to inhibit acetylcholinesterase and prolong epps at adult junctions were without effect.

Although the slow time course of synaptic potentials is most easily explained by a lack of functional acetylcholinesterase sites to terminate transmitter action, alternate or additional explanations should be kept in mind. It is quite possible, considering the simple morphology of early contacts and the probable absence of an external lamina and postsynaptic folds, that hydrolysis of acetylcholine is not the rate-limiting step in the termination of transmitter action. Diffusion of transmitter from simple disk-shaped synaptic clefts might halt the conductance change after 1–2 msec (170,171). The prolonged falling phase might simply reflect the long time constant characteristic of relatively small embryonic or cultured fibers (172) or denervated adult fibers (173). Alternatively prolonged synaptic potentials might be due to sustained transmitter release from newly functional nerve endings. Measurements of the true time course of synaptic current either by extracellular recording or voltage clamp are required.

Even though muscle cells that form in culture contain true acetylcholinesterase at a very early stage in differentiation, we have not been able to demonstrate active sites of the enzyme anywhere on the external surface of these cells. Acetylcholine potentials are not prolonged by conditioning pulses of edrophonium, a rapidly acting acetylcholinesterase inhibitor delivered from a second, immediately adjacent micropipette (164, c.f. 174,175). Preliminary attempts to localize the enzyme by the electron-microscopy methods described by Karnovsky (176) have revealed only occasional densities on the external aspect of the plasma membrane which may have been due to the diffusion of the reaction product (164).

Cultured muscle cells secrete a large amount of acetylcholinesterase into the bathing medium (177) as well as several other soluble proteins (178), and acetylcholinesterase activity can be removed from adult endplates by "mild" treatment with proteolyitc enzymes (179, 180,181). Thus catalytic sites of acetylcholinesterase molecules may not be an integral part of the postsynaptic membrane. After enzymatic digestion, the external lamina appears "thin and discontinuous" (182). It is possible that interaction between ectodermally derived nerve terminals and mesodermally derived muscle cells results in a unique external lamina that is capable of binding or trapping acetylcholinesterase active sites.

External Lamina

Although the extracellular matrix is thought to play a role in several aspects of morphogenesis and in the induction and maintenance of the differentiated state [see the paper by Balzas (183) and references cited by Cohen and Hay (184)], it would seem, at first glance, that an external lamina is not a prerequisite for nerve–muscle junction formation. Muscle fibers grown in fibroblast-free cultures are not covered by an amorphous coat, yet they can be innervated by spinal cord neurons. Most embryonic rat muscle fibers are "bare" when fine nerve fibers first become closely aposed to the muscle membrane (17,40). At this stage (16 days) an external lamina surrounds small bundles of myofibers. However, neither of these observations can be taken as strong evidence against a direct role of external lamina material in junction formation. It is now clear that neural tube cells as well as other epithelia can secrete collagen and form an amorphous extracellular matrix, so it is possible that small patches or amorphous material are present at newly formed contacts. To date, no electron micrographs of sections that are unambiguously through synaptic contacts have been published. At mature junctions the external lamina that covers the Schwann cell and nerve terminal appears to "fuse" with that surrounding the muscle fiber (122), but, in fact, the origin

and chemical composition of the amorphous material in the cleft and secondary postsynaptic folds are not known.

The external lamina may play a role in "cementing" established functional contacts. Motor nerve terminals can be removed from adult endplates after a relatively prolonged exposure to proteolytic enzymes (179–182, 185). The muscle and nerve terminal membranes are probably not markedly altered by the dissociation procedure: muscle membrane potential and acetylcholine sensitivity are not altered, and the free nerve terminal continues to conduct action potentials

SUMMARY

In response to the very general question, "What are the guiding forces that cause the axons of developing or regenerating nerve cells to grow, to travel long distances to their specific terminal stations and among millions of cells to make contact with only a selected few," Katz (1) said in 1966: "We can give no clearer answer . . . now than our predecessors could a hundred years ago." It should be apparent from this review that the situation has not changed in the last several years. It should also be apparent, however, that since 1966 there has been a renaissance of interest in synapse formation, that has yielded several interesting results and ideas, and, perhaps, the means to test them further.

REFERENCES

1. Katz, B., *Nerve, Muscle and Synapse*. McGraw-Hill, New York, 1966.
2. Eccles, J. C., *The Physiology of Synapses*, Academic Press, New York, 1964.
3. Gutmann, E., and Young, J. Z., *J. Anat.* **78**, 15 (1944).

4. Fex, S., and Thesleff, S., *Life Sci.* **6,** 635 (1967).

5. Fex, S., and Jirmanova, I., *Acta Physiol. Scand.* **76,** 257 (1969).

6. Harrison, R. G., *J. Exp. Zool.* **9,** 787 (1910).

7. Elsberg, C. A., *Science* **45,** 318 (1917).

8. Stuart, A. E., *J. Physiol.* **209,** 627 (1970).

9. Hamburger, V., *J. Comp. Neurol.* **88,** 221 (1948).

10. Fujita, S., *J. Comp. Neurol.* **122,** 311 (1964).

11. Langman, J., and Haden, C., *J. Comp. Neurol.* **138,** 419 (1970).

12. Lyser, K., *Dev. Biol.* **10,** 433 (1964).

13. Tello, J. F., *Trabajos Lab. Inv. Biol. Univ. Madrid,* V, **15,** 101 (1917).

14. Levi-Montalcini, R., and Visintini, F., *Boll. Soc. Ital. Biol. Sper.* **13,** 983 (1938).

15. Filogamo, G., and Marchisio, P., *Neurosci. Res.* **4,** 29 (1971).

16. Hamburger, V., *The Emergence of Order in Developing Systems.* 27th Symposium of the Society of Developmental Biology, M. Locke, ed., Academic Press, New York, 1968.

17. Kelly, A. M., and Zacks, S. I., *J. Cell Biol.* **42,** 135 (1969).

18. Ripley, K. L., and Provine, R. R., *Brani Res.* **45,** 127 (1972).

19. Provine, R., *Brain Res.* **41,** 365 (1972).

20. Crain, S., *J. Exp. Zool.* **173,** 353 (1970).

21. Kano, M., and Shimada, Y., *Brain Res.* **27,** 402 (1971).

22. Kano, M., and Shimada, Y., *J. Cell Physiol.* **78,** 233 (1971).

23. Robbins, N., and Yonezawa, T., *Science* **172,** 395 (1971).

24. Robbins, N., and Yonezawa, T., *J. Gen. Physiol.* **58,** 467 (1971).

25. Fischbach, G. D., *Science* **169,** 1331 (1970).

26. Fischbach, G. D., *Dev. Biol.* **28,** 407 (1972).

27. Cohen, M. W., *Brain Res.* **41,** 457 (1972).

28. Seecof, R. L., Teplitz, R. L., Gerson, I., Ikeda, K., and Donady, J. J., *Proc. Natl. Acad. Sci. U.S.* **69,** 566 (1972).

29. Fischbach, G. D., and Breuer, A., unpublished observations.

30. Chamley, J. H., Campbell, G. R., and Burnstock, G., *Dev. Biol.* **33,** 344 (1973).

31. Tennyson, V. M., *J. Cell Biol.* **44,** 62 (1970).

32. Yamada, K. M., Spooner, B. S., and Wessels, N. K., *J. Cell Biol.* **49,** 614 (1971).

33. Bunge, M. B., *J. Cell Biol.* **56,** 713 (1973).

34. Bray, D., *J. Cell Biol.* **56,** 702 (1973).

35. Bray, D., *Nature* **244,** 93 (1973).

36. Ingram, V. M., *Nature* **222,** 641 (1969).

37. Teichberg, S., and Holtzman, E., *J. Cell Biol.* **57,** 88 (1973).

38. Crain, S., and Peterson, E. R., *In Vitro* **6,** 373 (1971).

39. del Castillo, J., and Katz, B., *J. Physiol.* **124,** 560 (1954).

40. Kelly, A. M., and Zacks, S. J., *J. Cell Biol.* **42,** 154 (1969).

41. James, D. W., and Tresman, R. L., *Z. Zellforsch. Mikros. Anat.* **100,** 126 (1969).

42. Furshpan, E., and Potter, D., in *Current Topics in Developmental Biology,* Moscona, A. A., and Monroy, A., eds., Academic Press, New York, 1968, pp. 95–127.

43. Bennett, M. V. L., *Fed. Proc.* **32,** 65 (1973).

44. Peacock, J. H., and Nelson, P. G., *J. Neurol. Neurosurg. Psychiat.* **36,** 389 (1973).

45. Rash, J. E., and Fambrough, D., *Dev. Biol.* **28,** 242 (1973).

46. Katz, B., *The Release of Neural Transmitter Substances.* Sherrington Lectures, Liverpool University Press, England, 1969.

47. Miledi, R., *J. Physiol.* **154,** 190 (1960).

48. Dennis, M., and Miledi, R., *Nature New Biol.* **232,** 126 (1971).

49. Dennis, M., and Miledi, R., *J. Physiol.,* in press, 1974.

50. Dennis, M., and Miledi, R., *J. Physiol.,* in press, 1974.

51. Landmesser, L., *J. Physiol.* **213,** 707 (1971).

52. Bennett, M. R., Pettigrew, A. G., and Taylor, R. S., *J. Physiol.* **230,** 331 (1973).

53. Bennett, M. R., McLachlan, E., and Taylor, R. S., *J. Physiol.* **233,** 481 (1973).

54. Bennett, M. R., McLachlan, E., and Taylor, R. S., *J. Physiol.* **233,** 501 (1973).

55. Diamond, J., and Miledi, R., *J. Physiol.* **162,** 393 (1962).

56. Birks, R., Katz, B., and Miledi, R., *J. Physiol.* **150,** 145 (1960).

57. Miledi, R., and Slater, C., *Proc. Roy. Soc. (London)* **B169,** 289 (1968).

58. Harris, A. J., and Miledi, R., *J. Physiol.* **217,** 497 (1971).

59. Rahaminoff, R., and Yaari, Y., *J. Physiol.* **228,** 241 (1973).

60. Katz, B., and Miledi, R., *J. Physiol.* **195,** 481 (1968).

61. McArdle, J. J., and Albuquerque, E. X., *J. Gen. Physiol.* **61,** 1 (1973).

62. Martin, A. R., *Physiol. Rev.* **46,** 51 (1966).

63. Ginzborg, B. L., in *Excitatory Synaptic Mechanisms, Proc. Fifth Int. Mtg. Neurobiol.,* Andersen, P., and Jansen, J. K. S., eds., Universtets for laget, Oslo, 1970.

64. Kuno, M., *Physiol. Rev.* **51,** 647 (1971).

65. Hoel, P. G., *Introduction to Mathematical Statistics.* P. 68, Wiley, New York, 1954.

66. del Castillo, J., and Katz, B., *J. Physiol.* 630 (1956).

67. Kuno, M., Turkanis, S., and Weakly, J., *J. Physiol.* **213,** 545 (1971).

68. Ramon y Cajal, S., *Degeneration and Regeneration of the Nervous System,* May, R. M., transl., Hafner, New York, 1928.

69. Shimada, Y., and Fischman, D., *Dev. Biol.* **31,** 200 (1973).

70. Elmqvist, D., and Quastel, D. M. J., *J. Physiol.* **177,** 463 (1965).

71. Potter, L. T., *J. Physiol.* **206,** 145 (1970).

72. Collier, B., *J. Physiol.* **205,** 341 (1969).

73. Giller, E. L., Schrier, B. K., Shainberg, A., Fisk, R., and Nelson, P. G., *Science* **182,** 588 (1973).

74. Giacobini, G., *J. Neurochem.* **19,** 1401 (1972).

75. Giacobini, G., Filogamo, G., Weber, M., Boquet, P., and Changeux, J. P., *Proc. Natl. Acad. Sci. U.S.* **70,** 1708 (1973).

76. Robbins, N., and Fischbach, G. D., *J. Neurophysiol.* **34,** 570 (1971).

77. Snyder, D., Rifenberick, D., and Max, S., *Exp. Neurol.* **40,** 36 (1973).

78. Mallart, A., and Martin, A. R., *J. Physiol.* **196,** 593 (1968).

79. Katz, B., and Miledi, R., *J. Physiol.* **207,** 789 (1970).

80. Johnson, E. W., and Wernig, A., *J. Physiol.* **218,** 757 (1971).

81. Katz, B., and Miledi, R., *J. Physiol.* **224,** 665 (1972).

82. Potter, L. T., in *The Structure and Function of the Nervous System,* Bourne, E. H., ed., Academic Press, New York, 1972.

83. Weiss, P., and Edds, M. V., *J. Neurophysiol.* **8,** 173 (1945).

84. Steinbach, J. H., Harris, A. J., Patrick, J., Schubert, D., and Heinemann, S., *J. Gen. Physiol.* **62,** 255 (1973).

85. Nelson, P. G., personal communication.

86. Landmesser, L., *J. Physiol.* **220,** 243 (1972).

87. Langley, J. N., *J. Physiol.* **22,** 215 (1897).

88. Guth, L., *Physiol. Rev.* **36,** 441 (1956).

89. Vera, C. L., Vial, J. D., and Luco, J. V., *J. Neurophysiol.* **20,** 365 (1957).

90. Nelson, P. G., and Peacock, J. H., *Science* **177,** 1005 (1972).

91. Strom, T. B., Deisseroth, A., Morganroth, J., Carpenter, C., and Merrill, J. P., *Proc. Natl. Acad. Sci. U.S.* **69,** 2995 (1972).

92. Illiano, G., Tell, G. P., Siegel, M. I., and Cuatrecasas, P., *Proc. Natl. Acad. Sci. U.S.* **70,** 2443 (1973).

93. Byron, J. W., *Nature New Biol.* **241,** 152 (1973).

94. Sperry, R., *Quart. Rev. Biol.* **20,** 311 (1945).

95. Mark, R. F., *Brain Res.* **14,** 245 (1969).

96. Gaze, R. M., *The Formation of Nerve Connections,* Academic Press, London, 1970.

97. Mark, R. F., *Exp. Neurol.* **12,** 292 (1965).

98. Sperry, R., and Arora, H., *J. Embryol. Exp. Morphol.* **14,** 307 (1965).

99. Marotte, L. R., and Mark, R. F., *Brain Res.* **19,** 41 (1970).

100. Marotte, L. R., and Mark, R. F., *Brain Res.* **19,** 53 (1970).

101. Guth, L., and Bernstein, J. J., *Exp. Neurol.* **4,** 59 (1961).

102. Mark, R. F., Marotte, L. R., and Johnstone, J. R., *Science* **170,** 193 (1970).

103. Cass, D. T., Sutton, T. J., and Mark, R. F., *Nature* **243,** 201 (1973).

104. Feng, T. P., Wu, W. Y., and Yang, F. Y., *Scientia Sin.* **14,** 1717 (1965).

105. Hoh, J. F. Y., *Exp. Neurol.* **30,** 263 (1971).

106. Close, R., and Hoh, J. F. Y., *J. Physiol.* **198,** 103 (1968).

107. Miledi, R., Stefani, E., and Steinbach, A. B., *J. Physiol.* **217,** 737 (1971).

108. Buller, A. J., Eccles, J. C., and Eccles, R., *J. Physiol.* **150,** 417 (1960).

109. Weiss, P., and Hoag, A., *J. Neurophysiol.* **9,** 413 (1946).

110. Bernstein, J. J., and Guth, L., *Exp. Neurol.* **4,** 262 (1961).

111. Redfern, P. A., *J. Physiol.* **209,** 701 (1970).

112. Tello, F., *Trab. Lab. Inv. Biol. Univ. Madrid* **5,** 117 (1907).

113. Guth, L., and Zalewski, A., *Exp. Neurol.* **7,** 316 (1963).

114. Iwayami, T., *Nature* **224,** 81 (1969).

115. Lullman-Rauch, R., *Z. Zellforsch. Micros. Anat.* **121**, 593 (1971).

116. Zelena, J., Vyklicky, L., and Jirmanova, I., *Nature* **214**, 1010 (1967).

117. Hnik, P., Jirmanova, F., Vyklicky, L., and Zelena, J., *J. Physiol.* **193**, 309 (1967).

118. Miledi, R., *Nature* **199**, 1191 (1963).

119. Fex, S., Sonesson, B., Thesleff, S., and Zelena, J., *J. Physiol.* **184**, 872 (1966).

120. Jansen, J. K. S., Lømø, T., Nicolaysen, K., and Westgaard, R. N., *Science* **181**, 559 (1973).

121. Gwyn, D. G., and Aitken, J. T., *J. Anat.* **100**, 111 (1966).

122. Saito, A., and Zacks, S. T., *Pathol.* **10**, 256 (1969).

123. Guth, L., Albers, W., and Brown, W. C., *Exp. Neurol.* **10**, 236 (1964).

124. Pecot-Dechavassine, M., *Arch. Int. Pharmacodyn.* **176**, 118 (1968).

125. Kuffler, S. W., *J. Neurophysiol.* **6**, 99 (1943).

126. Miledi, R., *J. Physiol.* **151**, 1 (1960).

127. Albuquerque, E. Y., and McIsaac, R. J., *Exp. Neurol.* **26**, 183 (1970).

128. Fambrough, D. M., and Hartzell, H. C., *Science* **176**, 189 (1972).

129. Hartzell, H. C., and Fambrough, D. M., *J. Gen. Physiol.* **60**, 248 (1972).

130. Vogel, Z., Sytkowski, A. J., and Nirenberg, M. W., *Proc. Natl. Acad. Sci. U.S.* **69**, 3180 (1972).

131. Patrick, J., Heinemann, S., Lindstrum, J., Schubert, D., and Steinbach, J. H., *Proc. Natl. Acad. Sci. U.S.* **69**, 2762 (1972).

132. Holtzer, H., in *Cell Differentiation*, Schyeide, O., and de Vellis, J., eds., Van Nostrand Reinhold, New York, 1970, pp. 476–503.

133. Fambrough, D., and Rash, J., *Dev. Biol.* **26**, (1971).

134. Paterson, B., and Prives, J., *J. Cell Biol.* **59**, 241 (1973).

135. Harris, A. J., Heinemann, S., Schubert, D., and Tarakis, H., *Nature* **231**, 296 (1971).

136. Cohen, S. A., and Fischbach, G. D., *Soc. Neurosci. First Ann. Meeting*, 1971, p. 162A.

137. Fischbach, G. D., and Cohen, S. A., *Dev. Biol.* **31**, 147 (1973).

138. Sytkowski, A. J., Vogel, Z., and Nirenberg, M. W., *Proc. Natl. Acad. Sci. U.S.* **70**, 270 (1972).

139. Fischbach, G. D., and Cohen, S. A., unpublished observations.

140. Hartzell, C., and Fambrough, D., *Dev. Biol.* **30**, 153 (1973).

141. Lømø, T., and Rosenthal, J., *J. Physiol.* **221**, 493 (1972).

142. Drachman, D., and Witzke, F., *Science* **176**, 514 (1972).

143. Cohen, S. A., and Fischbach, G. D., *Science* **181**, 76 (1973).

144. Purves, D., and Sackmann, B., *J. Physiol.*, in press, 1974.

145. Barnard, E. A., Wieckowski, J., and Chiu, T. H., *Nature* **234**, 207 (1971).

146. Veneroni, C., and Murray, M., *J. Embryol. Exp. Morphol.* **21**, 369 (1969).

147. Fischbach, G. D., and Heuser, J., unpublished observations.

148. Fischbach, G. D., Henkart, M. P., Cohen, S. A., Brewer, A. C., Whysner, J., and Neal, F. M., in *Synaptic Transmission and Neuronal Interaction*, Bennett, M. V. L., ed., Raven Press, New York, 1974.

149. Ezerman, E. B., and Ishikawa, H., *J. Cell Biol.* **35**, 405 (1967).

150. Ishikawa, H., *J. Cell Biol.* **38**, 51 (1968).

151. Hourani, B. T., Torain, B. F., Henkart, M. P., Carter, R. L., Marchesi, V. T., and Fischbach, G. D., manuscript submitted for publication, 1974.

152. Beranek, R., and Vyskocil, F., *J. Physiol.* **188**, 53 (1967).

153. Feltz, A., and Mallart, A., *J. Physiol.* **218**, 85 (1971).

154. Hirano, H., *Z. Zellforsch. Mikros. Anat.* **79**, 198 (1967).

155. Altman, J., *Brain Res.* **30**, 311 (1971).

156. Palade, G., in *Subcellular Particles*, Hayashi, T., ed., Ronald Press, New York, 1959.

157. Hirano, H., Parkhouse, B., Nicholson, G. Lennox, E. S., and Singer, S. J., *Proc. Natl., Acad. Sci. U.S.* **69**, 2945 (1972).

158. Patrick, J., and Lindstrum, J., *Science* **180**, 871 (1973).

159. Kupfer, C., and Koelle, G. B., *J. Exp. Zool.* **116**, 397 (1951).

160. Engel, W. K., *J. Histochem. Cytochem.* **9**, 66 (1961).

161. Mumenthaler, M., and Engel, W. K., *Acta Anat.* **47**, 274 (1961).

162. Fluck, R. A., and Strohman, R. C., *Dev. Biol.* **33**, 417 (1973).

163. Lubinska, L., and Zelena, J., *Nature* **210**, 39 (1966).

164. Fischbach, G. D., Cohen, S. A., and Henkart, M. P., *Ann. N.Y. Acad. Sci.*, in press, 1974.

165. Tennyson, V. M., Brzin, M., and Slotwiner, P., *J. Cell Biol.* **51**, 703 (1971).

166. Guth, L., and Brown, W. C., *Exp. Neurol.* **12**, 329 (1965).

167. Eranko, O., and Teravainen, H., *J. Neurochem.* **14**, 947 (1967).

168. Lentz, T. L., *J. Cell Biol.* **42**, 431 (1969).

169. Bornstein, M. B., Iwanami, H., Lehrer, G. M., Brictbart, L., *Z. Zellforsch. Mikros. Anat.* **92**, 197 (1968).

170. Ogston, A. G., *J. Physiol.* **118**, 50 (1952).

171. Eccles, J. C., and Jaeger, J. C., *Proc. Roy. Soc. (London)* **B148**, 38 (1958).

172. Fischbach, G. D., Nemeroff, M., and Nelson, P. G., *J. Cell Physiol.* **78**, 289 (1971).

173. Nicholls, J., *J. Physiol.* **131**, 1 (1956).

174. Nastuk, W. L., and Alexander, J. T., *J. Pharmacol. Exp. Ther.* **111**, 302 (1954).

175. Katz, B., and Thesleff, S., *Brit. J. Pharmacol.* **12**, 260 (1957).

176. Karnovsky, M., *J. Cell Biol.* **23**, 217 (1964).

177. Wilson, B. W., Nieberg, P. S., Walker, C. R., Linkhart, T. A., and Fry, D. M., *Dev. Biol.* **33**, 285 (1973).

178. Schubert, D., Tarikas, H., Humphreys, S., Heinemann, S., and Patrick, J., *Dev. Biol.* **33**, 18 (1973).

179. Hall, Z., and Kelly, R., *Nature* **232**, 62 (1971).

180. Betz, W., and Sackmann, B., *Nature* **232**, 94 (1971).

181. Betz, W., and Sackmann, B., *J. Physiol.* **230**, 673 (1973).

182. McMahan, U. J., Spitzer, N. C., and Peper, K., *Proc. Roy. Soc. (London)* **B181**, 421 (1972).

183. Balzas, E. A., *Chemistry and Molecular Biology of the Intercellular Matrix. I. Collagen, Basal Laminae, Elastin*, Academic Press, London, 1970.

184. Cohen, A. M., and Hay, E. D., *Dev. Biol.* **26**, 578 (1971).

185. Peper, K., and McMahan, U. J., *Proc. Roy. Soc. (London)* **B181**, 431 (1972).

CHAPTER FOUR

Metabolic Cooperation in Cell Culture

A MODEL FOR CELL-TO-CELL COMMUNICATION

RODY P. COX

Departments of Medicine and Pharmacology
Division of Human Genetics
New York University School of Medicine
New York, New York

MARJORIE R. KRAUSS

Department of Medicine
Division of Human Genetics
New York University School of Medicine
New York, New York

M. E. BALIS

Sloan-Kettering Institute
for Cancer Research
New York, New York

JOSEPH DANCIS

Department of Pediatrics
Division of Human Genetics
New York University School of Medicine
New York, New York

The processes by which cells exchange information constitute one of the most intriguing areas of biology. As described in this monograph, cell-to-cell communication involves a number of fundamental biological processes that may also have relevance to human disease. Quantitative studies on cell communication are difficult, and many of the observations are necessarily descriptive rather than analytical.

Metabolic cooperation is the phenomenon, originally described by Subak-Sharpe, Bürke, and Pitts (1), in which the phenotype of mutant cells in tissue culture is corrected by intimate contact with normal cells. For example, fibroblasts deficient in the enzyme IMP:pyrophosphate phosphoribosyltransferase (EC 2.4.2.8), alternatively named hypoxanthine phosphoribosyltransferase (HPRT), incorporate, as detected by radioautography, much smaller amounts of tritiated hypoxanthine or guanine into cellular material than do normal cells. However, when certain HPRT-deficient (HPRT$^-$) cells are mixed and grown in close contact with normal fibroblasts, they incorporate these purines. Metabolic cooperation appears to be a relatively easily manipulated and analyzable model for cell communica-

tion. If the normal and mutant phenotype can be distinguished at the cellular
level, correction of the phenotype can
be determined. In this model system
modifications of the tissue culture milieu
and of the donor or recipient cell can
be carried out easily. Thus the effect of
environmental or cellular alterations on
cell communication can be evaluated.

CELL MARKERS FOR STUDIES ON METABOLIC COOPERATION

Prerequisites for the study of metabolic
cooperation are genetic markers capable
of being detected at the cellular level in
tissue culture. The enzymes HPRT,
adenine phosphoribosyltransferase
(APRT, EC 2.4.2.7), and thymidine
kinase (TK, EC 2.7.1.21) are examples
of markers whose activity is easily detected in individual cells by radioautography. The inability of mutant cells to
incorporate the specific substrates permits simple differentiation from wild-
type cells. Mutants can be selected from
established wild-type cell populations by
their resistance to purine or pyrimidine
analogs (6-thioguanine for HPRT⁻, 8-
azaadenine for APRT⁻, and 5-bromo-
deoxyuridine for TK⁻). Mutants are
available in established cell lines, and
similar mutations can be found in cells
derived from human subjects.

NATURE OF THE SUBSTANCE TRANSFERRED DURING METABOLIC COOPERATION

Skin fibroblasts derived from patients
with the Lesch–Nyhan syndrome are
deficient in HPRT (2) and therefore
incorporate much smaller amounts of
tritiated hypoxanthine or guanine into
cell nuclei and cytoplasm (Fig. 1a) than
do normal cells (Fig. 1b). However,

when Lesch–Nyhan cells are mixed with
equal numbers of normal human cells,
grown to confluency and labeled with
³H-hypoxanthine, nearly all cells will
incorporate label (Fig. 1c). Cell contact
apparently is required since HPRT⁻
cells growing close to, but not in contact
with, normal cells do not incorporate
³H-hypoxanthine. In theory the nature
of the substance that passes between
normal and mutant might be episomal
DNA, informational RNA, or a regulatory substance that endows the mutant
cell with the ability to synthesize or
activate a functional HPRT. It is also
possible that the normal HPRT⁺ cells
may provide HPRT⁻ mutant cells with
preformed enzyme. A final possibility is
that normal cells incubated with ³H-
hypoxanthine synthesize radioactive nucleotides or their derivatives, which are
then transferred to mutant cells. If the
last mechanism were operative, one
would predict that incorporation of
radioactivity into mutant cells would
cease promptly after separation of
HPRT⁻ cells from normal ones. If
either of the first two mechanisms were
operative, incorporation of radioactive
label into separated mutant cells should
continue as long as functional enzyme
persists.

Studies on the nature of the substance
transferred were carried out independently by somewhat different methods in
our laboratory (3) and in Pitts' laboratory in Glasgow (4) with quite similar
results. In our experiments monolayer
cultures of normal human fibroblasts,
Lesch–Nyhan fibroblasts, and 1:1 mixtures of normal and mutant cells were
grown in flat-bottomed glass bottles,
each bottle containing a small coverslip.
Prior to harvesting the confluent cell
monolayer the coverslips were removed
and placed in individual Leighton tubes
and incubated with ³H-hypoxanthine.
These coverslips were therefore repre-

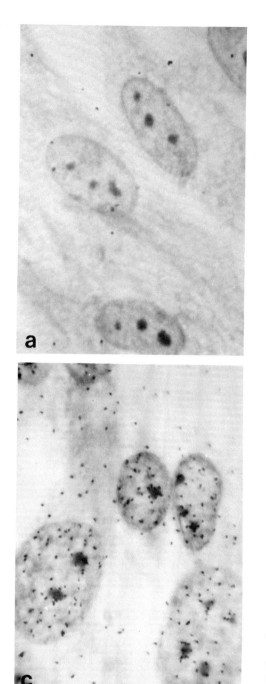

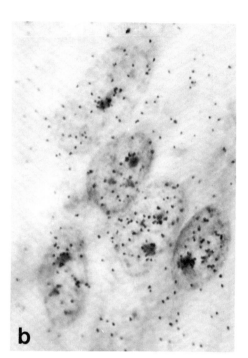

Fig. 1. Metabolic cooperation between normal human skin fibroblasts and cells from patients with the Lesch–Nyhan syndrome: radioautographs of human skin fibroblast monolayer cultures incubated for 3 hours with 100 μCi/ml of ^{3}H-hypoxanthine at 37°C. (*a*) Lesch–Nyhan fibroblasts; (*b*) normal fibroblasts; (*c*) 1:1 mixture of cocultured Lesch–Nyhan and normal fibroblasts, showing correction of the mutant phenotype.

sentative of the cell monolayer prior to separating the cells. The remainder of the cell monolayers in the bottles were then detached from the glass surfaces with a solution of trypsin and Versene (5), and the cell suspensions were diluted in fresh medium containing ^{3}H-hypoxanthine. Each of the cell suspensions was incubated in a siliconized Erlenmyer flask in a gyratory water bath at 36°C for 2 hours. The cells were prepared for radioautography. The incorporation of ^{3}H-hypoxanthine was determined by observing the number of silver grains over a large number of cell nuclei. The results of such grain counts are shown in Table 1. Cultures of HPRT$^-$ cells grown in the presence of ^{3}H-hypoxanthine, either in monolayer or suspension, incorporated little hypoxanthine as compared to normal fibroblasts. However, cells on coverslips removed from bottles with equal numbers of Lesch–Nyhan and normal cells were nearly all labeled, showing that there had been cell-to-cell interaction with resultant phenotypic modification of the HPRT$^-$ cells (top panel of Table 1). In contrast, the trypsinized mixture of normal and mutant cells that had been incubated with ^{3}H-hypoxanthine while in suspension (and therefore not in contact) exhibited a distribution approximating that to be expected if there were *no* cell interaction (bottom panel of Table 1) (3). The most likely explanation of this experiment is that metabolic cooperation between HPRT$^+$ and HPRT$^-$ cells involves transfer of a radioactive product of the normal enzyme from normal to mutant cells, unless the enzyme is very "unstable" and is rapidly degraded.

Pitts carried out similar experiments using HPRT$^-$ baby hamster kidney cells derived from the BHK$_{21}$ C$_{13}$ cell line (4). When mutant hamster cells were mixed with an equal number of wild-

type BHK fibroblasts and grown together in confluent culture, nearly all cells incorporated label. The proportion of HPRT$^-$ cells in an unlabeled duplicate cell mixture was determined by subculturing at a 1:00 dilution into medium containing ^{3}H-hypoxanthine. After 8 hours the cells were examined by radioautography, and single cells not in contact with other cells were scored as HPRT$^-$ or HPRT$^+$, depending on the grain count. The proportion of HPRT$^-$ cells in different experiments carried out by Pitts varied from 39 to 53%, indicating that the absence of cells with low grain counts in the confluent cell mixture was due to phenotypic masking of the mutant cells. Reconstruction experiments using prolonged radioautographic exposures further showed that if BHK–HPRT$^-$ cells in the sparse culture had incorporated ^{3}H-hypoxanthine at the wild-type rate for only 10 minutes, they could have been detected by the experimental method used.

The unlikely possibility that HPRT is extremely unstable was excluded in cell cultures by inhibiting protein synthesis with cycloheximide and determining the rate of fall of enzyme activity (Table 2) (3). Under these conditions HPRT fell only slightly over a period of 12 hours, whereas APRT, which was used as a control, was unstable, as shown by the marked reduction in the activity of this enzyme (Table 2). Since it is possible that cycloheximide might stabilize HPRT by inhibiting the synthesis of factors that normally degrade the enzyme and thereby abnormally prolong its survival, the experiment comparing cell monolayers and cell suspensions was repeated in the presence of cycloheximide. Under these conditions any enzyme already present in mutant cells should remain stable for at least 12 hours. However, this experiment also showed that in cell suspensions from

Table 1. Radioautographic Analysis of ³H-Hypoxanthine Incorporation into the Nuclei of Normal (HPRT+) and Mutant (HPRT⁻) Human Skin Fibroblasts and and 1:1 Mixtures[a]

Human Skin Fibroblast Strain	Grains per Cell Nucleus		
	<10	10–30	>30
Coverslip culture[b]:			
HPRT⁻	110	74	16
HPRT⁺	0	21	178
HPRT⁺ and HPRT⁻,			
1:1	10	116	74
Expected distribution			
if no cell interaction	55	47	97
Trypsinized cell suspension[c]:			
HPRT⁻	185	15	0
HPRT⁺	9	35	156
HPRT⁺ and HPRT⁻,			
1:1	110	32	58
Expected distribution			
if no cell interaction	97	25	78

Source. R. P. Cox, M. R. Krauss, M. E. Balis, and J. Dancis, *Proc. Natl. Acad. Sci. U.S.* **67**, 1573 (1970).

[a] A total of 200 cells were counted in each preparation.

[b] Monolayer cultures grown on coverslips were incubated with 100 μCi/ml ³H-hypoxanthine for 2 hours.

[c] Suspension cultures were incubated with 12.5 μCi/ml ³H-hypoxanthine for 90 minutes.

1:1 mixtures grown in monolayer approximately half the cells promptly reverted to the mutant phenotype. The results clearly support the view that metabolic cooperation is the result of transfer of enzyme product from normal to mutant cell.

We have carried out similar experiments using mixtures of APRT-deficient hamster cells and wild-type BHK cells (6), and Pitts has studied TK⁻ hamster cells and normal BHK cells (4). The results of the experiments indicate that enzyme product transfer is responsible for metabolic cooperation with these enzyme markers also.

It is noteworthy that Pitts has shown that the nucleotide TMP (thymidine 5′-phosphate) when added to medium is incorporated only very slowly (at less than 10% the rate of thymidine incorporation) into TK⁺ BHK cells. The small incorporation that is observed may not be due to the transport of TMP across the cell membrane but rather to the transport of degradation products of TMP formed in the medium (4). Subak-Sharpe (7) has also shown that adenosine 3′,5′-monophosphate (AMP) and guanosine 5′-monophosphate (GMP) cannot enter wild-type polyoma-virus-transformed hamster cells without loss of the phosphate group. Since transfer of nucleotide or nucleotide derivative appears to be the most reasonable basis for this form of cell-to-cell communication, the structures that allow this exchange must have different properties from the normal cytoplasmic membrane of cells not in physical contact with one another. In Chapter 1 of this volume N. B. Gilula presents evidence that the specialized membrane structure necessary for metabolic cooperation is the gap, or low-resistance, junction, and this structure also mediates ionic coupling. The production of a gap junction apparently requires close apposition of membranes of two cells. This requirement probably explains why intimate cell contact is needed for metabolic cooperation.

HETEROSPECIFIC CELL MIXTURES IN THE STUDY OF MECHANISMS OF METABOLIC COOPERATION

Investigations of the mechanisms responsible for metabolic cooperation are facilitated by the use of morphologically distinguishable donor and recipient cells. Cytological identification eliminates the need for statistical analysis and permits the assessment of metabolic cooperation on a cell-by-cell basis. Stoker (8) has demonstrated metabolic cooperation between morphologically similar cells by marking donor cells with ingested carbon or carmine granules to distinguish them from the mutant recipient. We have used heterospecific cell mixtures in which morphological differences between cells of different species permits identification of individual donor and recipient cells. Species differences do not affect the efficiency of cell communication (6). Therefore the use of heterospecific cell mixtures provides a good opportunity to study cell-to-cell interaction and the requirements for cell communication.

"NONCOMMUNICATING" CELL LINES

L Cells and Their Mutant Derivatives

Previous studies from several laboratories have demonstrated that certain cell lines are apparently unable to participate in metabolic cooperation when studied in the usual ways, that is, cocultured with competent recipient cells and labeled for 1½–3 hours with 35–100 μCi/ml of ^{3}H-hypoxanthine (4,6,9). The L cell (NCTC clone 929 L) and its sublines are the most completely studied representatives of these "noninteracting" cell lines. Gilula and associates (9) have shown that L cells apparently do not form gap junctions when cocultured with

Table 2. Effects of Cycloheximide on the Stability of HPRT and APRT in Human Fibroblast Cultures

Incubation Time (Hours) with Cycloheximide[a]	Specific Activity	
	HPRT[b]	APRT[b]
0	10.3	10.7
4	9.8	—
8	12.4	6.8
12	9.6	0.6

Source. R. P. Cox, M. R. Krauss, M. E. Balis, and J. Dancis, *Proc. Natl. Acad. Sci. U.S.* **67**, 1573 (1970).

[a] Cycloheximide (1.0 μg/ml) was added to cultures at time 0. Two replicate cultures were harvested at each time. This concentration of cycloheximide inhibited leucine-1-^{14}C incorporation by 88% in these cell cultures within 2 hours.

[b] Specific activity in millimicromoles of substrate reacted per minute per milligram of protein. Values are the average of duplicates that agreed within 20%. Cycloheximide was added to tissue cultures to inhibit the synthesis of new enzyme, permitting estimation of the survival of HPRT and APRT.

normal hamster fibroblasts, and this cell mixture does not exhibit metabolic cooperation or ionic coupling. The apparent failure of L cells to communicate may be due to an inability to form the specific cell junctions or a quantitative deficiency of them that impairs the detection of cell communication. These possibilities were studied by using a prolonged labeling with ^{3}H-hypoxanthine of cocultured HPRT$^+$ L cells and HPRT$^-$ human cells derived from a patient with the Lesch-Nyhan syndrome who had virtually complete absence of HPRT activity. Over 90% of the fibroblasts derived from the skin of this patient showed less than five grains of incorporated ^{3}H-purine on radioautography after 20 hours of labeling with 35 μCi/ml of ^{3}H-hypoxanthine (Fig. 2a). When freshly trypsinized mixtures of L cells and Lesch–Nyhan cells were inoculated into a medium containing 35 μCi of ^{3}H-hypoxanthine and cocultured for 20 hours in the presence of the label, the mutant human cells exhibited heavy labeling [>30 grains per nucleus on radioautography (Fig. 2c), demonstrating metabolic cooperation]. Replicate mixtures of L cells and Lesch–Nyhan fibroblasts that were cocultured for 20 hours and then labeled for 3 hours with ^{3}H-hypoxanthine showed little or no detectable transfer of label to the mutant human cell (Fig. 2b). Mixtures of L cells and Lesch–Nyhan fibroblasts cocultured for 30 hours and then incubated for 20 hours with ^{3}H-hypoxanthine also exhibited metabolic cooperation, indicating that the transfer of ^{3}H-nucleotide observed with freshly trypsinized cells is not the result of a trypsin-mediated modification of the cell surface. These results suggest that L cells have the capacity to communicate with a competent recipient but that the efficiency of communication

is greatly reduced. Pitts (10) has shown that BSC-1, an African green monkey kidney cell, and a HeLa cell line interact with HPRT$^-$ hamster fibroblasts with an intermediate efficiency in that, although ^{3}H-purine is transferred during a pulse-labeling experiment, the mutant receives between 50 and 20% of the label incorporated into the donor. These studies suggest that the cell interactions necessary for cell communication show quantitative differences between cells of different origins, but that so-called noncommunicating cell lines do interact with a reduced efficiency.

McCargow and Pitts (11) have shown that Sendai-virus-induced hybrids formed between the communicating BHK hamster fibroblasts and the noncommunicating L cell are able to function as donors in metabolic cooperation when studied by standard methods. The hybrid cell membrane contains both hamster and mouse specific antigens. These findings indicate that the capacity to form cell associations necessary for efficient cooperation is expressed in hybrid membranes.

Human Lymphocytoid Cell Lines

Human lymphocyte cultures capable of continuous cultivation *in vitro* can be established from normal subjects and patients with inborn errors of metabolism by exposing peripheral blood lymphocytes to Epstein–Barr virus. Many lymphocytoid cell lines have an apparently normal chromosomal complement, and those derived from patients with enzymatic deficiencies exhibit the inborn error (12,13). In stationary culture these lymphocyte suspensions clump together and grow as large aggregates. It is therefore possible to test metabolic cooperation between normal and Lesch–Nyhan

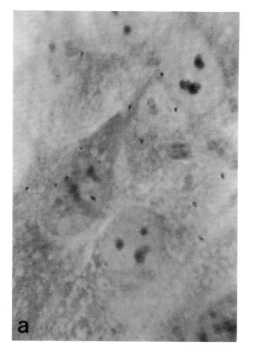

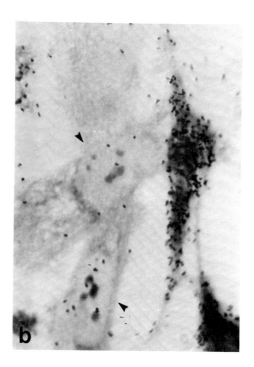

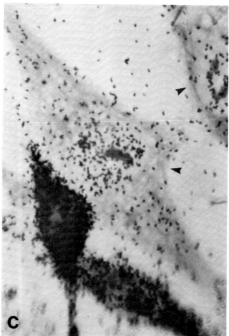

Fig. 2. Metabolic cooperation between putative "noncommunicating" L cells and HPRT⁻ human skin fibroblasts: radioautographs of cocultured Lesch–Nyhan cells and L cells labeled for varying times with 35 μCi/ml of ^{3}H-hypoxanthine at 37°C. The L cells are smaller, with darkly stained cytoplasm and nuclei. The HPRT⁻ human cells (marked by arrowheads), in cell mixtures, are lightly stained with nuclei containing three or four nucleoli. (*a*) Lesch–Nyhan cells incubated for 20 hours with ^{3}H-hypoxanthine show virtual absence of label incorporation and are a control for residual HPRT activity. (*b*) Cocultured 1:1 mixtures of L cells and Lesch–Nyhan fibroblasts incubated for 3 hours with ^{3}H-hypoxanthine show that L cells do not transfer detectable label to the mutant human cells during the 3 hour incubation. (*c*) Cocultured 1:1 mixtures of L cells and Lesch–Nyhan fibroblasts labeled for 20 hours with ^{3}H-hypoxanthine show transfer of label to the human cells.

lymphocyte cell lines by breaking the cell clumps and thoroughly mixing HPRT$^+$ and HPRT$^-$ cells together. Mutant and normal cells were agglutinated with one another and after being grown together for 48 hours were labeled with ^{3}H-hypoxanthine for 3 hours. As shown in Fig. 3, approximately half the cells retained the mutant phenotype despite close contact with HPRT$^+$ lymphocytes. These results indicate that permanent lymphocytoid cell lines do not form efficient cell associations necessary for metabolic cooperation.

Mouse Neuroblastoma Cell Lines

Mouse neuroblastoma cell lines have been the subject of intensive neurophysiological and neurochemical studies (14,15). Electrical coupling indicative of low-resistance junctions has been observed with certain clonal lines of these cells (16). We have studied interactions between two clonal lines of neuroblastoma cells that have been well characterized (A$_2$ and N-18) and Lesch–Nyhan fibroblasts, which are known to be efficient recipients in metabolic cooperation, in order to determine whether intercellular associations needed for efficient nucleotide transfer occur between these cells. Our results indicate that neither clonal line of neuroblastoma cells corrected the mutant phenotype of Lesch–Nyhan cells when these cell mixtures were labeled for 3 hours with 25 μCi/ml of ^{3}H-hypoxanthine. Experiments were also carried out in the absence of serum where neurite extension is prominent (17). Metabolic cooperation did not occur either in the presence or absence of serum, although apparent cell contact between neuroblastoma cell processes and Lesch–Nyhan fibroblasts was observed by light microscopy.

RATE OF FORMATION OF CELL ASSOCIATIONS NECESSARY FOR METABOLIC COOPERATION

Gap junctions have been implicated by Gilula and associates (9) as the structures necessary for both ionic coupling and metabolic cooperation. These low-resistance junctions are highly organized structures that contain tubular elements when viewed by electron microscopy. Their rate of formation can be estimated from the time required for demonstrating nucleotide transfer after trypsinized mixtures of normal and mutant cells have attached to glass and made contact in medium containing ^{3}H-hypoxanthine. A heterospecific cell mixture of normal hamster cells (HPRT$^+$) and human Lesch-Nyhan cells (HPRT$^-$), which can be distinguished cytologically, was inoculated into Leighton tubes. The cells attached to the glass substratum within 2–3 hours and began to elongate and become spindle shaped. When these cultures were grown with 35 μCi/ml of ^{3}H-hypoxanthine, transfer of label from normal hamster to HPRT$^-$ human cells was readily observed as soon as cell contact was made (Fig. 4a). The labeling of HPRT$^+$ hamster cells was only moderate, presumably because of the inhibition of nucleic acid synthesis after trypsinization; nevertheless, the normal cells transferred a portion of their label to the mutant cells. With increasing time the efficiency of cooperation increased, as shown in Fig. 4c. This may in part be attributed to an increase in nucleic acid synthesis in the hamster cells following recovery from trypsinization. Figures 4b and 4d show Lesch–Nyhan cells grown in a radioactive medium for the same length of time.

The possibility that the nucleotide was transferred through the medium was investigated and excluded in several

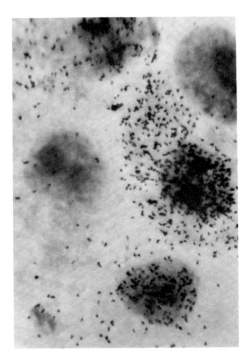

Fig. 3. Radioautograph of cocultured lympho-cytoid cell lines derived from a normal subject and from a Lesch–Nyhan patient and labeled for 3 hours with 50 μCi/ml of ^{3}H-hypoxanthine at 37°C. There was no detectable metabolic coop-eration since approximately half the cells are unlabeled.

ways. The nucleotide transfer required cell contact and was not the result of nucleotide leakage and uptake by recent-ly trypsinized cells since mutant cells that were not in contact with normal cells remained unlabeled. Moreover, transfer of radioactive medium from recently trypsinized BHK hamster onto recently trypsinized Lesch–Nyhan cells failed to increase the labeling of the mutant cells. Incubating coverslips of recently trypsinized BHK cells suspend-ed over recently trypsinized Lesch–Nyhan cells and filling the capillary space be-tween the coverslips with medium con-taining ^{3}H-hypoxanthine also failed to increase labeling of the mutant fibro-blasts.

INFLUENCE OF GROWTH CHARACTERISTICS AND CELL SHAPE ON METABOLIC COOPERATION

Effect of Contact Inhibition on Metabolic Cooperation

Surface interactions between potential donor and recipient cells are likely to be important in forming the intercellular associations needed for communication. Therefore the efficiency of metabolic cooperation was compared using as donors a cell line with a high degree of contact inhibition—cell line 3T3—or a polyoma-virus-transformed derivative of 3T3 with a marked reduction in its contact properties. As seen in Fig. 5, both the normal 3T3 cells (Fig 5a) and the polyoma-virus-transformed line (Fig. 5b) grown to confluency with HPRT⁻ human cells function with equal effici-ency in metabolic cooperation. It would appear that contact inhibition of growth is not related to the capacity of a cell to supply product in this form of cell com-munication. It should be noted that Pinto da Silva and Gilula (18) have recently identified gap junctions in cul-tures of normal chick embryo fibroblasts and those transformed by Rous sarcoma virus.

Effect of Cell Shape and Morphology on Metabolic Cooperation

Cell lines that are inefficient in cell com-munication (L cells, certain clonal lines of HeLa cells, and mouse neuroblastoma cell lines) have a common morphology in that they grow predominantly as rounded bipolar fibroblasts. Cells that effectively communicate have a more varied morphology—for example, elon-gated spindle-shaped or stellate cells (skin fibroblasts and BHK cells) or

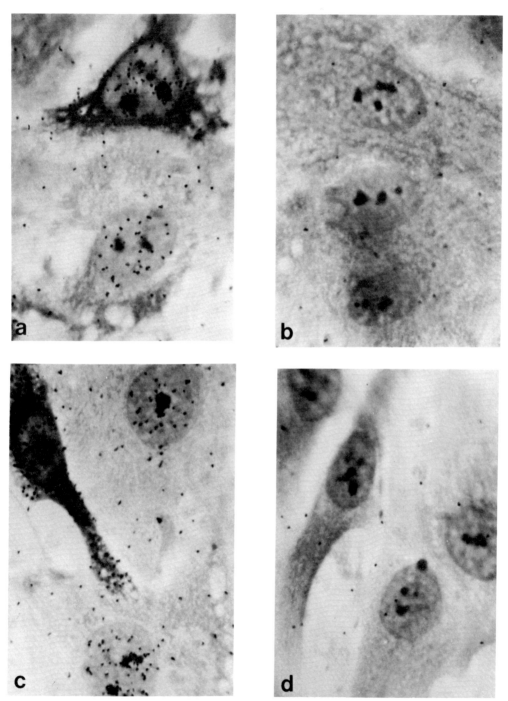

Fig. 4. Rate of formation of cell junctions necessary for metabolic cooperation: radioautographs of cocultured normal hamster cells and HPRT⁻ human skin fibroblasts grown in a medium containing 35 μCi/ml of ³H-hypoxanthine. Controls are Lesch–Nyhan fibroblasts incubated with ³H-hypoxanthine for the same time as the cell mixtures. (*a*) Cell mixtures 4 hours after inoculation; (*b*) Lesch–Nyhan fibroblasts 4 hours after inoculation; (*c*) cell mixture 5½ hours after inoculation; (*d*) Lesch–Nyhan fibroblasts 5½ hours after inoculation.

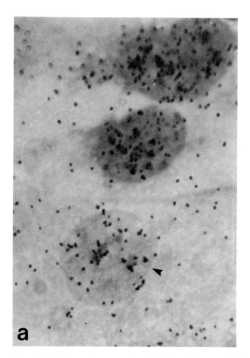
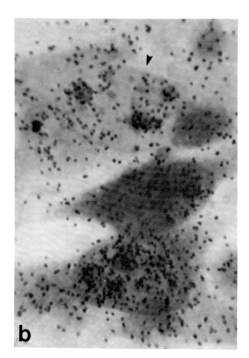

Fig. 5. Effect of contact inhibition characteristics of sublines of 3T3 mouse fibroblasts on metabolic cooperation: radioautographs of (*a*) mouse cell line 3T3 and (*b*) polyoma-virus-transformed 3T3 cocultured with Lesch–Nyhan skin fibroblasts (marked by arrowheads) and labeled for 3 hours with 35 μCi/ml of ^{3}H-hypoxanthine at 37°C. Both 3T3 and polyoma-virus-transformed 3T3 cells efficiently transfer label to mutant human fibroblasts, which indicates that metabolic cooperation is not affected by loss of contact inhibition or oncogenic virus transformation of 3T3 cells.

"cuboidal" cells (3T3 mouse fibroblasts and certain HeLa cell sublines). Since cell shape and mode of attachment to the substratum may affect cell associations and thereby influence cell communication, the effects on metabolic cooperation of altering the shape of cells with inefficient communication was investigated.

Both L cells and their derivatives (19) as well as certain clonal lines of HeLa cells (20), when grown in medium with cyclic AMP or its N^6, $O^{2'}$-dibutyryl derivative (DBcAMP), show a morphological alteration with a spreading and flattening of the cells, which become more adherent to the substratum. Hydrocortisone also alters the shape of HeLa and L cells producing a similar morpho-

logical change (20). Figure 6*a* shows a mixture of Lesch–Nyhan cells with a HeLa clone called HeLa$_{65}$ that exhibits inefficient communication, and Fig. 6*c* shows a different clonal line, HeLa$_{75}$—a cuboidal cell that exhibits efficient metabolic cooperation with human HPRT$^-$ cells. The observation that one HeLa cell alone is an effective donor in metabolic cooperation (Fig. 6*c*) while another is not (Fig. 6*a*) emphasizes the selectivity of cell communication. When HeLa$_{65}$ cells are grown in medium with hydrocortisone, they assume a cuboidal shape, adhere more tightly to the substratum, and morphologically resemble HeLa$_{71}$. However, as shown in Fig. 6*b*, HeLa$_{65}$ cells grown with hydrocortisone remain

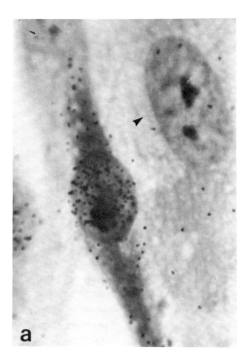

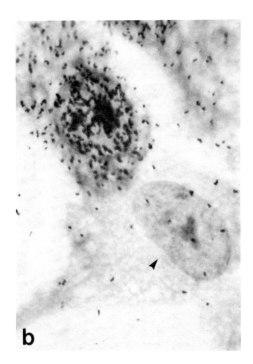

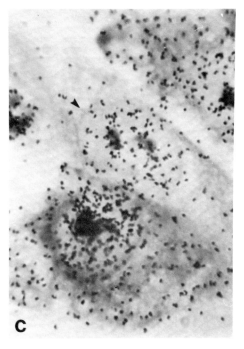

Fig. 6. Effect of cell shape and mode of binding to substratum on metabolic cooperation: radioautographs of HPRT⁻ human skin fibroblasts grown with (a) HeLa$_{65}$ clonal cell line, (b) Hela$_{65}$ cells grown in a medium containing 1 μM of prednisolone, and (c) HeLa$_{71}$ clonal cell line. Human Lesch–Nyhan cells are lighter stained, with nuclei having two or three nucleoli, and are marked with arrowheads. The HeLa$_{65}$ cells (a) grow as thick bipolar fibroblasts with large nucleoli and are more darkly stained than skin fibroblasts. They do not correct the mutant phenotype of HPRT⁻ fibroblasts. The HeLa$_{65}$ cells grown with prednisolone (b) become cuboidal and spread on the glass substratum; despite the morphological change they remain unable to donate label to Lesch–Nyhan fibroblasts. The cuboidal cells of HeLa$_{71}$ (c) exhibit metabolic cooperation with HPRT⁻ human fibroblasts. Cultures were incubated for 3 hours at 37°C with ³H-hypoxanthine: a and b, 50 μCi/ml; c, 35 μCi/ml.

79

unable to donate label to Lesch–Nyhan cells despite the morphological changes. Similar results were observed with L cells grown in medium with either DBcAMP or hydrocortisone, which produced a more spindle-shaped cell but did not promote metabolic cooperation with human HPRT$^-$ cells, an efficient recipient.

Nelson and Peacock (21) have reported that large nondividing L cells may be obtained by X-irradiation, and these cells are suitable for electrophysiologic studies. X-irradiated monolayers of L cells show that a hyperpolarization activation response can be elicited from cells adjacent to a stimulated cell, suggesting propagation of the response from one cell to another, not necessarily requiring cell junctions (21). Because of these relationships, metabolic cooperation was studied between X-irradiated giant L cells and HPRT$^-$ human cells. Figure 7a shows a mixture of L cells and Lesch–Nyhan fibroblasts that had been exposed to 35 μCi/ml of ^{3}H-hypoxanthine for 3 hours; there was no transfer of label from L cells to mutant human fibroblasts. Figure 7b shows a mixture of the giant irradiated L cells and Lesch–Nyhan fibroblasts that were similarly labeled. Despite the marked morphological changes and the evidence that similar preparations exhibited propagation of the hyperpolarizing activation response, there was no transfer of nucleotide to the HPRT$^-$ human cell under the conditions of this experiment.

Effect of Colcemid and Cytochalasin B on Metabolic Cooperation

Microtubules and microfilaments have been implicated in the maintenance of cell shape and cytoskeleton, and also in the transport and secretion of cellular materials. The importance of the integrity of the microtubule and the microfila-

ment structure on metabolic cooperation was studied by incubating communicating cell mixtures with either Colcemid (2.0 x 10^{-7}M), which disaggregates microtubules, or cytochalasin B (10 μg/ml), which breaks down microfilaments and also binds to cell membranes. Figure 8b shows a cocultured mixture of normal human fibroblasts and HPRT$^-$ hamster cells labeled with 35 μCi/ml of ^{3}H-hypoxanthine for 3 hours in the presence of Colcemid. The treated cells show retraction of cellular processes and vacuolization of the cytoplasm. The amount of label incorporated into normal human cells is reduced, but as long as contact is maintained between the normal human cell and the mutant hamster recipient, a proportionate amount of label is transferred, compared to replicate cell mixtures not treated with Colcemid (Fig. 8a). Figure 8c shows that the mutant HPRT$^-$ hamster cells do not metabolize ^{3}H-hypoxanthine. The integrity of the microtubular system apparently is not necessary for metabolic cooperation, and therefore this process differs from certain secretory activities that do require microtubules (22).

On the other hand, cytochalasin B treatment of communicating cell mixtures interferes with metabolic cooperation (Fig. 9b). Mixtures of normal BHK cells and Lesch-Nyhan fibroblasts treated with cytochalasin exhibit some retraction of cell processes, margination of the cell nucleus, and occasional binucleate cells. Cytochalasin-treated cell mixtures labeled for 3 hours with 35 μCi/ml of ^{3}H-hypoxanthine showed some heterogeneity with respect to metabolic cooperation in that most of the Lesch-Nyhan cells in intimate contact with BHK fibroblasts remain unlabeled (Fig. 9b). However, 10–15% of the Lesch–Nyhan cells were fed ^{3}H-purine by contact with BHK. The amount of label incorporated into the BHK cells was reduced in the presence of cytochalasin. However, experimental treatments of BHK–Lesch–Nyhan cell mixtures with other compounds—for example, colchicine or 450 mM sucrose solutions (see Fig. 10b)—reduced ^{3}H-hypoxanthine incorporation into BHK fibroblasts more than cytochalasin, but did *not* interfere with nucleotide transfer. Therefore microfilaments or cytochalasin-sensitive binding sites on plasma membranes appear to play a role in metabolic cooperation.

EFFECT OF ALTERATIONS IN MEDIUM COMPOSITION ON THE EFFICIENCY OF METABOLIC COOPERATION

The susceptibility of the intercellular structures that mediate metabolic cooperation to environmental alterations was studied in communicating cell mixtures by changing the composition of the medium.

Effect of Sodium Concentration on Metabolic Cooperation

The transfer of nucleotide or a derivative from one cell to another may involve a translocation system similar to active transport. The transport of certain compounds—such as potassium, sugars, and amino acids—frequently depends on the simultaneous transport of sodium ions (23). Therefore the role of sodium ions in metabolic cooperation was investigated in cocultured mixtures of normal human fibroblasts and HPRT-deficient hamster fibroblasts. An hour before labeling the cultures with ^{3}H-hypoxanthine, the medium in half the cultures was replaced by a MEM medium in which choline chloride substituted for NaCl (final Na$^+$ concentration 20 mM). In replicate cultures a medium containing a physiological amount of NaCl (145 mM Na$^+$) was added. During a subse-

quent 3 hour exposure to 35 μCi/ml of ^{3}H-hypoxanthine the amount of label transferred to mutant hamster cells from normal human donors was the same in both low-sodium and in normal-sodium media.

Effect of Divalent Ions on Metabolic Cooperation

Divalent ions, and in particular Ca^{2+} ions, have been shown to be necessary for ionic coupling in certain tissues—for example, insect salivary glands (24). The effect of divalent-ion depletion in tissue culture were studied by incubating cocultured normal hamster cells (BHK) and Lesch–Nyhan fibroblasts in a MEM medium without divalent ions and with 10% dialyzed fetal calf serum. The transfer of label from HPRT+ hamster fibroblasts to Lesch-Nyhan cells was as effective in the absence as in the presence of divalent ions. These results suggest that metabolic cooperation in cell cultures is not dependent on the presence of divalent ions in the medium.

Effect of Osmolarity on Metabolic Cooperation

Variations in the osmotic strength of the medium affect the permeability of cells and may also alter the organization of cell surfaces. In certain tissues ionic coupling is abolished by incubating the tissues in hypertonic solution (24). The effects on metabolic cooperation of increasing the osmolarity of the medium from 0.30 (Fig. 10a) to 0.45 (Fig. 10b) and 0.60 (Fig. 10c) was investigated by adding a sucrose-containing medium to cocultured mixtures of BHK hamster cells and Lesch–Nyhan fibroblasts 1 hour before labeling for 3 hours with ^{3}H-hypoxanthine. As shown in Fig. 10b, the labeling of normal hamster cells was reduced in

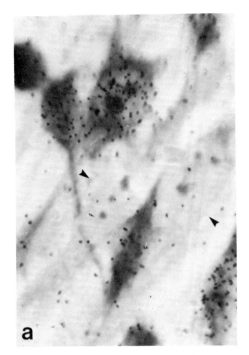

Fig. 7. Failure of X-irradiated L cells to undergo metabolic cooperation with Lesch–Nyhan cells: radioautographs of cocultured L cells and human skin Lesch–Nyhan cells. Human cells marked by arrowheads are lightly stained, whereas mouse L cells are darker. (*a*) Cocultured L cells and HPRT⁻ human cells; (*b*) giant L cells cocultured with HPRT⁻ human fibroblasts. The L cells were irradiated with 5000 rads and cultured for 10 days, becoming larger and more flattened. They were cocultured with HPRT⁻ human fibroblasts for 48 hours and exposed to 35 μCi/ml of ³H-hypoxanthine for 3 hours. There was no transfer of label from L cells or irradiated L cells to the mutant human fibroblasts.

a medium 0.45 os M with sucrose, but a proportionate amount of label was donated to the mutant human cells. At high sucrose concentrations cytopathic effects were observed and the HPRT⁺ hamster cells showed a marked reduction in the incorporation of ³H-hypoxanthine (Fig. 10*c*).

Effect of Serum Deprivation on Metabolic Cooperation

Serum is required for the growth of most cells in culture, and the removal of serum from the medium is associated with relatively prompt morphological alterations (cells, in general, become more spindle-shaped) and profound changes in cellular metabolism (marked reductions in DNA, RNA, and protein synthesis) (25). The effect of serum deprivation on the cell association necessary for metabolic cooperation was studied by incubating monolayers of normal BHK cells and Lesch–Nyhan fibroblasts in a serum-free medium for 30 minutes and then labeling them with 35 μCi/ml of ³H-hypoxanthine for 3 hours in the same medium. Cells in the serum-free medium became more spindle-shaped than replicate cultures incubated in a medium containing serum, but the efficiency of nucleotide transfer appeared to be unaltered.

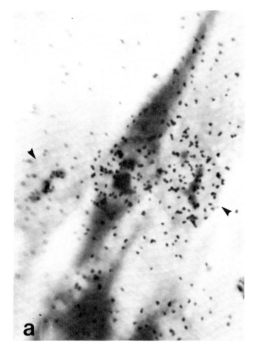

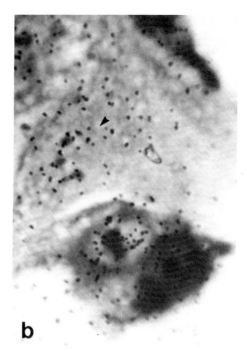

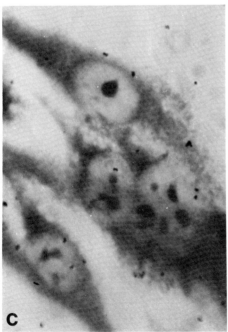

Fig. 8. Effect of Colcemid on metabolic cooperation: radioautographs of cocultured HPRT⁻ hamster and normal human skin fibroblasts. The hamster cell cytoplasm is darker staining and the nucleus has large nucleoli. The human cells (marked by arrowheads) are lighter staining. (*a*) Cocultured HPRT⁻ hamster cells and normal fibroblasts show metabolic cooperation. (*b*) Co-cultured HPRT⁻ hamster cells and normal human fibroblasts were treated with Colcemid (2.0×10^{-7} *M*) for 30 minutes and then labeled in the presence of Colcemid. Although there are morphological alterations induced by the drug and although the labeling of normal cells is reduced, the transfer of label to mutant hamster cells continued. (*c*) HPRT⁻ hamster cells incubated with ³H-hypoxanthine for 3 hours to show background labeling. Cultures were grown with 50 μCi/ml of ³H-hypoxanthine for 3 hours at 37°C.

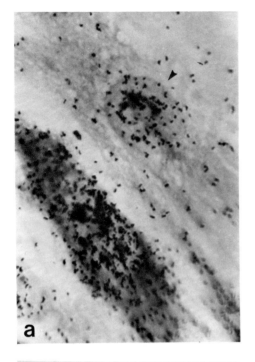

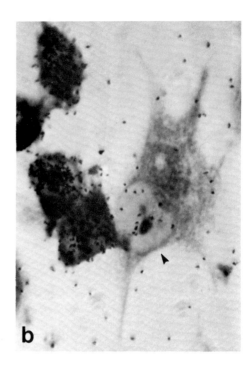

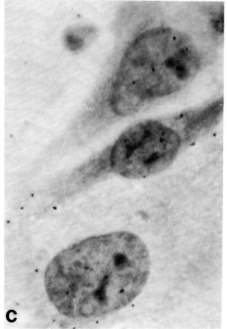

Fig. 9. Effect of cytochalasin B on metabolic cooperation: radioautographs of cocultured HPRT⁻ human fibroblasts (marked by arrowheads) and normal hamster (BHK) cells. (*a*) Cocultured BHK and Lesch–Nyhan fibroblasts show effective metabolic cooperation; (*b*) cocultured BHK and Lesch–Nyhan fibroblasts treated with cytochalasin B (10 μg/ml) for 30 minutes show the morphological alterations induced by the drug and a markedly reduced metabolic cooperation; (*c*) Lesch–Nyhan fibroblasts incubated with ³H-hypoxanthine for 3 hours show background labeling. Cultures were grown with 35 μCi/ml of ³H-hypoxanthine for 3 hours at 37°C.

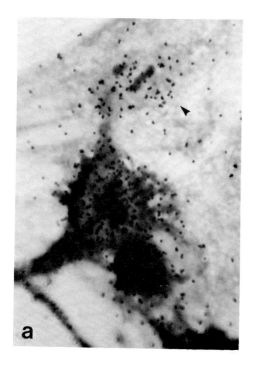

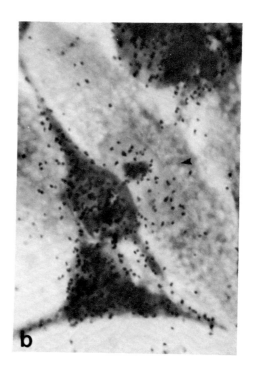

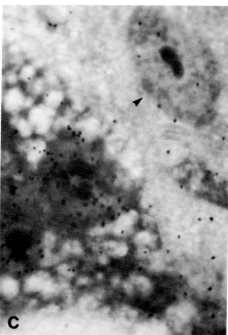

Fig. 10. Effect of osmolarity on metabolic cooperation: radioautographs of cocultured normal hamster (BHK) cells and Lesch–Nyhan fibroblasts. The hamster cells are dark-staining smaller cells which are feeding label to the paler and larger human cells marked by an arrowhead. (*a*) In a medium with an osmolarity of 0.30 os M (normal) metabolic cooperation between BHK cells and HPRT⁻ human fibroblasts was apparent. (*b*) In a medium 0.45 os M with sucrose there was reduced incorporation of label into BHK cells but a proportionate amount of label was transferred to the mutant human cells. (*c*) In a medium 0.60 os M with sucrose vacuolization of BHK and human cells, with marked reduction in the incorporation of ³H-hypoxanthine, was observed. Cultures were incubated for 30 minutes in media with different osmolarities and were then labeled for 3 hours with 35 μCi/ml of ³H-hypoxanthine at 37°C in the same medium.

EFFECT OF CELL SURFACE AND CYTOPLASMIC MEMBRANE MODIFICATON ON METABOLIC COOPERATION

Contact between donor and recipient cytoplasmic membranes is required for metabolic cooperation. A study of the effects of surface modification on metabolic cooperation was carried out to further define these cell interactions.

Effect of Neuraminidase Treatment on Metabolic Cooperation

The surfaces of mammalian cells contain sialic acid residues, which contribute to the electrical charge of the cell and also play a role in the biological properties of certain cells (26). For example, cells transformed by oncogenic viruses exhibit a marked increase in their membrane-associated sialic acid (27).

Since the chemical and biological properties of surfaces of mammalian cells may be important in the interactions required for metabolic cooperation, the effect of treating cells with neuraminidase was studied. Mixtures of normal hamster cells and HPRT$^-$ human fibroblasts were treated in a serum-free medium containing 50 units/ml of *Vibrio cholera* neuraminidase (Calbiochem Co.) for 30 minutes and then were labeled for 3 hours with 35 μCi/ml of ^{3}H-hypoxanthine in the presence of the enzyme. The cells exhibited the morphological effects similar to those of serum deprivation and became more spindle-shaped. However, there was no observable effect on metabolic cooperation.

Effect of Peroxides and Iodination of Surface Tyrosine Residues on Metabolic Cooperation

Apposition of cell membranes is required for metabolic cooperation (6). Recently a method for iodinating the surface tyrosine residues of cells has been described utilizing glucose, glucose oxidase, lactoperoxidase, and iodine (28). The iodination has been shown to be confined to surface structures (29). The chemical reaction responsible for iodination is not established. However, generation of peroxide and gluconolactone from glucose by glucose oxidase and the subsequent reduction of the peroxide with lactoperoxidase has been postulated to convert iodide to "nascent" iodine, which reacts with tyrosyl residues on the cell surfaces. The effect of iodination of BHK cells on metabolic cooperation with Lesch–Nyhan fibroblasts was investigated as described in the caption of Fig. 11. The iodination of BHK cells prevented the transfer of nucleotide to human HPRT$^-$ fibroblasts (Fig. 11e). The effect was not, however, the result of iodination of surface tyrosine residues since treatment of cells with glucose and glucose oxidase without added lactoperoxidase or iodine also inhibited metabolic cooperation (Fig. 11b). Figure 11a shows cooperation between a BHK and Lesch–Nyhan cell mixture treated by the same methods as in Figs. 11b and 11c except that glucose oxidase was omitted from the reactants. Nucleotide transfer occurs with the usual efficiency in this control (Fig. 11a). The compound responsible for interfering with metabolic cooperation was demonstrated to be hydrogen peroxide. The incubation of BHK cells with 1×10^{-5} M hydrogen peroxide in Hanks' solution, under conditions similar to those used for surface iodination (see caption of Fig. 11), prevented metabolic cooperation between treated BHK and Lesch–Nyhan cells. The incubation of BHK cells, under similar conditions, with $10^{-3}M$ gluconolactone did not inhibit or affect cooperation with Lesch–Nyhan fibroblasts.

When BHK cells were treated with either glucose and glucose oxidase (Fig.

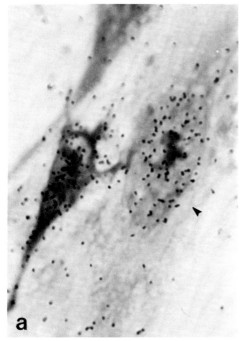

a

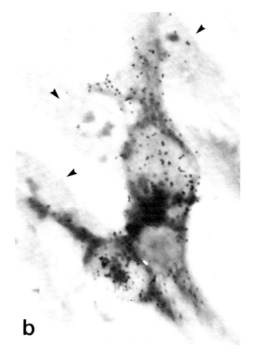

b

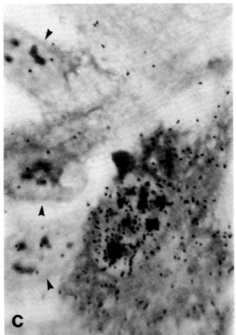

c

Fig. 11. Effect of peroxides and iodination of surface tyrosine residues on metabolic cooperation: radioautographs of cocultured normal hamster cells (BHK) treated as described below and Lesch–Nyhan skin fibroblasts (marked with arrowheads). BHK cells were detached from the glass substratum by 0.02% Versene in 0.1% dextrose and 0.05 *M* sodium phosphate buffered saline (0.15 *M*), pH 7.3 (PBS). The cells were pelleted by centrifugation at 800 *g*, washed once

in Hanks' balanced salt solution, and resuspended in 2 ml of Hanks' solution. The complete reaction mixture consisted of 0.5 ml of 1% glucose; 0.2 mg of lactoperoxidase in 0.2 ml, with an optical density ratio at 412 to 280 mm of 0.6–0.7; 0.2 ml of glucose oxidase containing 30 μg of protein and 0.05 ml of NaI (3.5×10^{-10} *M*). The coupled iodination reaction is the sum of the following two reactions:

1. glucose + O_2 $\xrightarrow[\text{oxidase}]{\text{glucose}}$ H_2O_2 + gluconolactone

2. H_2O_2 + I^- + protein $\xrightarrow{\text{lactoperoxidase}}$ iodoprotein + other products

As controls for the iodination reaction replicate BHK-cell suspensions were handled similarly, but various components of the complete reaction mixture were omitted.

(*a*) BHK cells treated in suspension with glucose alone and then cocultured with human HPRT⁻ fibroblasts for 18 hours prior to labeling with ³H-hypoxanthine. Efficient metabolic cooperation occurs in this control. (*b*) BHK cells treated with the H_2O_2-generating system (reaction 1) and then cocultured with human HPRT⁻ fibroblasts for 18 hours prior to labeling with ³H-hypoxanthine. BHK cells were unable to transfer label to Lesch–Nyhan fibroblasts. (*c*) BHK cells treated with the complete reaction mixture (reactions 1 and 2) and then cocultured with human HPRT⁻ fibroblasts for 18 hours prior to labeling with ³H-hypoxanthine. BHK cells were unable to transfer label to Lesch–Nyhan fibroblasts.

All cultures were labeled for 3 hours with 35 μCi/ml ³H-hypoxanthine at 37°C.

11b) or with hydrogen peroxide, a reduced incorporation of ^{3}H-hypoxanthine
was observed. However, the labeling of
treated BHK donor cells in these experiments was greater than or equal to that
observed in cultures incubated with
either Colcemid (Fig. 8b) or sucrose
solutions (Fig. 10) in which metabolic
cooperation was not blocked. Therefore
the reduced incorporation of ^{3}H-hypo-
xanthine into peroxide-treated BHK cells
probably cannot explain the failure of
nucleotide transfer. Peroxides may react
with surface structures required for effective cell interaction and block metabolic cooperation.

Modification of Cell Surfaces by Urea

In the presence of urea, contact inhibition of movement and growth of normal
fibroblasts is reduced; this alteration is
reversible, but reversal requires protein
synthesis (30). Cells treated with urea
are agglutinable by concavalin A, and
evidence suggests that urea has removed
from cell surface components a non-
dialyzable factor that may play a role in
the social behavior of cells (30). Because
of these relationships the effect of urea
on metabolic cooperation was studied in
communicating (BHK and Lesch–Nyhan) and in inefficiently communicating
(L cells and Lesch–Nyhan) cell mixtures. Cocultured cell mixtures were
incubated for 18 hours with 200 mM
urea solutions in complete Waymouth's
medium and were then labeled for 3
hours with ^{3}H-hypoxanthine in the same
medium. Radioautographs showed that
in communicating cell mixtures of BHK
fibroblasts and Lesch–Nyhan cells urea
treatment did not affect nucleotide transfer. Moreover, in inefficiently communicating cell mixtures of L cells and
HPRT$^-$ human fibroblasts urea also
did not promote nucleotide transfer.

Urea did, however, alter the morphology
of all three cell lines. In general, cells
grown in medium with 200 mM urea
tend to become cuboidal and are flattened and more adherent to the substratum.

Effect of Proteolytic Enzymes on Metabolic Cooperation

Since the surfaces of mammalian cells
are likely to play an important role in
determining the cell interactions necessary for communication the effects of
proteolytic enzymes on metabolic cooperation were investigated. Cocultured
mixtures of normal human fibroblasts
and HPRT$^-$ hamster cells were incubated with 0.8% crude trypsin solution
in Puck's saline A (1:250 trypsin,
GIBCo) for 2–5 minutes and were then
labeled for 2 hours with 50 μCi/ml ^{3}H-
hypoxanthine in Waymouth's medium.
The transfer of label from donor to recipient cell did not appear to be affected
by proteolytic enzymes as long as cell
contact was maintained. In areas of the
culture where the trypsin solution
caused cells to retract their processes and
separate, the transfer of label did not
occur. It would appear that proteolysis
does not affect the specific cell interactions until it causes the cells to separate
or that the structures necessary for
metabolic cooperation are rapidly re-
synthesized when cells are returned to
complete medium.

STUDIES ON THE TRANSLOCATION MECHANISMS INVOLVED IN METABOLIC COOPERATION

The exchange of nucleotides by cells in
contact has been shown to be reciprocal.
When HPRT$^-$ hamster cells were cocultured with APRT$^-$ hamster cells and

replicate cultures were exposed to either ^{3}H-hypoxanthine or ^{3}H-adenine, all of the cells in both preparations were labeled (31). The mechanisms responsible for the transfer of nucleotides and perhaps other metabolites from cell to cell by intimate contact are not known, although they may be regarded as a special type of transport. The energy requirements and possible importance of sulfhydryl-containing molecules in the translocation process were investigated.

Effect of Metabolic Inhibitors on Metabolic Cooperation

An uncoupler of oxidative phosphorylation, dinitrophenol (DNP) at a concentration of 5 mM, reduced the incorporation of ^{3}H-hydroxanthine into HPRT$^+$ cells, but the transfer of label to mutant cells still occurred despite the reduced labeling of donor cells. Two potent inhibitors of both glycolysis and oxidative metabolism, potassium azide (5 mM) and sodium fluoride (2 mM), markedly reduced the labeling of HPRT$^+$ cells, but the transfer of label to mutant cells occurred. These results suggest that energy is required for the incorporation of radioactive purines into cellular material, but the transfer of label from normal to mutant cells may be less energy dependent (6).

Effect of Sulfhydryl-Blocking Agents on Metabolic Cooperation

Sulfhydryl-containing proteins have been implicated in the translocation mechanisms of several transport systems (32, 33). Therefore the effects on metabolic cooperation of sulfhydryl-blocking agents were investigated. Confluent mixtures of normal hamster cells and Lesch–Nyhan cells were labeled with radioactive hypoxanthine in the presence or absence

of 0.1 mM p-chloromercuribenzoate. This concentration had previously been shown to markedly inhibit cation transport in mammalian cell cultures (32). However, under the conditions of the present experiments, there was some reduction of incorporation of label into HPRT$^+$ cells, but the amount of label transferred to human HPRT$^-$ cells was not significantly affected. It would appear that the process responsible for the transfer of nucleotide does not involve susceptible sulfhydryl groups.

EFFECT OF SUBCELLULAR LOCALIZATION OF LABELED PURINES IN DONOR CELLS ON EFFICIENCY OF METABOLIC COOPERATION

Subcellular compartmentalization of labeled purine derivatives may influence the efficiency of transfer from donor to recipient cell. In all of the previously presented experiments pulse labeling with ^{3}H-hypoxanthine was carried out; under these conditions label is primarily localized in the nucleus of cocultured donor and recipient cells, and the efficiency of transfer is high. In the present experiment the efficiency of metabolic cooperation was studied by comparing pulse-labeled heterospecific cocultured cell mixtures, in which normal donor and mutant recipient could be identified cytologically, with similar mixtures in which ^{3}H-purine was chased from labeled donor HPRT$^+$ cells during the 4 or 5 hours required for them to make contact with HPRT$^-$ fibroblasts. Figure 12a is a radioautograph of cocultured HPRT$^+$ hamster fibroblasts and HPRT$^-$ human cells labeled for 3 hours with ^{3}H-hypoxanthine. Most of the grains in the radioautograph of this labeled cell mixture are nuclear and nucleolar in both donor and recipient, and transfer has occurred with high efficiency. However, as seen in Fig. 12b, when the normal hamster cells were first labeled and then inoculated onto monolayers of Lesch–Nyhan fibroblasts, the transfer of nucleotide did *not* occur during the subsequent 18 hours of cocultivation. The label in these donor BHK cells, after the chase, is primarily cytoplasmic, and little label was transferred to the mutant human cells (Fig. 12b).

When the reciprocal cell mixture was studied (i.e., HPRT+ human cells as donors and HPRT− hamster cells as recipients), label was primarily nuclear, and nucleolar and nucleotide transfer occurred efficiently in cocultured cell mixtures during the three hour incubation with ³H-hypoxanthine (Fig 12c). Figure 12d shows a radioautograph of a similar cell mixture in which the HPRT+ human cells were labeled with ³H-hypoxanthine for 3 hours and then the label was chased as these cells were inoculated onto monolayers of HPRT− hamster cells. The cell mixture was cocultured for 18 hours and examined by radioautography. The label in human cells remains primarily nuclear following the chase, although some increase in cytoplasmic labeling is also seen (Fig. 12d). The transfer of label to HPRT− hamster cells occurred, although it appears less efficient than in labeling experiments where cell contact was established prior to adding ³H-hypoxanthine.

These findings raise the possibility that ³H-nucleotide or its derivatives may be transferred primarily from nucleus of donor cell to recipient or that in hamster cells that have been labeled and chased the ³H-purine has been metabolized to a product that cannot be transferred. The latter possibility cannot be excluded. However, as shown in Table 3, after labeling and a chase, both hamster and human fibroblasts contain moderately high levels of ³H-compounds soluble in trichloroacetic acid (TCA); this suggests that precursor molecules are still available. The proportion of TCA-soluble radioactivity to TCA-insoluble radioactivity in BHK cells after a labeling and chase is less than in human cells, but the amount of total soluble label is greater. At this time it is not possible to assess the significance of these differences in metabolic cooperation.

That the nuclear localization of labeled purine in donor cells seems to correlate with the efficiency of transfer of ³H-purine to recipient suggests that subcellular compartmentalization may play a role in metabolic cooperation. It is noteworthy that Pitts (4) has shown metabolic cooperation between normal hamster cells and those deficient in thymidine kinase when cocultured cell mixtures are pulse labeled with tritiated thymidine. Since the thymidine nucleotide pools are located predominantly in the nucleus (34), it suggests transport of labeled material from cell to cell via internuclear exchange or a rapid equilibrium of the transferred nucleotide between nuclear and cytoplasmic pools. In this regard Bendich, Vizoso, and Harris (35) have presented evidence of internuclear bridges between cells.

NATURE OF INCORPORATED ³H-LABEL IN DONOR AND RECIPIENT CELLS AS DETECTED BY ENZYMATIC DIGESTION

The fate of radioactive label incorporated into donor and recipient cells following growth in medium with ³H-hypoxanthine can be approximated by studying the susceptibility of the label to hydrolytic enzymatic digestion in fixed preparations. The assumption is

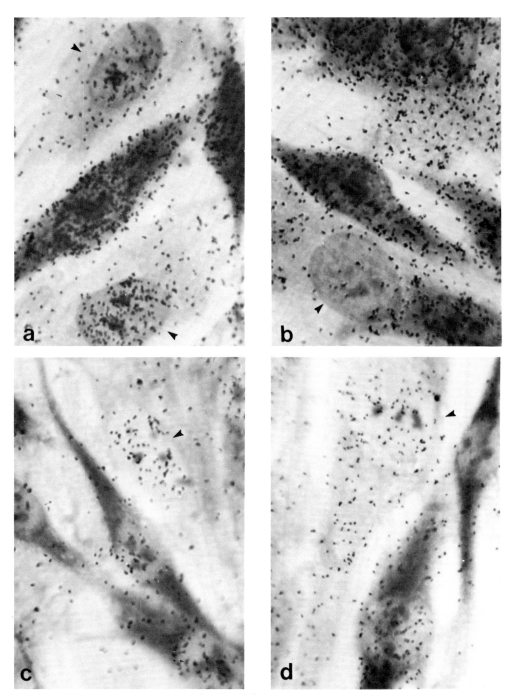

Fig. 12. Effect of subcellular localization of labeled purines in donor cells on the efficiency of metabolic cooperation: radioautographs of human and hamster (BHK) fibroblast cultures comparing metabolic cooperation in cocultured cell mixtures labeled with ³H-hypoxanthine to cell mixtures in which the donor cells were labeled with ³H-hypoxanthine and then the label was chased as donor cells made contact with potential recipients.

(a) Cocultured HPRT⁺ hamster cells and Lesch–Nyhan fibroblasts (arrowhead) labeled for 3 hours with 35 μCi/ml ³H-hypoxanthine. Label is predominantly nuclear, and there is efficient metabolic

(Continued to page 92)

Table 3. Comparison of TCA-Soluble and TCA-Insoluble Radioactive Label in Labeled and in Labeled-Chased Human and Hamster Fibroblasts Incubated with ^{3}H-Hypoxanthine

Experiment	Radioactivity (cpm $\times$ 10[-3])[a]	
	TCA-Soluble	TCA-Insoluble
Human fibroblasts[b,c]:		
Labeled (8)	8.42 $\pm$ 0.77	1.71 $\pm$ 0.7
Labeled and chased (8)	4.23 $\pm$ 0.83	3.67 $\pm$ 0.27
Hamster fibroblasts (BHK)[b,c]:		
Labeled (4)	113.65 $\pm$ 6.29	14.44 $\pm$ 3.19
Labeled and chased (4)	44.81 $\pm$ 1.12	29.85 $\pm$ 2.60

[a] The mean and standard deviations are shown.

[b] Replicate coverslip preparations were set up. Human and hamster fibroblasts were incubated for 3 hours with ^{3}H-hypoxanthine, 50 and 30 μCi/ml, respectively. The labeled coverslips were washed three times in ice-cold Hanks' buffer and placed in vials containing Bray's Scintillant. The labeled and chased coverslips were washed with buffer after labeling and then incubated in a non-radioactive medium for 18 hours prior to harvesting. They were harvested by washing three times in ice-cold Hanks' buffer and placed in vials containing Bray's Scintillant. Following counting both the labeled and the labeled and chased coverslips were removed from scintillant, washed in buffer, and fixed in ice-cold 5% TCA for 20 minutes. The coverslips were rinsed in buffer and placed in vials containing Bray's Scintillant and recounted. The differences between the first and second counts represent the TCA-soluble fraction.

[c] The number of determinations is shown in parentheses.

that extensive cleavage of a labeled macromolecule by a specific enzyme will release ^{3}H-purine, which diffuses out of the cells, reducing the grains in subsequent radioautographs. By correlating the specificity of the enzyme used with the reduction in grains and the subcellular distribution of the remaining grains it is possible to determine the macro-molecular species into which the label was incorporated. Heterospecific cell mixtures in which the normal cell can be cytologically distinguished from the mutant permit the comparison of the nature of incorporated label in donor and in recipient cells following metabolic cooperation.

Figure 13a shows that the amount and

(Continued from page 91)

cooperation. (b) HPRT[+] hamster cells were labeled for 3 hours with 35 μCi/ml ^{3}H-hypoxanthine; then the cells were inoculated onto Lesch–Nyhan fibroblasts (arrowhead), and the label was chased as BHK cells made contact with Lesch–Nyhan fibroblasts. Radioactive label in BHK cells is predominantly cytoplasmic, and little label is transferred to HPRT[-] human cells. (c) Cocultured HPRT[+] human cells (arrowhead) and HPRT[-] hamster fibroblasts labeled for 3 hours with 50 μCi/ml ^{3}H-hypoxanthine. Label is predominantly nuclear, and there is efficient metabolic cooperation. (d) HPRT[+] human fibroblasts (arrowhead) were labeled for 3 hours with 50 μCi/ml ^{3}H-hypoxanthine; then the cells were inoculated onto HPRT[-] hamster cells, and the label was chased as human fibroblasts made contact with mutant hamster fibroblasts. Radioactive label is predominantly nuclear in human cells, although some increase in cytoplasmic labeling is also seen. Metabolic cooperation occurs with moderate efficiency between the HPRT[+] human donor and the HPRT[-] hamster recipient.

Cell cultures were labeled for 3 hours with 35 μCi/ml ^{3}H-hypoxanthine at 37°C as described.

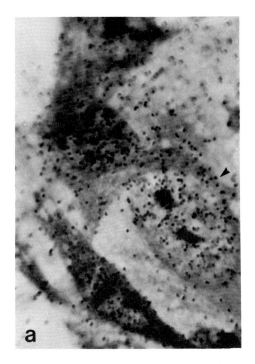

a

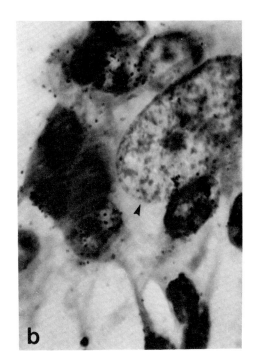

b

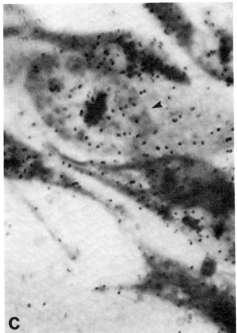

c

Fig. 13. Comparison of RNase and DNase treatment with respect to release of incorporated ³H-purine from fixed preparations of cocultured HPRT⁺ hamster donor and HPRT⁻ human recipient fibroblasts (marked by arrowheads) after metabolic cooperation: cultures were incubated with 20 µCi/ml ³H-hypoxanthine for 3 hours at 37°C followed by fixation in methyl alcohol overnight. (*a*) Control—no enzyme treatment; (*b*) RNase 1.7 µg/ml for 30 min; (*c*) DNase 2 µg/ml for 30 min. The enzyme preparations were in 0.04 *M* Tris–0.9% saline containing 4 m*M* $MgCl_2$ at pH 7.2. From R. P. Cox, M. R. Krauss, M.E. Balis, and J. Dancis, *Exp. Cell Res.* **74**, 251 (1972).

distribution of grains from incorporated ^{3}H-hypoxanthine are the same in normal hamster fibroblasts and Lesch-Nyhan cells following metabolic cooperation. The label appears to be incorporated into similar macromolecules in the donor and recipient cell, as shown by the effects of RNase (Fig. 13b) and DNase (Fig. 13c) on replicate cell mixtures. Incubation of fixed coverslip preparations in solutions containing RNase markedly decreased the number of grains in both cytoplasm and nucleus of donor and recipient cells following metabolic cooperation. The basophilic staining characteristic is also decreased in these cells, supporting the loss of nuclear RNA in the treated cells. The incubation of similar preparations with solutions of DNase caused only moderate diminution of labeling in both donor and recipient nuclei. The similarity of subcellular distribution and reduction of radioactive label in HPRT[+] hamster cells and HPRT[−] human cells probably indicates that ^{3}H-hypoxanthine or its transferred derivative was incorporated into similar macromolecules.

CONCLUSIONS AND SUMMARY

Metabolic cooperation is a form of cell communication in which cells in contact exchange metabolites and other molecules. Intercellular communications exist between most cells that are in contact in tissue culture. Similar interactions occur *in vivo*, so that the phenomenon is not merely a tissue culture phenomenon (10,36).

From an evolutionary view cell-to-cell interactions provide multicellular organisms with an important mechanism for homeostasis and control of metabolic activity. The interconnected cell systems, rather than individual cells, may therefore constitute the functional unit.

Direct cell contact has been implicated in regulating growth, differentiation, and embryonic development. Multicellular organisms may have developed metabolic cooperation to circumvent the permeability barriers imposed by cell membranes. The development of specialized structures or junctions in areas of cell apposition provide channels for exchanging regulatory substances. Present studies on metabolic controls in eukaryotic cells have emphasized the importance of cyclic nucleotides as regulators of metabolism as well as the mediators of certain hormonal effects. The exchange of these regulatory nucleotides would ensure that interconnected cells would respond in a coordinate manner to a stimulus. The studies presented in this chapter indicate that the transfer mechanism is very resistant to perturbations of many types. Such results might be expected of a process of fundamental importance to survival in multicellular organisms. Current evidence also indicates that there is a selectivity in the specialized cell junctions with respect to the molecules transferred.

The capacity to form effective junctions for communication is an intrinsic property of the cell and is not altered by changes in cell shapes, modes of binding to the substratum, or marked alterations of the extracellular milieu. Cells of different origins exhibit quantitative differences in the efficiency of metabolic cooperation and certain lines that were formerly classified as "noncommunicating" have been shown to exchange nucleotides with a reduced efficiency. Although certain neoplastic cells fail to show electrical coupling when in contact with homologous cells or with normal cells (37,38,39), it is difficult to relate the malignant process to the lack of communication *in vitro* since in other studies tumor cells and cells transformed by oncogenic virus do show coupling (18,39,40) and metabolic cooperation (6). Perhaps differences in response to regulatory substances are responsible for the varied behavior of neoplastic cells.

Studies on the mechanism of metabolic cooperation in cell culture are likely to yield information on more general aspects of cell communication. To date most of the studies have been descriptive, but they also provide information for more analytical evaluations that may help elucidate certain fundamental processes involved in cell communication.

References

1. Subak-Sharpe, J. H., Burke, R. R., and Pitts, J. D., *J. Cell Sci.* **4**, 353 (1969).

2. Seegmiller, J. E., Rosenbloom, F. M., and Kelley, W. N., *Science* **155**, 1682 (1967).

3. Cox, R. P., Krauss, M. R., Balis, E. M., and Dancis, J., *Proc. Natl. Acad. Sci. U.S.* **67**, 1573 (1970).

4. Pitts, J. D., in *Growth Control in Cell Culture*, Ciba Foundation Symposium, Wolstenholme, G. E. W., and Knight, J., eds., Churchill & Livingstone, London, 1971, p. 89.

5. Cox, R. P., and MacLeod, C. M., *J. Gen. Physiol.* **45**, 439 (1962).

6. Cox, R. P., Krauss, M. R., Balis, M. E., and Dancis, J., *Exp. Cell Res.* **74**, 251 (1972).

7. Subak-Sharpe, J. H., in *Homeostatic Regulators*, Ciba Foundation Symposium, Wolstenholme, G. E. W., and Knight, J., eds., Churchill & Livingstone, London, 1969, p. 276.

8. Stoker, M. G. P., *J. Cell Sci.* **2**, 293 (1967).

9. Gilula, N. B., Reeves, O. R., and Steinbach, A., *Nature* **235**, 262 (1972).

10. Pitts, J. D., in *Third Lepetit Colloquium on Cell Interactions*, Silvestri, L. G., ed., North-Holland, Amsterdam, 1972, p. 277.

11. McCargow, J., and Pitts, J. D., *Biochem. J.* **124**, 48P (1971).

12. Choi, K. W., and Bloom, A. D., *Science* **170**, 89 (1970).

13. Streeter, S., Spector, E., and Bloom, A., *Birth Defects* (Original Article Series) **9**, 138 (1973).

14. August-Tocco, G., and Sato, G., *Proc. Natl. Acad. Sci. U.S.* **65**, 311 (1969).

15. Nelson, P., Ruffner, W., and Nirenberg, M., *Proc. Natl. Acad. Sci. U.S.* **64**, 1004 (1969).

16. Nelson, P. G., Peacock, J. H., and Amano, T., *J. Cell Physiol.* **77**, 353 (1971).

17. Seeds, N. W., Gilman, A. G., Amano, T., and Nirenberg, M. W., *Proc. Natl. Acad. Sci. U.S.* **66**, 160 (1970).

18. Pinto da Silva, P., and Gilula, N. B., *Exp. Cell Res.* **71**, 393 (1972).

19. Johnson, G. S., Friedman, R. M., and Pastan, I., *Proc. Natl. Acad. Sci. U.S.* **68**, 425 (1971).

20. Ghosh, N. K., Griffin, M. J., and Cox, R. P., manuscript in preparation, 1973.

21. Nelson, P. G., and Peacock, J. H., *Science* **177**, 1005 (1972).

22. Rasmussen, H., *Science* **170**, 404 (1970).

23. Crane, R. K., *Fed. Proc.* **24**, 1000 (1965).

24. Lowenstein, W. R., *Ann. N.Y. Acad. Sci.* **137**, 441 (1966).

25. Dulbecco, R., *Nature* **227**, 802 (1970).

26. Weiss, L., in *The Cell Periphery, Metastasis and Other Contact Phenomena*, Wiley, New York, 1967, p. 267.

27. Warren, L., Fuhrer, J. P., and Buck, C. A., *Proc. Natl. Acad. Sci. U.S.* **69**, 1838 (1972).

28. Schenkein, I., Levy, M., and Uhr, J. W., *Cell Immunol.* **5**, 490 (1972).

29. Baur, S., Vitetta, E. S., Sherr, C. J., Schenkein, I., and Uhr, J. W., *J. Immunol.* **106**, 1133 (1971).

30. Weston, J. A., and Hendricks, K. L., *Proc. Natl. Acad. Sci. U.S.* **69**, 3727 (1972).

31. Bürk, R. R., Pitts, J. D., and Subak-Sharpe, J. H., *Exp. Cell Res.* **53**, 297 (1968).

32. Fox, C. F., and Kennedy, E. P., *Proc. Natl. Acad. Sci. U.S.* **54**, 891 (1965).

33. Cox, R. P., *Mol. Pharm.* **4**, 510 (1968).

34. Adams, R. L. P., *Exp. Cell Res.* **56**, 49 (1969).

35. Bendich, A., Vizoso, A. D., and Harris, R. G., *Proc. Natl. Acad. Sci. U.S.* **57**, 1029 (1967).

36. Frost, P., Weinstein, G. D., and Nyhan, W. L., *J. Am. Med. Assoc.* **212**, 316 (1970).

37. Lowenstein, W. R., and Kanno, Y., *Nature* **209**, 1248 (1966).

38. Borek, C., Higashino, S., and Lowenstein, W. R., *J. Membrane Biol.* **1**, 274 (1969).

39. Sheridan, J. D., *J. Cell Biol.* **45**, 91 (1970).

40. Johnson, R. G., and Sheridan, J. D., *Science* **174**, 717 (1971).

CHAPTER FIVE

Transfer of Macromolecules Between Cells in Contact

GERALD M. KOLODNY

Radiology Research Laboratory
Massachusetts General Hospital
Harvard Medical School
Boston, Massachusetts

Examples of cell-to-cell communication exist in various situations in animal biology. Such processes as embryonic development, wound healing, antibody responses, and contact inhibition show interactions between cells *in vivo* and *in vitro*. It is possible that some or all of these processes are mediated by macromolecules that are transferred between cells and provide the signals for activation or inhibition of cellular activities.

TRANSFER OF MACROMOLECULES BETWEEN CELLS

Ideally, to investigate transfer of macromolecules between cells, one would like to have a method that would allow for biochemical analysis of the molecules transferred. Radioautographic techniques such as those used so successfully by Subak-Sharpe and collaborators (1) and others (2–4) to demonstrate metabolic cooperation have the limitation that the molecules transferred cannot be isolated and biochemically examined.

Method

Methods have now been developed to biochemically examine macromolecules transferred between cells (5). A recipient population of cells is cocultivated with a donor population of cells containing radioactively labeled macromolecules. After varying periods of cocultivation, the two populations of cells are separated and the recipient cells are examined for labeled macromolecules that have passed over from the donor cells. The two populations of cells are separable by virtue of one population being heavier than the other and thus separating on centrifugation. Cells are made heavier by having them phagocytize particles of metallic tantalum powder prior to cocultivation.

Figure 1 illustrates the techniques involved. Tantalum particles (2 μ average diameter) are placed on confluent cultures of cells. Over the next 48–72 hours the cells phagocytize these particles. These cells are then cleansed of extracellular tantalum particles by washing them over a petri dish that has multiple score lines on its surface to catch the excess tantalum. The donor and recipient cells are cocultivated for 5–20 hours during which time the tantalum-containing donor cells and the recipient cells become confluent in intimate contact with one another. The coculture is then trypsinized, and the cells are centrifuged on a 0–17% Ficoll gradient on top of a 17% Ficoll buffer. The heavy donor cells containing tantalum pass through the Ficoll to the bottom of the tube, whereas the light recipient cells float within the Ficoll gradient.

Donor Cells

Donor cells are prepared from confluent cells to which tantalum powder is added. Confluent cells are used to ensure that there will be a minimal amount of cell division since cell division leads to a decrease in tantalum particles per cell.

Subconfluent cells grown in the presence of tantalum particles will, however, grow normally as seen by light phase microscopy, with normal generation time, morphology, and contact inhibition. The tantalum is not seen in the nucleus but is restricted to the cytoplasm.

Heavy cells are used as donor, rather than recipient, cells. Donor cells, when put in mixed culture, even when cleansed of the majority of excess extracellular powder, will still have some small amount of excess extracellular powder associated with them. The recipient cells in mixed culture could take up this excess powder, become heavy, and sediment with the donor cells. If the light cells were the donor cells, these light cells taking up powder would seriously contaminate the recipient cells. On the other hand, the loss of light recipient cells to the sedimenting heavy cells would decrease rather than increase the apparent macromolecular transfer.

Control experiments show that during separation on Ficoll about 15% of the light cells sediment with the heavy cells or attach to the side walls of the tube, indicating an 85% recovery of light cells from the floating band within the Ficoll gradient. About 1–2% of the heavy cells appear with the floating cells.

Cells that ingest tantalum powder retain the particles. If they did not do so, one could object that any apparent molecular transfer observed could be explained by a loss of particles from heavy donor cells and thereby an increased contamination of light recipient cells. If heavy tantalum-containing cells are placed in dilute suspension in fresh dishes, there is no visible migration of tantalum out of the cells. Moreover, if heavy cells are mixed in culture with light fresh cells for periods of up to 2 weeks, there is no visible increase in the percentage of cells containing particles, indicating that these particles do

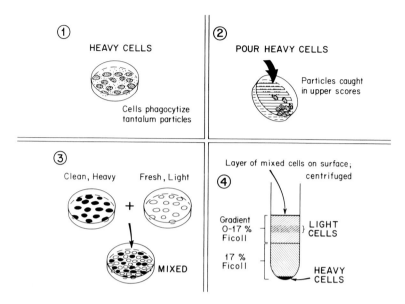

Fig. 1. Techniques for separating two populations of cells in combined culture. (*1*) Tantalum particles (1 mg/ml) placed on confluent culture, which phagocytizes particles. (*2*) To eliminate excess extracellular tantalum, culture is placed on scored plate. Excess particles are caught in scorings when the plate is tipped and washed. (*3*) Heavy cells mixed in culture with fresh cells and cocultured for 4–20 hours. (*4*) Light and heavy cells separated by centrifugation on 0–17% Ficoll gradient on top of 17% Ficoll buffer.

not pass freely between cells. Controls have also been done with heavy cells incubated in the absence of excess tantalum particles for varying periods of up to 2 weeks and then mixed with fresh light cells, the mixture being immediately separated on Ficoll. Contamination of light cells by heavy cells has always remained at about 1.5%. If the heavy cells were losing tantalum, the percentage of contamination would have increased. In addition, a doubling of tantalum concentration during preparation of donor cells increases the average number of particles per cell but does not alter the apparent percentage of macromolecular transfer observed below.

Using the above methods to cocultivate and separate two populations of cells in culture, recipient cells have been cocultured with radioactively prelabeled donor cells and then separated from the donor cells and analyzed for labeled material. The experiments to be described here on the transfer of macromolecules between cells do not reveal the quantity of cell protein or RNA or what percentage of total cell protein or RNA may be transferred intercellularly but only the percentage of these macromolecules labeled during a brief labeling period that may be transferred.

RNA Transfer

For all the biological processes exhibiting cell-to-cell interactions the information passed between the cells appears to be highly specific. Since nucleic acids with their highly specific base sequences are known to be informational molecules, the techniques described in the preceding section were used to obtain evidence of the transfer of RNA between cells in culture (5). The RNA of heavy, tantalum-containing donor cells was labeled with radioactive uridine before coculture with the recipient cells. The donor cells were then cocultured together with nonlabeled recipient cells in the presence of a large excess of unlabeled uridine. After separation of the donor and recipient cells, radioactively labeled RNA was found in the recipient cells. After a 4 hour coculture period, about 10% of the donor cell radioactivity precipitable in trichloroacetic acid (TCA) appeared in the recipient cells. After 18–20 hours, about 20% of the radioactivity in the donor cells appeared in the recipient cells. These results suggest that a significant amount of labeled macromolecular RNA was transferred from the prelabeled heavy cells to the light cells.

Hydrolysis of RNA by treatment of the donor and recipient cells with hot (90°C) TCA for 30 minutes resulted in a reduction of all counts to less than twice the background, indicating that the radioactive counts in both the donor and recipient cells were located primarily in RNA.

Cell-to-Cell Transfer of Macromolecular RNA

In several cell culture systems (6,7) "pulse-chase" experiments with uridine have been shown to be less than ideal. Although the incorporation of labeled uridine into acid-precipitable products is greatly suppressed after the addition of a large excess of nonlabeled uridine, there is still a detectable increase in incorporation. Hence the possibility still existed that the radioactive RNA seen in the recipient cells represented RNA synthesized in those cells. Experiments were therefore done in an attempt to rule out the possibility that the radio-

activity found in the recipient cells represented synthesis of RNA in those cells rather than macromolecular transfer during coculture.

ACID-SOLUBLE POOLS. If radioactive RNA seen in the recipient cells represented RNA synthesized in those cells, uridine pools in the donor and recipient cells must have been exchangeable; that is, ^{3}H-uridine in the donor cell pools must have been transferred directly into the recipient cell pools without going into the extracellular medium and being diluted there. Examination of the acid-soluble pools in both donor and recipient cells after coculture and cell separation showed radioactivity to be at the background level. It would therefore seem unlikely that these pools could account for the labeled RNA seen in the recipient cells. The background level of radioactivity in the uridine pools of the donor and recipient cells after coculture argues against the possibility that radioactive uridine which remained in these pools after donor cell labeling and which could not be diluted by excess cold uridine was responsible for the label in recipient cell RNA.

ACTINOMYCIN. Actinomycin was used to suppress RNA synthesis in recipient cells. The presence of actinomycin had no significant effect on the apparent transfer of RNA. The recipient cells in the coculture experiments had about 12% of the radioactivity seen in the donor cells after a 5 hour coculture period. To determine the maximum amount of RNA synthesis that could possibly take place in the recipient cells in the presence of actinomycin, control cultures consisting of separate cultures of donor and recipient cells were established during the time of coculture. Both of these separate cultures were maintained in the presence of actinomycin and ^{3}H-uridine equal in concentration to that used to label the donor cells. This concentration was of course many magnitudes higher than that to be found in the uridine pools in cells in the experimental cocultures. At the time of separation of donor and recipient cells from the cocultures the separate control cultures were trypsinized, mixed, and put immediately on Ficoll separation gradients. The recipient cells in these control experiments contained only 3% of the radioactivity seen in the donor cells. These experiments therefore also indicated that the labeled RNA seen in the recipient cells was not synthesized in those cells but was transferred into those cells after synthesis.

It is interesting that treatment with actinomycin does not prevent the cells from settling on the culture surface and spreading in a morphologically normal manner, indicating that for this cell function apparently little or no new DNA-dependent RNA synthesis is necessary (8).

URIDINE/CYTIDINE RATIOS. Because uridine is converted to cytidine intracellularly, RNA synthesized in the presence of ^{3}H-uridine will show a decrease in the ^{3}H-uridine/^{3}H-cytidine ratio as synthesis proceeds (9,10). If the labeled RNA seen in the recipient cells represented RNA synthesized in those cells from precursor nucleotides from the donor cells during coculture, the uridine/cytidine ratio of the RNA would be expected to be different in the donor and recipient cells. These ratios, however, were very similar, suggesting again that labeled RNA seen in the recipient cells does not represent RNA synthesized in the recipient cells *de novo*.

METHYL LABELING. The RNA of donor cells was also labeled by incubating the cells with ^{3}H-methyl methionine in a methionine-deficient medium, containing formate to prevent one carbon transfer into purine. The donor and recipient cells were cocultured for 9 hours in an excess of methionine. From 10 to 13% of the ^{3}H-methyl radioactivity in the donor cell RNA extracted by hot phenol appeared with the recipient cell RNA.

The extent of methyl label seen in the recipient cells after donor cell labeling also indicates that RNA was presumably transferred intact from donor cells. The washing of the cells followed by a cold methionine "chase" would be expected to prevent any further significant incorporation of methyl label. In addition, had there been breakdown of RNA in the donor cells, the methylated bases would not be expected to appear in newly synthesized recipient cell RNA. This is because the methylation of RNA occurs at the polymer, rather than the nucleotide, level and methylated bases are not used in the synthesis of methylated RNA (11–14).

NO INTERCELLULAR DNA TRANSFER. Although RNA appears to be exchanged between cells, these techniques with ^{3}H-thymidine label have not yielded any evidence of DNA exchange. The absence of DNA exchange makes it unlikely that the apparent RNA exchange could be due to phagocytosis or fusion with whole living or dead donor cells by recipient cells. It also makes it unlikely that some sort of aggregation of donor and recipient cells, not seen in the controls, took place. In addition, the loss of tantalum particles by donor cells, not seen in controls, and their subsequent loss of weight so that they float with the recipient cell fraction during separation on Ficoll, cannot be taken as an explanation of the results with RNA. If these spurious explanations were the cause of the apparent transfer of RNA, one would expect to have seen DNA transfer as well.

Other Considerations. There are several theoretically possible ways in which macromolecular RNA could be exchanged between cells in culture. The RNA could be transferred by cell fusion or recipient cell phagocytosis of dead donor cells. However, in these experiments, if donor and recipient cells had fused, or had recipient cells phagocytosed dead donor cells, the tantalum from the donor cells would have caused the fused cells to sediment with the donor cells, and no increase in radioactivity would have been noted in the recipient cells. Also spontaneous cell fusion is a rare event and probably could not explain the degree of RNA transfer that apparently takes place.

It is unlikely also that small bits of cytoplasm from donor cells removed at trypsinization could be transferred to recipient cells. This is because the donor cells, after being cleansed of excess tantalum powder, are sedimented before addition to coculture at 140 rpm for 1 minute. No cytoplasmic particles would sediment at such low G forces and thus be carried into coculture. Moreover, as will be seen, nucleases in medium would rapidly degrade cytoplasmic RNA. Radioautographic experiments have shown transfer of radioactively labeled donor cell RNA to the interior of recipient cells, eliminating the possibility that the apparent RNA observed is due to extracellular tags of donor cell cytoplasm adherent to recipient cells.

RNA Transfer Between Transformed Cells

Since it is possible that intercellular RNA exchange might be related to growth control in contact-inhibited cell cultures, it is of interest to know whether a transformed cell line will show cell-to-cell transfer of RNA. Loewenstein (15) and coworkers have demonstrated the absence of ionic communication between cancer cells in isolated tumor explants whose normal tissue counterparts show ionic coupling. On the other hand, Potter and associates (16) found normal electrical coupling between transformed 3T3 fibroblasts in culture.

Using some variations in the previously described techniques, transfer of RNA between transformed cells has been investigated (17). These variations required that the donor cells not be dividing during the time of tantalum uptake. If the donor cells are dividing, the tantalum powder will be diluted among daughter cells, so that cell separation will be inefficient. As soon as enough powder is ingested by a donor cell to allow it to be efficiently separated from recipient cells on Ficoll, the cell divides to yield only half as much powder per cell. Also, if the cells continue to divide after a monolayer is formed, so that they pile on top of one another, only the top cell in the pile will ingest the particles.

In order, therefore, to examine for RNA exchange between transformed cells by the above techniques, conditions were used in which the donor cell population did not divide during the time of phagocytosis of tantalum. In the first series of experiments nontransformed contact-inhibited 3T3 cells were used as the donor cells and SV-40 virus–transformed 3T3 cells were used only as the recipient cells. A second series of experiments used excess thymidine to inhibit cell division (18) in SV-40 virus–transformed 3T3 cells so that these cells would efficiently phagocytize tantalum particles. X-Irradiation was also used in a third set of experiments to inhibit cell division in transformed cells so that the cells would efficiently phagocytize particles.

Under these conditions SV-40 virus–transformed 3T3 cells accepted RNA from other donor SV-40 virus–transformed 3T3 cells treated so as to increase their efficiency of tantalum phagocytosis. The amount of RNA transferred, about 10% of the donor cell labeled RNA, was very similar to that exchanged between nontransformed cells.

It is interesting that X-irradiated transformed cells transferred less RNA than other cells in these experiments. Either X-irradiation alters the process of preparing these macromolecules for transfer or the anatomy or physiology of the cell surface juncture is altered by X-ray treatment.

Although RNA exchange between transformed cells similar to that seen in nontransformed cells does occur, this does not rule out the possibility that macromolecules exchanged between cells may be important in growth regulation. It may be that growth-regulating RNAs represent a very small part of the total RNA exchanged between nontransformed cells. Although these growth-regulating molecules may not be among those exchanged between transformed cells, they represent too small a fraction of the total to be detected by the methods discussed here. Alternatively other exchangeable molecules may act as growth regulators, and their exchange may be inhibited in transformed cells.

Intercellular Transfer of Proteins

The transfer of cytoplasmic proteins can also be demonstrated using similar techniques to those used to observe inter-cellular RNA transfer (19). Donor cells have been labeled with radioactive leucine, then placed in coculture with unlabeled recipient cells and a 1000-fold excess of unlabeled leucine. After a 5 hour period of coculture, the recipient and donor cell populations were separated on Ficoll gradients, and TCA-precipitable radioactivity in each group of cells was determined. Control experiments were also done in which recipient and donor cells were mixed and immediately separated on Ficoll. In the control experiments about 1% of the radioactivity seen in the donor cells appeared in the floating "recipient" cell fraction. In the experimental cocultures, however, about 5% of the radioactivity seen in the donor-cell fraction appeared in the recipient cell fraction, thus indicating significant intercellular protein transfer.

After trypsinization of donor cells and cleansing them of excess extracellular tantalum particles prior to coculture for examination of protein transfer, the cells are sedimented at 140 rpm for 1 minute, as in the techniques described for observing intercellular RNA transfer. This procedure prevents the addition to the coculture of any labeled cellular particles or debris that could be taken up by the recipient cells. It is possible that some of the apparent intercellular protein transfer was the result of fragments of cell membranes adhering to neighboring cells on trypsinization and cell dispersion of the cocultures prior to separation on Ficoll. However, it is unlikely that this could account for very much transfer inasmuch as these proteins lying on the cell surface would be expected to be digested by the trypsin treatment. Also the presence of these adhering proteins would not account for the intercellular nuclear protein and histone transfer, as described below, that takes place in cocultures.

Transfer of Proteins to Recipient Cell Nuclei

Various studies have been reported on the effects of both acidic nuclear proteins and histones on DNA transcription (20–24). Salas and Green (25) have on the basis of their studies suggested that proteins binding to DNA regulate cell growth. Bonner and associates (24) believe that histones in conjunction with "chromosomal" RNA act in gene regulation. Others (20,21) have shown evidence of changes in nuclear acidic proteins during changes in the patterns of DNA transcription. Specific differences in the nonhistone nuclear proteins and their tissue-specific interaction with DNA have also been reported (26–28).

If nuclear proteins do in fact regulate gene transcription, evidence of the transfer of proteins to recipient cell nuclei would suggest the possibility that cells could influence their neighbors in processes requiring changes in patterns of gene expression by cell-to-cell transfer of proteins to cell nuclei.

Examination of recipient cell nuclei from coculture experiments has shown that proteins are also transferred to cell nuclei. About 0.52% of the radioactivity in the heavy donor cell fraction was seen in the recipient cell nuclei, whereas in the control experiments only 0.18% of the heavy cell radioactivity was seen in the nuclei from the "recipient" cell fraction of the Ficoll gradients.

Examination of the proportion of label in the nucleus compared to the cytoplasm showed a difference in the donor cells as compared to the recipient cells. In the donor cell population about 14% of the labeled proteins were located in the nuclei. In the recipient cells a smaller percentage, about 11% of the radioactive proteins, appeared in the nuclei, suggesting a proportionately greater intercellular transfer of proteins to the cytoplasm than to the nucleus.

The nuclear proteins were further examined to see whether any particular nuclear protein fraction was transferred to a greater or lesser extent than any other fraction. Donor cell proteins were labeled with ^{3}H-leucine and recipient cell proteins with ^{14}C-leucine. Nuclear proteins were then extracted from recipient cell nuclei after a 5 hour coculture. These proteins were fractionated, and the ratio of ^{3}H/^{14}C was determined for each of the nuclear protein fractions. The ratio of ^{3}H-labeled histones to ^{14}C-labeled histones within the recipient cell nuclei was found to be greater than the ^{3}H/^{14}C ratios of any of the other nuclear protein fractions.

These results suggest that histones were transferred into recipient cell nuclei to a greater extent than other proteins. Although it is possible that nuclear proteins are transferred from one nucleus to another, it would seem more likely that donor cell nuclear proteins were synthesized in the donor cell cytoplasm, then transferred to recipient cell cytoplasm and from there entered recipient cell nuclei.

To analyze the extent of transfer of the histone subfractions, the histones within the recipient cells have been analyzed on acrylamide gels. Donor cells were labeled with ^{3}H-lysine and recipient cells with ^{14}C-lysine in lysine-free media. After coculture for 5 hours, the donor and recipient cells were separated and the recipient cell nuclei were isolated. Histones were extracted from recipient cell nuclei, and the dual labeled histones were subfractionated on acrylamide gels. Radioactive gel electrophoretograms were obtained after slicing the gels and plotting the ^{3}H and ^{14}C dpm. Peaks representing all of the histone subfractions were seen, indicating transfer of all the histone subfractions. Plots of ^{3}H and ^{14}C dpm across the gels also paralleled one another. No major differences existed therefore in the proportion of each subfraction present in the donor cells that was transferred. The same percentage of each labeled histone subfraction present in the donor cells was transferred.

The nuclear extraction procedures used results in nuclei that under phase microscopy appear clean of cytoplasmic tags. However, it is still possible that some of the proteins extracted from these nuclei represent cytoplasmic proteins adhering to the nuclear membrane. It is unlikely, however, that these proteins represent a very large portion of the proteins designated nuclear proteins because the nuclei were cleansed until no cytoplasmic tags were visible. Moreover, labeled histones can be identified in discrete bands on the acrylamide gels, which would not be expected were there extensive cytoplasmic contamination.

These studies, except for the case of histones, reveal little about the molecular size of the transferred proteins. Histones from 3T3 cells are presumably like all the other histones so far studied: small molecules with a chain length of 100–225 amino acids (29). It is not surprising therefore that among the nuclear proteins the greatest amount of transfer was found in the histone fraction, which, because of its relatively small molecular size, would be expected to show greater diffusion than other larger protein molecules.

INTERCELLULAR MACROMOLECULAR TRANSFER VIA MEDIUM

Medium Protein

Experiments have been done to determine what portion if any of the labeled protein that is transferred between cells in culture could be transferred by first being secreted into the medium and then taken up from the medium by recipient cells, rather than being directly transferred through intercellular contacts. Cells were labeled overnight, and the medium was removed; these donor cells were then trypsinized and replated in fresh media to simulate the conditions of coculture. After 5 hours of incubation, the fresh media were removed and transferred to new recipient cells that had just been trypsinized and replated. Five hours later the TCA-precipitable radioactivity in the recipient cells was determined. About 0.64% of the radioactivity seen in the donor cells was present in the recipient cells, far less than that seen to be transferred between cells in contact. Several experiments showed no loss of TCA-precipitable radioactivity in 5–10 hours in media taken off prelabeled cultures, indicating that no significant breakdown of protein occurred in the media. Since they do exhibit stability in the medium, the lack of appreciable protein transfer via the medium cannot be attributed, in those 10 hour control experiments, to degradation of labeled protein that would not have been seen in the 5 hour coculture experiments.

Labeled proteins found in the media could be metabolic waste products, proteins important for such cell functions as adhesion (8) or tissue construction, or could represent macromolecules important in intercellular interactions in spite of the fact that they represent such a small percentage of the total intercellular protein transfer. For example, Fratantoni et al. (30,31) found that defects in Hunter and Hurler cells in culture could be corrected by media taken from normal cultures. The corrective factor appears to be a protein (32). In studies on the correction of C-4 deficiency (33) the corrective factor within the medium is not as yet as well characterized.

The source of proteins found in the media could be from active cellular secretion, leakage from dead or dying cells, or due to cell damage caused by trypsinization.

Medium RNA

It is possible that some of the RNA exchanged between cells could pass first from donor cells into the medium and then be taken up from the medium by recipient cells. Cocultures established at one-half, one-fourth, and one-eight the usual cell density show, however, a progressively decreased rate of RNA transfer per cell (Table 1), indicating that the major portion of intercellular RNA transfer probably occurs between cell contacts, and not via the medium.

Medium RNA has also been studied directly (34). Cells were incubated in a medium containing ^{3}H-uridine for 24 hours. After discarding the labeling medium and adding new medium for 8 hours, the RNA from the fresh medium was precipitated with TCA. About 0.65% of the radioactivity in the cells was present in the medium. This concentration did not vary appreciably as the serum concentration was varied from 0 to 10% calf serum. The RNA was also precipitated from the labeling medium itself after a 24 hour labeling period, and this medium also contained about 0.75% of the TCA-precipitable radioactivity seen in the cells incubated in that medium.

To check for the stability of medium RNA, medium was removed from labeled cells and incubated in the absence of cells. At intervals over the succeeding 48 hours aliquots of media

Table 1. Transfer of TCA-Precipitable Radioactivity

Cell Dilution	Radioactivity (cpm)		cpm Light/Heavy (%)
	Light Cells	Heavy Cells	
None	660	6643	10.0
1:2	565	6867	8.3
1:4	614	9070	6.8
1:8	119	5463	2.2
Control	98	6753	1.4

NOTE: Donor cells were labeled for 5 hours with 5 μCi/ml ^{3}H-uridine and then placed in coculture with unlabeled recipient cells. Cocultures were established using cells trypsinized from one confluent dish of recipient cells mixed with donor cells cleaned of excess tantalum after being trypsinized from one confluent dish (no dilution). Similar mixtures containing the same number of cells were each also divided to make two (1:2) cocultures, four (1:4) cocultures, and eight (1:8) cocultures. After incubation for 5 hours cocultures were separately pooled from the 1:2, 1:4, and 1:8 cocultures, and the donor and recipient cells were separated in each group. The control group consists of donor and recipient cells maintained separately instead of cocultured and mixed just prior to separation on Ficoll. As the cell density in the cocultures decreases the extent of transfer of TCA-precipitable radioactivity also decreases, suggesting that the majority of transfer occurs through cell-to-cell contacts.

were taken, the RNA was precipitated with TCA, and the radioactivity was determined. There was no decline in the radioactivity in each of these aliquots over a 48 hour period, indicating that medium RNA is quite stable.

No differences were found in the amount, stability, methylation, or size range of medium RNA from normal 3T3 cells in comparison with virally transformed 3T3 cells. The morphologic and biochemical changes associated with vrial transformation cannot therefore be linked to any properties of medium RNA.

Figure 2 is a sucrose density gradient profile of the phenol-extracted RNA from the media. Media from confluent and subconfluent 3T3 and SV-40 virus–transformed 3T3 cells all contained RNA that gave a similar density profile. As can be seen from this figure, the RNA from medium overlying labeled cell cultures sediments predominantly in a region containing carrier 4-5S RNA.

Medium RNA is fractionated into three components on Sephadex G-100. The major component is in the size range 5S RNA with a smaller amount in the range 4S. In addition a minor component of medium RNA is greater than 5S and is in the size range 7–8S RNA.

Because of the possibility that the RNA in the medium might represent degraded ribosomal RNA, ribosomes have been isolated from cells prelabeled

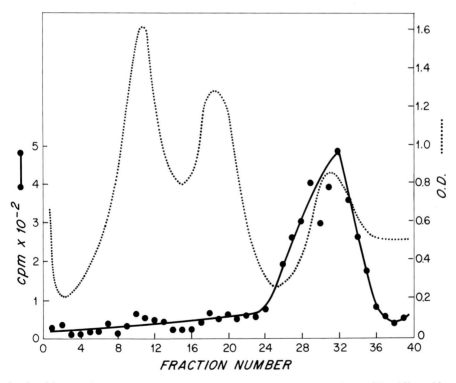

Fig. 2. In this experiment 3T3 cells in culture were labeled with 5 μCi/ml ³H-uridine. After 16 hours the RNA in the medium was extracted with carrier cellular RNA and sedimented on a 5–30% sucrose gradient. Radioactivity and optical density at 260 mμ of gradient fractions (0.2 ml) are plotted. Medium RNA sediments in the low-molecular-weight region of the gradient.

with ^{14}C-uridine. These ribosomes were incubated for 4 hours in a medium that had been taken from cell cultures labeled with ^{3}H-uridine. The RNA from the medium containing the labeled ribosomes was then extracted with phenol and placed on sucrose gradients. The ^{14}C radioactivity from the ribosomal RNA incubated in medium cosedimented with the ^{3}H radioactivity of medium RNA, which is consistent with the possibility that RNA in the medium may be degraded ribosomal RNA.

Medium RNA labeled with ^{3}H-methyl also cosediments with ^{14}C-uridine-labeled RNA in the medium. Medium RNA has a much greater extent of methylation than the major intracytoplasmic RNA species. Medium RNA is methylated 14 times more than 28S RNA and 4–6 times more than t-RNA.

One approach to identifying the source of medium RNA is to identify the methylated components of this RNA and compare it with the methylated components of known cellular t-RNA and r-RNA. It has previously been shown that t-RNA is rich in methylated guanine and contains very little methylated ribose (35); likewise, 5-methylcytosine and thymidine are found primarily in t-RNA (35). Ribosomal RNA, on the other hand, is very rich in methylated ribose and contains 6-methyladenine and 6-dimethyladenine, which are not major components in t-RNA (36).

The methylation pattern of medium RNA is quite simple in comparison with the complex patterns seen in t-RNA and r-RNA. Instead of a whole series of peaks on chromatography, as seen with intracellular RNA, only two major peaks are observed with medium RNA. The first peak is assumed to be a methylguanine isomer since methylated guanines are the only major components of

either t-RNA or r-RNA that migrated in that region of the chromatogram. The second peak has not yet been characterized.

These methylation studies suggest that the source of medium RNA is not t-RNA or r-RNA. On the other hand, as mentioned above, incubation of ribosomes in medium leads to degradation of r-RNA to low-molecular-weight RNA that cosediments with medium RNA. It is known that phosphate bonds adjacent to methylated nucleotides are resistant to the action of ribonucleases (37–40). It is possible therefore that medium RNA represents r-RNA degraded in a fashion that would degrade all the methylated regions of RNA except those containing those two methylated nucleotides seen in the methylation patterns of medium RNA.

It is possible that medium RNA represents unique species of low-molecular-weight RNA. Several studies have been reported on low-molecular-weight RNA from cell nuclei (41–43), including the nuclei of 3T3 cells (44). These species of nuclear RNA, like medium RNA, are reported to be extensively methylated. Slavkin et al. (45) have described several species of low-molecular-weight methylated RNAs in the intercellular matrix of embryonic rabbit tooth primordia. They describe four distinct species varying from 2 to 7S in size. Several investigators have also found evidence of RNA associated with cell surfaces (46–48). It is possible that medium RNA is derived from this RNA or that membrane RNA represents medium RNA in the process of secretion.

The role of medium RNA is unknown. It could be intracell surplus or merely conveniently sized packages of degraded RNA for disposal outside the cell. Another alternative is that it represents informational molecules in transit from one cell to another. Studies have shown that extracellular RNA can produce changes in cell functions (49-51).

It is unlikely, however, that medium RNA is responsible for any large fraction of the RNA that has been found to be transferred by cells in contact as described here. Medium RNA is not present in sufficient quantity to account for any large portion of the RNA transferred during coculture. In addition, representatives of all major species of RNA have been found to be transferred intercellularly, whereas medium RNA contains only low-molecular-weight species. Nevertheless, it is still possible that significant processes in contact inhibition, embryonic development, and the like are mediated by medium RNA.

CONTAMINATION

In all the studies on macromolecule transfer between cells and in the studies on macromolecules in the culture media, repeated tests were done for contamination by PPLO, mouse leukemia virus, or other microorganisms. Contamination by mycoplasma was checked for both by the culture method of Dienes (52) and by a biochemical assay using radioautography to assay for ^{3}H-thymidine-labeled mycoplasma within the cytoplasm or at the cell periphery. Assays at periodic intervals failed to detect the presence of these contaminants. Therefore the observed macromolecules within the media and those exchanged between cells cannot be attributed to contamination of the cell cultures.

CELL JUNCTIONS

Macromolecules could pass between cells by going through membrane junctions between the cells. The sites of junctions between cells have been studied by several laboratories (53–56) using the electron microscope. Tight junctions, intermediary junctions, and desmosomes have been identified, but it is difficult to be certain where along the intercellular junctions exchange may be taking place.

The electron-microscopy studies of Bullivant and Lowenstein (57) and Revel and Karnovsky (55) have shown junctions between cells with channels of about 90 Å in diameter between closely opposed cell membranes. It is possible that macromolecules exchanged between cells traverse these channels. Bendich, Vizoso, and Harris (58) have presented histologic evidence of intercellular connections between cells with, in some cases, long processes extending from one cell to the next. These processes are shown by radioautography to contain ^{3}H label from thymidine and are considered to be transferring cytoplasmic and nuclear material from cell to cell. Although these authors present their microscopic data as evidence of DNA transfer between cells, the other studies outlined here show no evidence of DNA transfer.

SIGNIFICANCE

The transfer of macromolecules between cells indicates that cells may be open to the intercellular passage of material both structural and informational. Effects initiated in one cell could possibly pass to neighboring cells and an entire tissue. The behavior of a cell within a tissue may not necessarily be that of a cell acting alone in tissue culture without contact with its neighbors. Also the behavior of an entire tissue may not necessarily be the sum behavior of individual cells acting entirely alone. Macromolecules exchanged between cells might be important mediators in the related phenomena of contact inhibition of movement and division seen when cells are in close contact, wound healing, cellular interactions seen during development, and antigen–antibody interactions during cellular immune processes.

REFERENCES

1. Subak-Sharpe, H., et al., *J. Cell Sci.* **4,** 353 (1969).
2. Cox, R. P., et al., *Proc. Natl. Acad. Sci. U.S.* **67,** 1573 (1970).
3. Ashkenazi, Y. E., and Gartler, S. M., *Exp. Cell Res.* **64,** 9 (1971).
4. Fumimoto, W. Y., and Seegmiller, J. E., *Proc. Natl. Acad. Sci. U.S.* **65,** 577 (1970).
5. Kolodny, G. M., *Exp. Cell Res.* **65,** 313 (1971).
6. Cooper, H. L., *J. Biol. Chem.* **243,** 31 (1965).
7. Warner, J. R., et al., *J. Mol. Biol.* **19,** 349 (1966).
8. Kolodny, G. M., *Exp. Cell Res.* **70,** 196 (1972).
9. Hurlbert, R. B., and Potter, V. R., *J. Biol. Chem.* **195,** 257 (1952).
10. Scott, J. F., *Biochim. Biophys. Acta* **61,** 62 (1962).
11. Borek, E., *Cold Spring Harbor Symp. Quant. Biol.* **28,** 139 (1963).
12. Borek, E., and Srinivasan, P. R., *Ann. Rev. Biochem.* **35,** 275 (1966).
13. Hurlwitz, J., et al., *J. Biol. Chem.* **240,** 1256 (1965).
14. Srinivasan, P. R., and Borek, E., *Prog. Nuclear Acid Res.* **5,** 157 (1966).
15. Loewenstein, W. R., *Dev. Biol. Suppl.* **2,** 151 (1968).
16. Potter, D. D., et al., *Proc. Natl. Acad. Sci. U.S.* **55,** 328 (1966).
17. Kolodny, G. M., *J. Cell Phys.* **79,** 147 (1972).
18. Xeros, N., *Nature* **194,** 682 (1962).
19. Kolodny, G. M., *J. Mol. Biol* **78,** 197 (1973).
20. Rovera, G., and Baserga, R., *J. Cell Phys.* **77,** 201 (1971).
21. Baserga, R., and Stein, G., *Fed. Proc.* **30,** 1752 (1971).
22. Kamiyama, M., and Wang, T. Y., *Biochim. Biophys. Acta* **228,** 563 (1971).
23. Allfrey, V. G., et al., *Proc. Natl. Acad. Sci. U.S.* **49,** 414 (1963).
24. Bonner, J., et al., *Science* **159,** 47 (1968).
25. Salas, J., and Green, H., *Nature New Biol.* **229,** 165 (1971).
26. Gilmour, R. S., and Paul, J., *FEBS Letters* **9,** 242 (1970).
27. Wang, T. Y., *Exp. Cell Res.* **69,** 217 (1971).
28. Spelsberg, T. C., and Hnilica, L. S., *Biochem. J.* **120,** 435 (1970).
29. DeLange, R. J., and Smith, E. L., *Ann. Rev. Biochem.* **40,** 279 (1971).
30. Fratantoni, J. C., et al., *Science* **162,** 570 (1968).
31. Fratantoni, J. C., et al., *Proc. Natl. Acad. Sci. U.S.* **64,** 360 (1969).
32. Barton, R. W., and Neufeld, E. F., *J. Biol. Chem.* **246,** 7773 (1971).
33. Colten, H. R., *Proc. Natl. Acad. Sci. U.S.* **69,** 2233 (1972).
34. Kolodny, G. M., et al., *Exp. Cell Res.* **73,** 65 (1972).
35. Iwanami, Y., and Brown, G. M., *Arch. Biochem. Biophys.* **124,** 472 (1968).
36. Iwanami, Y., and Brown, G. M., *Arch. Biochem. Biophys.* **126,** 8 (1968).
37. Bayev, A. A., et al., *Biokhimiya* **28,** 931 (1963).
38. Gray, M. W., and Lane, B. G., *Biochim. Biophys. Acta* **134,** 243 (1967).
39. Michelson, A. M., and Pochon, F., *Biochim. Biophys. Acta* **114,** 469 (1966).
40. Brownlee, G. G., et al., *J. Mol. Biol.* **34,** 379 (1968).
41. Dingham, C. W., and Peacock, A. C., *Biochemistry* **7,** 659 (1968).
42. Nakamura, T., et al., *J. Biol. Chem.* **243,** 1368 (1968).
43. Weinberg, R. A., and Penman, S., *J. Mol. Biol.* **38,** 289 (1968).
44. Rein, A., and Penman, S., *Biochim. Biophys. Acta* **190,** 1 (1969).
45. Slavkin, H. C., et al., *Dev. Biol.* **23,** 276 (1970).
46. Beierle, J. W., *Science* **161,** 798 (1968).
47. Bennett, M., et al., *J. Cell Physiol.* **74,** 183 (1969).
48. Snow, C., and Allen, A., *Biochem. J.* **119,** 707 (1970).
49. Amos, H., *Biochem. Biophys. Res. Commun.* **5,** 1 (1961).

50. Galand, P., and Ledoux, L., *Exp. Cell Res.* **43,** 391 (1966).

51. Niu, M. C., et al., *Proc. Natl. Acad. Sci. U.S.* **47,** 1689 (1961).

52. Madoff, S., *Ann. N.Y. Acad. Sci.* **79,** 383 (1960).

53. Farquahar, M. G., and Palade, G. E., *J. Cell Biol.* **17,** 375 (1963).

54. Fawcett, D. W., *Exp. Cell Res. Suppl.* **8,** 174 (1961).

55. Revel, J. P., and Karnovsky, M. J., *J. Cell Biol.* **33,** 7 (1967).

56. Gilula, N. B., et al., *Nature* **235,** 262 (1972).

57. Bullivant, S., and Loewenstein, W. R., *J. Cell Biol.* **37,** 621 (1968).

58. Bendich, A., et al. *Procl. Natl. Acad. Sci. U.S.* **57,** 1029 (1967).

CHAPTER SIX

Cellular Interactions in Skin

C. N. D. CRUICKSHANK

M.R.C. Unit on Experimental Pathology of Skin
The Medical School
The University of Birmingham
Birmingham, England

The skin is an ideal tissue for studying cellular interactions. The epidermis contains cells that have three distinct origins and functions, and interact with each other. Moreover, interactions can be shown to take place between the epidermis and the underlying dermis. In the epidermis the principal cell type is the keratinocyte, which has the primary function of producing the hard, relatively impervious outer horny layer, hair, and nails. Keratinocytes are derived from the embryonic ectoderm. Under normal conditions divisions of these cells in the basal layer give rise to daughter cells, one of which progresses to the surface, during which time it undergoes the various changes associated with keratinization.

The melanocytes are derived from the neural crest and migrate into the epidermis in early embryonic life. Their function is to manufacture pigment (melanin) granules, which they impart to keratinocytes to afford protection from solar radiation. In lightly pigmented species they can be distinguished by the DOPA reaction.

The origin of the third type—the Langerhans cells, or "high level clear cells"—is in doubt. Because of their resemblance to the cells of the tumor histiocytosis X, they are regarded as probably being of mesenchymal origin. Their characteristic histochemical reaction (in the human) is the ability to stain with 5-nucleotidease (ATPase). Their function has not yet been adequately defined, but the most likely function seems to be an involvement in the control of keratinization. Another ingenious theory is that, if they are of mesenchymal origin, they may be mediators of dermal influences within the epidermis.

The three cell types in the epidermis have well-defined spatial relationships. The melanocytes are normally interspersed with the cells of the basal epidermal layer and are in contact with the basal lamina. Unlike the keratinocytes, which develop hemidesmosomes, they have no such distinctive junction points. Their dendrites ramify between cells of the basal and spinous layers making contact with a number of keratinocytes that varies from site to site but is on the order of 15–20. This has become known as the epidermal melanin unit (1).

The Langerhans cells are interspersed between cells of the spinous layer and are oriented vertically. They too have no distinctive attachment points with neighboring cells, and their dendrites spread laterally but mainly in a vertical plane reaching from the basal layer to the stratum granulosum.

Recently it has been demonstrated that under normal circumstances the keratinocytes, particularly of the granular and horny layer, are arranged in orderly columns in most body sites and that the apparent disarray has been due to histological artifact (2,3). Something on the order of four cells of the basal layer cover the equivalent lengths of the cells of the granular and horny layers.

The skin is not a homogeneous tissue but a collection of organs. Apart from the different cell lineages, the epithelial cells have been modified to form hair follicles, sebaceous glands, and sweat glands. The skin is readily accessible for observation and experimentation *in vivo* and can be studied by organ culture and monolayer culture *in vitro*. The following account of cellular interaction will be confined to adult skin. In embryo skin the process of development complicates intercellular relationships. Much work of importance, however, has been done on cell interactions in embryo skin (4,5). A distinction too must be drawn between (a) cellular changes brought about by "embryonic inducers" (and their equivalents in the adult), which are lasting, and (b) modulations such as may be caused by hormones and vitamins, which require the continued presence of the stimulant.

DERMAL–EPIDERMAL INTERACTIONS

The commonest example of the influence of the dermis on epidermal cells is encountered in the healing of superficial abrasions and burns. If the original epidermis is completely lost as a result of trauma, epithelial cells derived mainly from hair follicles but also from the necks of the sebaceous gland migrate onto the cut surface of the dermis and eventually form a confluent sheet. The cells originally devoted to the production of hair and sebum now proceed in their new location to behave exactly as the original keratinocytes of the epidermis and produce normal surface keratin. If the lesion is deeper than the bottoms of the hair follicles, repair in this way can occur from sweat gland epithelium, but in this instance the procedure takes considerably longer both in the migratory stage and in the final formation of a horny layer.

Much light has been thrown on dermal–epidermal interactions by the recombination grafting experiments of

Billingham and Silvers (6,7). By separating epithelium from such body sites as the trunk and sole, and recombining with dermis from differing regions and implanting on a "neutral" site, they showed that the dermis acted as the determinant of the behavior pattern of the overlying epidermis. For example, ear epidermis, which is normally thin and lightly keratonized, converted to a thick, heavily keratinized epithelium with the histological characteristics of the sole when recombined with sole dermis. Reversing the orientation of the dermis produced the same effect, indicating that the effector agent was not confined to the superficial regions.

However, corneal epithelium grafted onto the trunk retains its original characteristics. This observation indicates that there are limits to the effectiveness of the dermal influence, presumably if the epithelium has become too highly specialized.

One of the most interesting studies on epithelial mesenchymal interactions has been carried out by Oliver (8–12) using the rat vibrissa. These have a regular arrangement that makes individual follicles readily definable. Initial experiments showed that, after removal of the dermal papilla from the hair bulb by microdissection, new dermal papillae were regenerated and that several generations of normal whiskers grew. Even after removal of the whole of the lower third of the vibrissa follicle, dermal papillae were regenerated and the growth of short whiskers occurred. If more than the lower third was removed, the remaining length was unable to induce papillae, and whisker formation was not obtained. If dermal papillae removed from other follicles were implanted into the cut ends of these remaining lengths of follicle tubes, the epithelial cells previously unable to produce whiskers now did so.

In a further series of experiments various recombinations of dermal papillae and follicle tubes were made under the ear skin of the same animal. Segments of plucked follicle wall consisting of the outer root sheath and the mesenchymal layer derived from the lower third of the follicle regenerate papillae and whiskers in the ear skin. It was found that if the follicle were slit longitudinally to form a sheet, the dermal papilla could not induce whisker formation. Instead in some long-term experiments some of the flattened sheets had become rounded, often keratinizing into an eccentrically placed lumen. This suggests that some spatial relationship is required for the induction of whisker growth.

Implants of ear epidermis combined with dermal papillae provided no convincing evidence of the induction of whiskers. More conclusively, combination of dermal papillae with scrotal epidermis, which is non-hair-bearing, was negative. However, in both instances evidence was obtained that the dermal papillae induced a reorganization of the epidermis into irregular "matrices" of cells around them.

Apart from the obvious problem of spatial relationships, it is possible that either ear and scrotal epidermis is less susceptible to the influence of the dermal papilla or that possibly the local dermis exerted an overriding influence. Although some level of organization could be achieved, the more specialized function of whisker production failed.

Most of these studies suggest that the dermis has the dominant role, but the induction of new dermal papillae indicates that in some situations there is a two-way system in which the epithelium can cause local modification of the dermal structures. The dependence of epidermis on the dermis is emphasized by the failure of pure epidermis to survive except for very short periods in organ culture (personal observation) and when implanted on the chorioallantoic membrane of the developing chick (13).

KERATINOCYTE INTERACTIONS

Keratinocytes comprise the main bulk of the epidermis and form a self-regulating system in which dead squamous cells are shed continuously and are replaced by cells arriving at the surface as a result of divisions in the basal layer, thus maintaining a constant thickness in any one site. Bullough (14) has provided good evidence that a chemical mediator he has named chalone controls this process. It is epidermis specific but not species specific, and small amounts have been extracted from fish and pig skin, and characterized as probably a small protein or glycoprotein. Added to slices of mouse skin incubated *in vitro* it inhibits mitosis for 5–9 hours. The effect wears off but can be activated by a wash with epinephrine in amounts that would not normally have any effect. Thus the general hypothesis is that chalone, along with the cofactor of epinephrine, acts as a negative feedback system—chalone being released by the keratinocytes at some stage during their maturation. Release ceases at some stage when they start on the process of keratinization, and the level of chalone concentration regulates the number of mitoses occurring in the basal layers. Tissue-specific chalones that inhibit the growth of melanocyte tumors have also been isolated.

Bullough's concept was originally invoked to explain the burst of mitoses that occurs after injuring epidermis. It is my view that one is on less sure ground here. The postulate is that injury causes inhibition of chalone and thus an increased mitotic rate. There are two main arguments against this.

Hell and Cruickshank (15) have provided evidence that injury induces a point stimulus and synchronization of cell cycles that is more compatible with the positive release of a "wound hormone," and later Hell (16) showed that extracts of skin can induce *in vitro* an increase in the proportion of cells in DNA synthesis on the order of that occurring *in vivo*. The extracted material is very unstable and so far has not been identified. It is possible that there are two mechanisms: one for normal regulation of growth and the other an overriding one to combat injury.

CELLULAR INTERACTIONS IN ORGAN CULTURE

Tissue culture affords an excellent method for studying cellular interactions. In 1948 Medawar (17) described a simple method of organ culture of adult skin that has been modified in various minor ways by several workers. Essentially a thin slice of skin is floated dermal side down on the surface of a culture medium and incubated at 37°C. Under these conditions the process of wound healing is mimicked exactly. Cells from the cut ends of the epidermis migrate across the surface of the cut dermis. These cells are augmented by cells from any hair follicle epithelium in the vicinity to give rise to a continuous sheet on the undersurface of the dermis. This thickens up and eventually keratinizes.

The ultrastructural appearances have been described by Sarkany and Gaylarde (18) and by Constable (19). The cells of original epidermis appear normal, but keratinocytes that had migrated over the dermis are elongated and heaped in several layers even near to the leading edge of the migrating sheet. It appears as if the cells are moving as a group rather than as a monolayer. The nuclei and organelles also conform to the general shape of the cell. The tonofilaments, however, are disorganized and appear as clumps within the cytoplasm. Desmosomes are seen between the cells, and they have clumps of associated tonofilaments. Hemidesmosomes are seen some 30–40 μ from the leading edge, and the development of a basal lamina is first seen subjacent to these. The new basal lamina does not appear to be formed systematically, and although it is most consistently present at the edges where the epidermal cells had been longest in contact with the new dermis, there are stretches where gaps are present. Desmosomes are not numerous, but there are many areas of interdigitating processes (Fig. 1). This again again suggests a more massive mobilization of existing epidermal cells than one sees by light microscopy. Desmosome formation and interdigitations are not mutually exclusive, but what determines the nature of the junction between two cells remains unknown. It is clear that the basal lamina and associated hemidesmosome formation are produced as a dermal–epithelial cell interaction because they are formed only when dermis and epithelium have been in contact for a certain time, but again the nature of the stimulus is unknown.

Both melanocytes and Langerhans cells apparently migrate with the keratinocytes, although they have not as yet recognized special contact areas with keratinocytes.

CELLULAR INTERACTIONS IN MONOLAYER CULTURE

Keratinocyte Interactions

A method has been described (20) for studying interactions of cells from the epidermis for prolonged periods by phase contrast and time-lapse cinephotomicrography. Recently minor improvements and further observations have been made which will be described elsewhere (21). In brief, however, thin slices of guinea pig ear skin containing a minimal amount of dermis are mildly trypsinized to enable the epidermis to be separated from the demis, and a suspension is made of the epidermal cells in a tissue culture medium. These cells are incubated on the cover glass of a specially devised chamber suitable for long-term culture. Cultures have also been studied by transmission and scanning electron microscopy (22).

When the epithelial cells are dispersed in tissue culture medium, they are usually seen by phase contrast as individual rounded cells, although occasionally small clumps are seen.

The first stage in the formation of the culture is the reaggregation of the individual cells into clumps of between 6 and 10 cells. Initially this appears to be a random process caused by chance contact, but once the cells meet, they tend to adhere to each other. These clumps, which are in three dimensions, then develop an adherence to the glass. Movement of the clumps occurs by ruffling of the cytoplasm of those cells at the periphery and in direct contact with the glass. When clumps join, the active cells establish contact, but this is usually temporary, and they retreat from each other. This stage lasts for about 18 hours, and in the ensuing few days two types of event may occur. The clumps may lose their attachment to the glass and fall off, or individual cells may crawl out of the clumps, leaving dead cells behind.

Those keratinocytes that flatten on the glass start to divide in about 3 days and eventually form areas, roughly circular, in monolayers.

After division the cells maintain con-

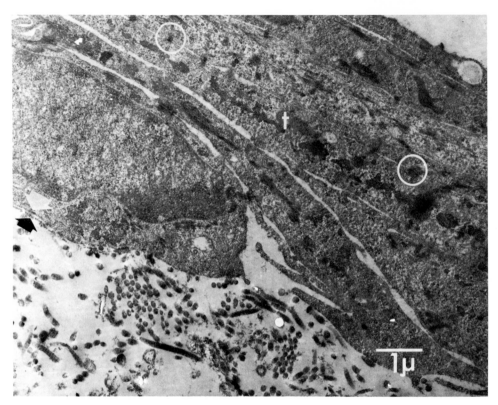

Fig. 1. Epiboly in organ culture. The circles and dagger indicate desmosomes and disorganized tonofilaments, respectively; the arrow shows basal lamina formation.

tact with each other. In fact during mitosis contact is never completely lost as the rounded premitotic cell remains attached to its neighbors by very fine filaments of cytoplasm. In the initial stages of colony formation the cells exhibit some contact inhibition; moreover, adjacent clones appear to move directionally toward each other as if by some chemotactic influence. It is interesting, however, that epithelial cells readily lose contact inhibition and readily heap up in layers and clumps long before confluence is reached. Even in the early stages of colony formation the contact inhibition is not absolute in that using scanning electron microscopy it could be seen that, as in contact-inhibited fibroblasts (23), overlap occurred to a

considerable extent (Fig. 2). This took the form of large areas of cytoplasm imposed on another cell and a multitude of ramifying cytoplasmic process above and below several other cells.

By electron microscopy it can be shown that changes in intercellular junction areas were detectable in the intact skin slice as soon as 2 minutes after incubation in trypsin (Fig. 3). These consisted of a loss of intercellular contact layer between opposing halves of the desmonsomal junctions and a slipping or frank separation of these halves. There was some patchy attenuation of the basal lamina. These changes became more prominent until after 11 minutes the basal lamina was largely absent and the cells had drawn apart and become

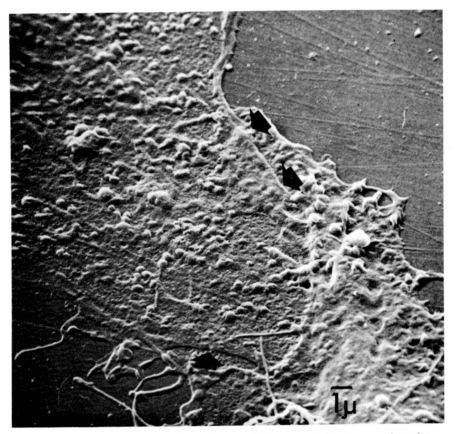

Fig. 2. Scanning electronmicrograph of keratinocytes. The arrows indicate areas of overlap.

rounded. It is of interest that, after the parting of the two halves of the desmosomes, the tonofilaments became withdrawn from their insertion in the attachment plaques. (Fig. 4). Many microvilli were seen extending from these otherwise rounded cells and in some instances were in contact with microvilli from other cells. Many cells had thickened plasma membranes. These observations indicate that keratinocyte adherence *in vivo* is maintained not only by an intercellular cement material at the desmosome but by interdigitating villous processes.

As can be seen from Fig. 5, there is a surprising variety of boundaries and junctions in culture, including fully formed desmosomes [macula adhaerens of Farquar and Palade (24)] and tight junctions [fascia adhaerens of Farquar and Palade (24)]. Since desmosomes (previously discussed in relation to their associated tonofilaments) are prominent as much as 12 days after starting culture and the half-desmosomes caused by trypsinization are not seen in culture at all, it seems that the desmosomes are formed *de novo* in culture, and presumably such a function involving two cells results as a consequence of cell boundary interaction. These desmosomes are apparently fully developed, having tonofilaments inserted into the attachment plaque and an intercellular contact layer. Membranous appositions and intermediate-

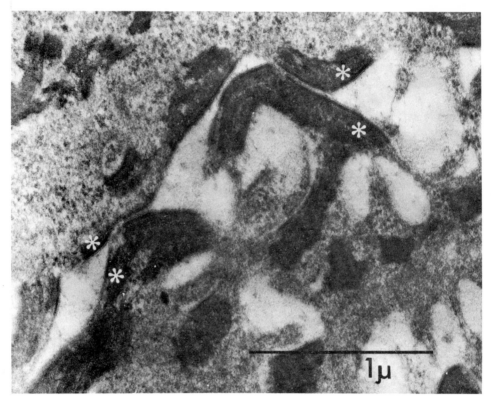

Fig. 3. Early effects of trypsin: loss of intercellular contact layer; slipping and parting of desmosomes. (indicated by asterisks).

type junctions are also seen. These intermediate-type junctions are like the zonulae adhaerens of Farquar and Palade (24), which, in the gut epithelia they were using, are a separate type of junction. They are also like the "early desmosomes" of Breathnach (25), which develop into desmosomes with their characteristic structure, and so it is possible that the membranous appositions or intermediate junctions represent early stages of desmosome formation.

Cup-shaped indentations as described by Abercrombie, Heaysman, and Pegum (23) in chick heart fibroblasts were seen. Abercrombie et al. have suggested that (a) these indicate points of cell–substrate adhesion and (b) they and the

microfilaments sometimes seen associated with them are connected with the locomotion of the cells. Carter's (1967) view, however, is that the movement of cells is passively controlled by the interaction of substrate, cell, and culture medium; he has called this mechanism haptotaxis. By stereo electron microscopy the keratinocytes appear to "wet" the substrate. However, the fact that along the side of a single cell there may be spreading parts and parts raised up from the substrate indicates, as of course does time-lapse cinephotomicrography, that the cell is ruffling (i.e., the boundary changing its physical properties). Haptotaxis alone is probably insufficient to explain cell movement. Abercrombie's suggested mechanism is one that would

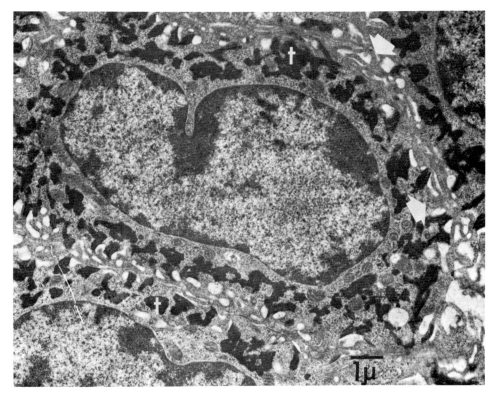

Fig. 4. Early effects of trypsin: withdrawal of tonofilaments (t); numerous microvilli.

fit into this gap, Although there is no further evidence from the epidermal cells in culture to support it, there is also nothing to contradict or exclude it.

Electron microscopy has shown that the majority of the cells that have formed rounded clumps become keratinized and give a picture identical with that of the stratum corneum *in vivo* (Fig. 6). A form of keratinization may take place in flattened cells, but there is the strong impression that organization in three dimensions promotes keratinization.

This concept would seem to be confirmed by the studies of Karasek and Charlton (27), who cultivated adult cells on the surface of a collagen gel. The early behavior of the cells was similar to that on other substrates except that the tendency to clumping and aggregation in the first 24 hours did not occur to the same extent. This is presumably because the cell–cell adhesive forces are greater than the cell–glass forces and the cell–collagen adhesion must more nearly approximate the cell–cell adhesive force. However, following attachment, the cells reassociate into a monolayer colony. The interesting phenomenon, however, is that in about 14 days under these conditions of culture the cells multiply both horizontally and vertically to the surface to form a three-dimensional structure with cells in varying stages of maturation, including the formation of a loose keratin layer. In relation to dermal–epidermal interactions it is of interest that inclusion of viable skin fibroblasts or fibroblast-conditioned medium en-

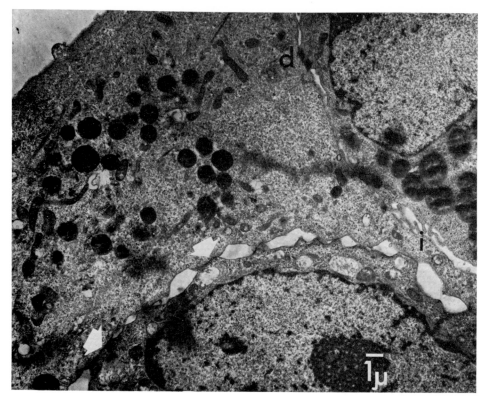

Fig. 5. Keratinocytes in monolayer—cell junctions, including newly formed desmosomes (d), inter-digitations (i), and membranous apposition (arrows).

hanced the growth and survival of skin epithelial cells.

Keratinocyte–Melanocyte Interactions

In the culture system described melanocytes also survive and proliferate, enabling one to study their relationships with the epidermal cells and the process of pigment transfer. The concept of the epidermal melanin unit is reinforced by the fact that pigment donation may occur simultaneously between one melanocyte and several keratinocytes (28). The mechanism is variable: sometimes the end of a dendrite is approximated to the keratinocyte mebrane at right angles. The end of the dendrite splays out, and, as judged by the rapid dancing

movement of the granules, the cytoplasm becomes more fluid. Coincidentally, in the region of the dendrite the cytoplasm of the epithelial cell becomes extremely active, exhibiting rapid ruffling movements. This is particularly marked when the dendrite attempts to withdraw. Long tongues of cytoplasm reach out after the dendrite until the only connection left is a very fine strand that finally snaps.

Sometimes the dendrite lies along the surface of the keratinocyte (Fig. 7). The granules move toward the end of the dendrite, and a clump of granules is similarly released. The digested granules are originally surrounded by their cytoplasm, but this does not persist, and in the Caucasian human and the guinea pig the granules end up as a nuclear cap.

The importance of these observations

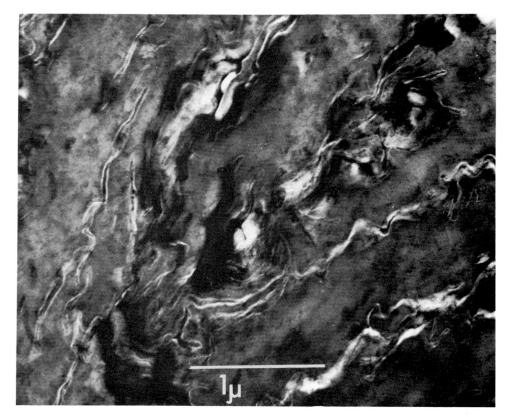

Fig. 6.   Keratinization in dispersed cell culture.

lies not in the demonstration of pigment transfer but in indicating that the process is not "cytocrine"; rather the epidermal cell plays an active part, in effect phagocytozing part of the living dendrite. It is also of interest that contact between a keratinocyte and a dendrite did not inevitably result in pigment transfer. Which cell said "yes" is unknown, but once the decision had been made, the recipient seemed the more active partner.

Cohen and Szabo (29) confirmed these findings, but in addition produced cinefilm of keratinocytes or melanocytes showing that the melanocytes exhibited no contact inhibition toward a monolayer of epithelial cells but wandered freely over it. They also demonstrated melanocytes in mitosis.

Melanocyte Interactions

In most cultures melanocytes tend to move more actively than keratinocytes, extending and retracting their dendrites among them. In some cultures, however, colonies of melanocytes develop and remain as a relatively constant group within which they extend and retract processes with each other. The apposition at times is extremely close, and time-lapse film gives the impression that material may be passing from cell to cell. So far we have been unable to obtain any confirmation of this possibility at the ultrastructural level. Several melanocytes

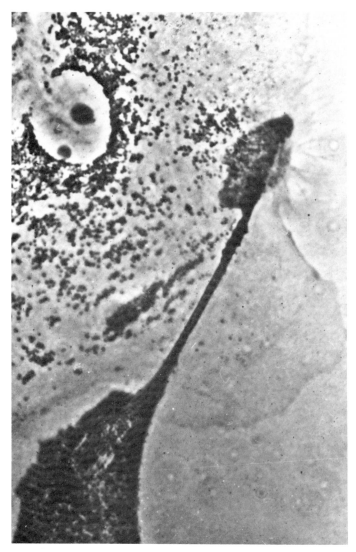

Fig. 7. Pigment donation in culture. Phase contrast, ×1600.

may be interlocked with each other so closely it is difficult to be sure where one dendrite ends and the other begins. The same melanocyte may be indulging in this behavior and donating pigment to a keratinocyte at the same time (Fig. 8).

Langerhans Cell Interactions

Until recently Langerhans cells in cultures have not been recognized. Fritsch

and Diem (30), using our culture system followed by ATPase staining, described positively staining dendritic cells in young primary cultures and also obtained electron-microscopic confirmation. We have also obtained evidence of these cells (Fig. 9) and by comparing ATPase with phase-contrast photographs can now identify them in time-lapse film and describe their relations with keratinocytes. This will be published in detail elsewhere. In effect, however, they

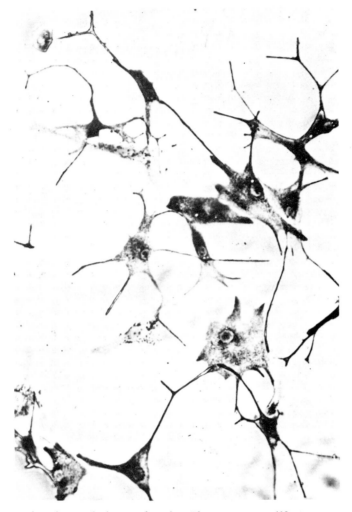

Fig. 8. Melanocyte junctions and pigment donation. Phase contrast, ×400.

appear to have some organizing ability over keratinocytes. When cultures aggregate, in the early stages (3–12 days) Langerhans cells are frequently found in the middle of the clump. When monolayers persist, the Langerhans cells are usually found in the middle of a lacuna. They extrude their dendrites in all directions, at times giving the impression of pushing the keratinocytes apart. In this situation rounded ends resembling podia appear on the tips of the dendrites. These "podia" often overlay the cytoplasm of the keratinocyte and are certainly in close enough apposition to suggest the possible transmission of material from cell to cell. The contact period, however, is relatively brief: a few minutes as opposed to 1–2 hours in the case of pigment donation.

CONCLUSION

Apart from the demonstration of chalone, the interactions described are essentially inferences based on studies of cell and tissue behavior. The method of transfer of information is quite unknown. Cavoto

Fig. 9. Langerhans cell in culture ATPase reaction. ×600.

and Flaxman (31) have shown by inserting microelectrodes that epidermal cells in close apposition will transmit current and thus are electrically coupled, but the relevance of this to behavior is somewhat dubious. It seems more likely that a study of the outer plasma membrane coat might be the most profitable line of investigation.

REFERENCES

1. Fitzpatrick, T. B., and Breathnach, A. S., *Dermatol. Wochenschr.* **147,** 481 (1963).

2. McKenzie, I. C., *Nature* **222,** 881 (1969).

3. Christophers, E., *J. Invest. Dermatol.* **56,** 165 (1971).

4. Wessels, N. K., *Dev. Biol.* **4,** 87 (1962).

5. Moscona, A. A., and Garber, B. B., in *Epithelial-Mesenchymal Interactions*, Fleischmajor, R., and Billingham, R. E., eds., Williams and Wilkins, Baltimore, 1968, p. 230.

6. Billingham, R. E., and Silvers, W. K., *J. Exp.* **125,** 429 (1967).

7. Billingham, R. E., and Silvers, W. K., in *Epithelial-Mesenchymal Interactions*, Fleischmajor, R., and Billingham, R. E., eds., Williams and Wilkins, Baltimore, 1968, p. 252.

8. Oliver, R. F., *J. Embryol. Exp. Morphol.* **15,** 331 (1966).

9. Oliver, R. F., *J. Embryol. Exp. Morphol.* **16,** 231 (1966).

10. Oliver, R. F., *J. Invest. Dermatol.* **47,** 496 (1966).

11. Oliver, R. F., *J. Embryol. Exp. Morphol.* **17,** 27 (1967).

12. Oliver, R. F., in *Epithelial-Mesenchymal Interactions*, Fleischmajor, R., and Billingham, R. E., eds., Williams and Wilkins, Baltimore, 1967, p. 267.

13. Briggaman, R. A., and Wheeler, C. E., *J. Invest. Dermatol.* **51,** 454 (1968).

14. Bullough, W. S., *Sci. Journal*, April 1969.

15. Hell, E., and Cruickshank, C. N. D., *J. Invest. Dermatol.* **31,** 128 (1973).

16. Hell, E., *Br. J. Dermatol.* **83,** 632 (1970).

17. Medawar, P. B., *Quart. J. Microsc. Sci.* **89,** 187 (1948).

18. Sarkany, I., and Gaylarde, P. M., *Br. J. Dermatol.* **83,** 572 (1970).

19. Constable, H., MsSc Thesis Univ. of Birmingham (1972).

20. Cruickshank, C. N. D., Cooper, J. R., and Hooper, C., *J. Invest. Dermatol.* **34,** 339 (1960).

21. Cruickshank, C. N. D., et al., manuscript in preparation.

22. Constable, H., *Br. J. Dermatol.* **86,** 27 (1972).

23. Abercrombie, M., Heaysman, J. E., and Pegum, J. S., *Exp. Cell Res.* **67,** 359 (1971).

24. Farquar, M. G., and Palade, G. E., *J. Cel Biol.* **17,** 357 (1963).

25. Breathnach, A. S., *Ultrastructure of Human Skin*, J. & A. Churchill, London, 1971, p. 10.

26. Carter, S. B., *Nature* **213,** 361 (1967).

27. Karasek, M. A., and Charlton, M. E., *J. Invest. Dermatol.* **56,** 205 (1971).

28. Cruickshank, C. N. D., and Harcourts. *J. Invest. Dermotl.* **42,** 183 (1964).

29. Cohen, J., and Szabo, G., *Exp. Cell Res.* **50,** 418 (1968).

30. Fritsch, P., and Diem, D., *Arch. Dermatol. Forsch.* **243,** 364 (1972).

31. Cavoto, F. V., and Flaxmann, B. A., *J. Inv. Dermatol.* **59,** 370 (1972).

CHAPTER SEVEN

Regulation of Animal Cell Growth*

HARRY RUBIN

Department of Molecular Biology
University of California
Berkeley, California

Differentiated animal cells have characteristic growth rates in their normal organ and tissue environment—rates that are related to their function and need for replacement. The spectrum ranges from nonmultiplying cells, such as nerve and muscle, to cells multiplying rapidly and continuously, such as intestinal epithelium and hematopoietic tissue. In between lie a number of variations, which have been considered and categorized by Goss (1).

The characteristic growth rates of course are subject to alteration. This occurs in the normal course of events through hormonal stimulation and other physiological changes. It occurs in pathology with the appearance of malignancy. Moreover, it occurs in response to wounding and other types of tissue damage. The paradigm for wound healing is the repair of skin epidermis (2). Here we have a tissue in which a measured slow replacement by basal cells of differentiating and ultimately dying superficial cells is continuously occurring. On removal of some of the epidermal tissue through wounding, a train of events is set in motion that elicits migration into the wound by ranks of adjacent cells and an enhanced rate of multiplication to replace the damaged tissue. Variations on this theme have been demonstrated in

* The term "cell growth" will be used here as equivalent to "cell multiplication."

various epithelial tissues of the body. The phenomenon of wound healing has been under investigation by pathologists for many years. Research in the field has spawned a number of theories about the types of controls and stimuli that exist in normal and damaged tissues. Work in the intact organism is, however, faced with formidable difficulties. The damaged tissue is complex. It contains not only ordinary epidermis but also hair follicles, basement membrane, several layers of diverse connective tissue, intervening ground substance, and blood vessels. Bleeding is an unavoidable accompaniment of the wound. It is difficult to quantitate the cell numbers and biochemical processes involved. For these reasons the advent of the technique of tissue culture at the beginning of the century was regarded as providing the seeds of solution of the problem. Since the earliest days of tissue culture its practitioners have been inevitably concerned with the problems of cell growth and its regulation, and more specifically with the problems of wound healing, as far as they could be approached in culture.

The work on regeneration of tissues in culture during the first half of the century is thoroughly reviewed in Fischer's classic treatise *The Biology of Tissue Cells* published in 1945 (3). Up to that historical juncture, the object of study was the tissue explant embedded in plasma clots. Explants consisted of thousands of cells that initially retained their original topographical relationships to one another. It was noted early on that, for the most part, the only cells that multiplied were those that migrated from the bulk explant into the plasma clot. When the migrating region became crowded, cell growth slowed down, but it could be restimulated by wounding the tissue. Although these studies were largely descriptive and the measurements relatively imprecise, the evidence was

convincing that cell migration was a factor in stimulating growth and that the topographical relationships of the cells were important in determining their growth rate.

The second half of the century saw the introduction of techniques for dispersed cell culture, including the monolayer technique (4). This involved the separation of cells from one another, their attachment to a solid substratum, and their multiplication thereon. The monolayer technique permitted the continuous observation of all the cells in the culture. It provided a uniformity of environment for the cells and permitted precise techniques for quantitating cell numbers and a variety of biochemical processes. It provided a method for rapidly labeling the cells with radioisotopes. It should be noted at the outset, however, that up to now the great bulk of work in this field has been done with fibroblasts. These cells most readily survive the dispersal techniques and multiply freely in the dispersed state, whereas epithelial cells rarely survive dispersal into individual units, and when they do, cannot be grown with facility (5).

The general observation was made that fibroblasts multiplied rapidly until they formed a confluent sheet, whereupon their growth rate slowed down. This observation seemed to bear some relationship to the "monolayering tendency" described by Abercrombie in his studies of cell movement (6). He coined the term "contact inhibition" to describe the failure of cells to move over one another when attached to a solid substratum. The term was expropriated to apply to the limitation on cell multiplication seen in confluent sheets (7). Unfortunately the term carries with it the connotation that cell multiplication, like cell locomotion, is inhibited immediately on contact between two cells. Since this is manifestly incorrect in the organism, where most migrating and multiplying cells in a wound remain completely surrounded by contiguous cells, the term "contact inhibition" will be dropped here in favor of the less committal term "density-dependent inhibition (8). The facts about growth and its regulation in monolayer cultures will first be examined, and an attempt will be made to bring the diverse facts together in a coherent working hypothesis.

GROWTH CURVES AND "SATURATION DENSITIES" OF ANIMAL CELLS IN CULTURE

Sparse cultures of fibroblasts (i.e., 10^3 cells per square centimeter) grow exponentially in the appropriate medium. Cells are sensitive to variations in their environment at these low densities. This is particularly true of primary or secondary cultures recently derived from the animal. It is frequently necessary therefore to use medium conditioned by large numbers of cells (9) or to use X-irradiated feeder layers of cells to obtain exponential growth when the cell density is less than 10^3 cm^{-2} (10). Growth of the cells depends not only on the micronutrients and ions of the medium but also on macromolecular constituents of serum. The quantitative requirements for serum vary with different cell types and cell lines; for example, sparse cultures of chicken embryo fibroblasts can grow at near maximal rate in the presence of as little as 0.5% serum (11), whereas chicken embryo epithelial cells require 10% or more, as do certain established lines of mouse cells. Indeed, too high a concentration of serum may be toxic to low densities of fibroblasts in primary cultures (11).

Because of the highly asymmetric shape of fibroblasts attached to a planar surface, they start making contact with one another long before they form a confluent sheet. In this preconfluent state they form a sparse network of cells in point contact with one another. Thus, when chicken fibroblasts cover only about one-fourth the surface area of a petri dish, about 80% of the cells are touching other cells (12). It is at this juncture that the particular concentration of serum present in the medium begins to loom large in determining the growth rate of the cells. If the serum concentration is low, the growth rate of the cells may decrease before the cells reach confluency. If it is high, exponential growth may continue beyond confluency (11).

Cells can continue to increase in number after they have achieved confluency and yet maintain themselves in a monolayer. The increase in cell number is achieved at the expense of the cross-sectional surface area of the cells. They are squeezed closer together; since their volume is not decreased in proportion to the decrease in area covered, a compensatory increase in height occurs.

Close inspection of cultures that appear to be monolayered reveals that there is much overlapping of the cytoplasms of cells, although the degree of nuclear overlap may be small (13). However, it is possible to achieve frank multilayered growth for most cell types by using high serum concentrations and frequently changing the medium (9). An apparent exception to this generalization is the 3T3 line of mouse embryo fibroblasts, which is obtained by procedures favoring fairly sharp limitation of growth at the monolayer stage. Even in those cultures that achieve multilayered growth the multiplication of cells slows down some time after confluency is reached.

The extent of the slowdown with confluency depends on the type of cell used. Chick embryo fibroblasts never stop growing completely as long as glucose, and amino acids, remain in ample supply (11). Their growth rate may slow down as much as tenfold at confluency, but their numbers continue to creep up. The number of 3T3 cells in a culture may become constant when confluency is reached (14). Constancy of number, however, it not always assurance that the growth rate equals zero, since some cell death and replacement occur with all cell lines.

These observations point up some of the ambiguities of terminology in the field of cell growth regulation. The term "contact inhibition" was originally devised to describe the paralysis of cell movement that occurs almost immediately after the leading lamella of one cell contacts the surface of another (6). When applied to the problems of cell multiplication, it carries with it the implication that a similar type of minimal contact causes inhibition of growth, when in fact such minimally contacted cells may be multiplying at their maximal rate (10). Indeed, under conditions of ample serum and nutrient supply, the growth rates of most cells do not decrease until the cells are entirely surrounded by contiguous cells (15). Because the term "contact inhibition" carries with it implications from its origin in studies of locomotion, other terms have been devised to describe the multiplicative behavior of cells. One such term is "density-dependent inhibition" (8), and another is "topoinhibition" (16). The first will be used here because it carries with it no preconceptions about the relative roles of diffusible inhibitors versus topographical arrangement.

RELATION BETWEEN CELL MOVEMENT AND MULTIPLICATION

In the early days of tissue culture it was observed that cell multiplication was largely restricted to the zone of cell migration, and it was proposed that movement and multiplication were related (3). The question of whether such a relation exists has been debated ever since. As so frequently happens, the debate has taken on categorical aspects: that is, is there a hard and fast relationship between cell movement as such and cell multiplication? It might be more fruitful to consider that movement itself is a manifestation of a more generalized activity of the cell membrane that provides the stimulus for multiplication. This attitude makes sense when we consider that malignant cells, which are not subject to the regulatory effects of population density, are *slower* in movement than normal cells, when that movement is measured as lateral displacement of the entire cell (17). The malignant cells exhibit a more aimless movement to and fro, and show a high degree of membrane motility when viewed by phase-contrast microscopy. Among normal cells of a given type, however, the extent of lateral cell displacement is probably a fair measure of membrane motility.

The relationship of cell contact to cell movement has been precisely quantitated by Abercrombie and associates in a series of classic papers (6,18). He established that the rate of movement of a cell varies inversely with the number of cells it contacts. Thus, when cells become crowded, their rate of movement decreases. The leading lamella of the cells, which undergoes continuous ruffling movements when moving into free space, ceases moving when it contacts another cell. When the cell is bounded along its entire periphery by other cells, it is reduced to gradual oscillatory movements in concert with its neighbors.

Not only is movement restricted by population density but so is the overall surface area of the cell. When a cell is in the isolated state, it is spread to the maximum extent, and the surface-to-volume ratio is maximized. As cells are forced closer, their surface areas shrink while volumes remain relatively constant. In very crowded cultures even fibroblasts tend to assume a quasi-cuboidal appearance and approach a minimum surface-to-volume ratio.

MEDIUM DEPLETION AND THE WOUND-HEALING EXPERIMENT

One of the problems that has plagued the study of density-dependent regulation of growth is that of medium depletion. When a culture gets very crowded, it rapidly depletes the medium of glucose and amino acids (19). This problem cannot be obviated by large increases in concentration of medium constituents because excessive concentrations cause deleterious side effects. Excessive glucose is metabolized to lactic acid, which decreases the pH of the medium and probably of the interior of the cells. Some amino acids in excessive amounts are toxic to cells. Even daily replacement of medium is not absolute insurance against depletion (20). One approach to coping with this problem is the continuous perfusion of medium. Kruse and Miedema (20) found it necessary to perfuse crowded cultures of some cell lines with the equivalent of about 60 medium changes in a day to maintain a reasonably constant nutritional environment. Since most workers in the field find them to be too troublesome and expensive for the advantage gained, the perfusion techniques have not been widely adopted.

The problem of medium depletion is most severe when a high serum concentration or a high pH has been used since both conditions increase the utilization of glucose and possibly other nutrients and thereby hasten depletion (11,13). In any study of density-dependent regula-

tion the trivial effect of medium depletion must be ruled out. It is surprising how infrequently this is done. Severe depletion can be readily tested by removing medium from crowded cultures and measuring its capacity to support the growth of small numbers of cells. If the medium supports the growth of sparse cultures at a more rapid rate than the crowded culture from which it was derived, it is resonable to suppose that the reduced growth rate of the latter is not due to medium depletion.

A more unequivocal demonstration of the effect of population density as opposed to medium depletion on growth rate is the wound-healing experiment, which, in effect, measures the growth rate of sparse and crowded cells sharing the same medium in the same dish (21, 22). The technique is a simple one. Cells are grown to confluency. The medium is removed, and a swath of cells is scraped from the dish. The culture is washed to remove the cells, and the original medium is replaced. Several variations on this theme have been described. After allowing variable time periods for migration of cells into the denuded region, ^{3}H-thymidine is added to the medium for incorporation into DNA. The cells are prepared for autoradiography. If the ^{3}H-thymidine is present for a short period, only a fraction of the cells will be labeled, and that fraction is proportional to the growth rate of the cells.

Within minutes after wounding, cells along the edge of the wound assume an orientation normal to the wound axis and start to migrate into the denuded region. After a period characteristic of the cell type (about 4 hours in chick embryo fibroblasts and 20 hours in 3T3 cells), the fraction of labeled cells among those that have migrated into the denuded region begins to increase over the fraction of labeled cells in the confluent region. The difference between the two regions reaches a maximum a few hours later and is sustained as long as the difference in population density remains. The increase in the fraction of cells synthesizing DNA is followed in a few hours by an increase in the fraction of cells in mitosis. The length of the delay in the onset of DNA synthesis and the occurrence of mitosis only after DNA synthesis indicate that most of the cells in the crowded culture were in the Gl period of the mitotic cycle.

The results of the wound-healing experiment leave no doubt that population density is a crucial determinant of cell growth rate. They make it difficult to sustain the contention that diffusible inhibitors play a dominant role in regulating growth of crowded cultures, since the crowded and sparse cells are within a few micrometers of one another. If diffusible inhibitors were to play a dominant role in growth regulation, one would have to contend that they were very large, slowly diffusing molecules or that they were highly unstable and could act only in the immediate vicinity of the cells producing them. While there has been a report that an inhibitor is released by cultured cells (23), it is described as being of low molecular weight; and it is apparently stable since it withstands an arduous purification procedure. The rationalization offered by the authors to explain the wound-healing experiment is that the inhibitor is active only on crowded cells. Thus, even if the putative inhibitor were to have physiological significance, the nature of the cellular response would be dependent on population density.

In any case, inhibitors of physiological significance must be distinguished from toxic materials arising as artifacts of the tissue culture method. It is relevant to note that the most inhibitory media are derived from malignant cultures, which are not themselves subject to population

density effects (24). Such media are inhibitory to normal and malignant cells alike and can in no way be considered significant in growth regulation.

In summary, the evidence from wound-healing experiments, the negative findings of many who have searched for inhibitors, and the variety of interpretations possible for the positive findings do not support the existence of diffusible growth regulators in culture.

SERUM AND GROWTH REGULATION

Before monolayer cultures were available, serum components were ubiquitous constituents of the medium, since the tissue explants were embedded in plasma clots (3). It was only necessary to add embryo extract to initiate cell proliferation (but see ref. 25). With the elimination of the plasma clot, it became necessary to add serum to the medium as well as micronutrients in order to sustain multiplication. Numerous attempts have been made down through the years to isolate the constituents of serum required for multiplication; to date no unique component has been found that by itself is capable of promoting multiplication. More recently it was discovered that the maximal population density attained in a culture is a function of the serum concentration present (26). Indeed, if the depletion of micronutrients is circumvented, it is possible to stimulate multiplication in slow-growing, crowded cultures by simply adding fresh serum to the medium.

Holley and coworkers have fractionated the macromolecular constituents of serum to determine their respective contributions to cell multiplication (27). The fractions are assayed for their effects on cell movement, DNA synthesis, mitosis, and other functions. The fractions are added to medium already containing low concentrations of whole serum since no single fraction will by itself support cell multiplication. This is an indication that serum, as a whole, is performing complex functions. Fractions that support multiplication of one cell line are ineffective for another cell line. Again it seems that there is no unique fraction of serum that can be deemed *the* growth-promoting agent.

There has been some contention about whether cell population density or serum concentration is the primary regulatory factor for multiplication. It is apparent that cells migrating into a wound multiply faster than their confluent counterparts on the same dish, so population density is dominant in this situation. It is just as obvious that crowded populations can be made to multiply by adding more serum, so one can adopt the attitude of its dominance here. The argument seems futile. It can be agreed that both factors determine the extent of cell multiplication in culture. Carried to an extreme, however, one might contend that the supply of glucose or amino acids is limiting and therefore growth regulating. We must then ask what are the growth-regulating situations in real life in the animal. It is most unlikely that the supply of glucose and amino acids is limiting: the concentrations in the blood are relatively constant and are maintained by a variety of homeostatic mechanisms—that is, there is no counterpart in the animal to medium depletion in culture (28). By the same token, there is no evidence that the macromolecular constituents of serum are ever depleted in the animal (29) as they are in culture, except conceivably in the case of liver regeneration, where the main source of serum proteins is itself involved. But certainly in the cases of wound healing in various surface epithelial tissues there is ample evidence of change in cell shape, movement, and population density (30).

These changes mimic in some respects the cellular changes seen in wound healing in culture and help us focus on the topographical state of the cell as a primary determinant in growth regulation.

It may well be, however, that serum and population density are acting at the same effector site in the cell. In both cases the cell surface appears as a prominent candidate for that site. It is most unlikely that the large, stimulatory molecules of serum penetrate the cell; that is, their primary action is likely to be at the cell surface. The most obvious effect of altering population density is to change the motility of the cell surface and the surface-to-volume ratio. This suggests that the cell surface acts as a transducer that responds to external effectors, to entrain a reaction chain that terminates in cell division.

OTHER GROWTH-STIMULATING SUBSTANCES

The addition of serum to a density-inhibited culture is the most effective way of stimulating DNA synthesis by medium alteration. There are a number of other substances, however, that are capable of stimulating DNA synthesis. Trypsin when added to the medium of density-inhibited chick embryo cultures at concentrations too low to detach the cells stimulates the synthesis of DNA, which is followed by mitosis (31). The experiments are successful only if chicken serum is the sole serum source in the medium, since it has little trypsin-inhibitory activity. The trypsin must remain for many hours to produce an observable effect. Pronase and ficin cause the same effect.

A similar effect has been reported in mouse cells (32), but it is claimed that a 30 minute exposure is sufficient to produce a maximal stimulation. Such short exposures produce only a minimal response in the chicken cells.

A group of Soviet workers has reported that DNA synthesis can be stimulated by a diverse group of substances, including digitonin, ribonuclease, and hyaluronidase (33). The authors speculated that all these substances were surface-active agents at the cell membrane.

If DNA synthesis is inhibited by depriving chick embryo cells of serum, it can be restored to its previous rate by adding sublethal concentrations of either cadmium, manganese, or zinc (34). Other metals that are not toxic in the concentration range tested do not stimulate DNA synthesis. Those concentrations that stimulate DNA synthesis coincide with those that stimulate the uptake of 2-deoxy-D-glucose. Neither response occurs in the absence of the other, although the increased uptake of 2-deoxy-D-glucose precedes that of increased DNA synthesis by several hours.

A similar effect was obtained in serum-deprived cultures by the addition of the highly carcinogenic hydrocarbon 9,10-dimethyl 1,2-benzanthracene (DMBA). Other carcinogenic hydrocarbons, such as methylcholanthrene and benzpyrene, are ineffective. Of the three, only DMBA is damaging to the cells, and it also stimulates 2-deoxy-D-glucose transport, so the effect seems analogous to that caused by the metals. The coincidence of stimulated DNA synthesis and 2-deoxy-D-glucose transport in all cases suggests a primary membrane effect by this diverse group of agents.

pH AND GROWTH STIMULATION

The pH of the blood of birds and mammals is about 7.4, and tissue culture media are usually adjusted to this pH. Lowering the pH of the medium of dense cultures is inhibitory to both DNA synthesis and 2-deoxy-D-glucose transport (13). In sparse cultures and in cultures of transformed cells it is necessary to reduce the pH below 7.0 to produce a similar inhibition. Increase of pH above 7.4 is stimulatory to DNA synthesis in dense cultures. Increases above pH 7.8 cause damage to the cells. The differential effect of pH on dense cultures as opposed to sparse and to transformed cultures suggests that pH in combination with cell density might be a regulatory factor for cell multiplication. This matter will be discussed further in a later section.

Ceccarini and Eagle (35) reported that cell lines have differing pH optima for multiplication. They found that the saturation density of cultures could be increased by using organic buffers to stabilize the pH and speculated that fluctuations in pH might be responsible for early contact inhibition. It should be noted that they used relatively high concentrations of serum, which promote rapid utilization of glucose, lactic acid formation, and marked lowering of pH. They also used a closed-bottle system for culture, which permits the accumulation of carbon dioxide and thereby speeds up the lowering of pH. When low serum concentrations are used in an open system, pH changes are minimized, and it can be seen that density-dependent inhibition is not just a response to fluctuating pH (13). It can also be seen that the optimal pH for growth is not fixed for any cell line, but varies proportionately with population density.

FACTORS THAT INHIBIT
CELL GROWTH

Removal of serum and lowering of pH have been shown to inhibit the growth rate of cells. Of course, there are many substances that are toxic to cells and thereby inhibit their multiplication. We are here interested only in those substances that *reversibly* inhibit growth. Among such compounds are the sulfated polysaccharides occurring in agar (11). The basis for their effect is not known, but it has been suggested that these highly acidic macromolecules might adsorb to the cell surface and thereby lower the pH in the immediate vicinity of the cell surface.

An inhibitory material that has been intensively investigated in recent years is cyclic 3′,5′-adenosine monophosphate (cAMP). This substance is considered to be the second messenger in the action of many polypeptide hormones. When added to fibroblastic cells in culture in the form of its dibutyryl derivative, cyclic AMP inhibits cell multiplication (36). [Note, however, that it purportedly stimulates DNA synthesis in cultured thymocytes (37).] It also makes transformed cells look somewhat like normal cells. The concentration of cyclic AMP is lower in transformed than in normal cells (38) and is also lower in fast-growing than in slow-growing cells (39). The effects of cyclic AMP appear in some experiments to occur in the G1 period (40, 41) and in others to occur in the G2 period (42). This series of observations has led some to the conclusion that cyclic AMP is a key controlling element in natural growth regulation (40).

This conclusion can be questioned on a number of *a priori* grounds, one of which is the stimulatory effect of cyclic AMP on the growth of some cell types, such as thymocytes (37), and its failure to affect at all the growth of certain other cell types (43). Another ground for skepticism about the role of cyclic AMP is the peculiar kinetics of its effect on DNA synthesis. Thus it is only inhibitory if present throughout the entire period of preparation for DNA synthesis in 3T3 cells (40). If added 3 hours after the cells are seeded, it has no effect on DNA synthesis, although the synthesis does not start for many hours. It has been reported that in cultures treated with dibutyryl cyclic AMP, which show no net growth, there are as many cells as in normal cultures that are synthesizing DNA and undergoing mitosis, but that many of the cells die or are unable to undergo cytokinesis (42,43). Therefore no convincing case can be made at present for cyclic AMP as an intermediary in normal growth regulation.

Another factor that has been claimed to enhance density-dependent inhibition is concanavalin A (44). This lectin has attracted much attention because it differentially agglutinates transformed as opposed to normal cells (45). It has been reported that monovalent concanavalin A reversibly inhibits the growth of transformed cells; that is, it restores them to a state of density-dependent inhibition. The conclusion from this work is that the exposure of cell sites for agglutination directly underlies the escape from density-dependent inhibition and that covering these sites restores normal control. This conclusion is thrown under a cloud by the more recent demonstration that concanavalin A binds to normal cells to the same extent as to transformed cells (46,47), that is, the same reactive sites exist on the surface of normal cells. Resolution of this matter awaits further confirmation of the single claim of the growth-inhibiting effects of monovalent concanavalin A.

The growth of cells can be inhibited by depriving them of zinc (48). This is accomplished by adding the chelating agent ethylenediamine tetraacetate (EDTA) to the medium. The amount added must be less than the concentrations of calcium and magnesium in the medium to avoid toxic effects. Under these conditions there is a fivefold to tenfold inhibition of DNA synthesis, and this inhibition can be reversed only by zinc. The concentration of zinc required for reversal is much less than that of the EDTA added to the medium, which indicates that there are materials in the cell that bind zinc even more firmly than does EDTA. The synthesis of RNA is only slightly inhibited by zinc deprivation, and protein synthesis is unaffected. In these respects zinc deprivation causes an effect like that of density-dependent inhibition. There is at least one distinctive difference though. Zinc deprivation does not cause an inhibition in the transport of glucose (49). It therefore appears not to affect the membrane properties of the cell. It has been noted that zinc is required for the initiation of nucleic acid synthesis by several purified polymerases, including the DNA polymerase of Kornberg (50), RNA polymerase (from *Escherichia coli*) (51), and the terminal nucleotidyl transferase of Bollum (52). It is also required for a number of other enzyme activities, especially some of those of the glycolytic and respiratory pathways that require nicotinamide–adenine dinucleotide (NAD) as a cofactor (53). Any one, or all, of these enzymes could be involved in the inhibition of growth by EDTA. It remains to be ascertained whether the activity of zinc-dependent enzymes is altered in density-dependent inhibition before its role in normal growth regulation can be evaluated.

PERMEABILITY STUDIES OF CELLS IN CULTURE

The permeability of cultured cells has been examined under various conditions of growth using the uptake of radioactive materials as a measure. The uptake of glucose and its analog 2-deoxy-D-glucose is increased when density-inhibited cells are stimulated to multiply by adding serum (54). The increase begins within minutes after the addition of the serum and reaches its peak within 6 hours. The increase in the rate of uptake requires protein synthesis,* but not RNA or DNA synthesis. A marked difference in the rate of glucose uptake has been found to exist between normal cells and those transformed by infection with RNA-containing sarcoma viruses. The rate of uptake can be as much as 10 or more times as great in the transformed as in the normal cells (55). Unlike the serum stimulation of normal cells, the increased uptake of glucose by the transformed cells persists as long as the cells remain transformed (56).

The rate of uptake of a number of other substances is also increased after serum stimulation and in the malignant transformation, but the extent of this increase is less than that for glucose. Inorganic phosphate, uridine, and thymidine (57–59) fall into this category. There is some difference of opinion as to whether the uptake of amino acids varies with the growth rate of cells (54,60,61). It appears that the uptake of some amino acids is somewhat affected, while that of others is not. Even for those that are affected, it is necessary to deplete existing amino acid pools to demonstrate the effect (61).

The rates of thymidine and uridine uptake could hardly be expected to be important influences in normal growth regulation since the presence of these substances in the medium is not required for growth by the cells. However, glucose is the major nutritional substrate used by

* We have recently found that the initial increase in the rate of uptake of 2-deoxy-D-glucose does not require protein synthesis, although the sustained increase over a period of hours does. (H. Rubin and D. Fodge, 1974, in *Control of Proliferation in Animal Cells*, B. Clarkson and R. Baserga (eds.), Cold Spring Harbor Laboratory, in press.)

cells and is the source of the energy and of many of the building blocks for macromolecular synthesis. It is also readily depleted from the medium (62). It is unlikely, however, that variations in the rate of glucose uptake determine the growth rate of cells. For one thing, the cells that most rapidly deplete the medium of glucose are just those least likely to be density inhibited. For another, prolonged starvation for glucose leads not to cell regulation but to cell death. Addition of excessive concentrations of glucose to the medium of density-inhibited cells does not increase their growth rate, nor does a fivefold to tenfold reduction in the concentration of glucose in the medium of sparse cultures inhibit their growth (11). Indeed, it has been reported that the initiation of DNA synthesis requires only the presence of monovalent ions, phosphate, and bicarbonate (63). I have recently found, however, that a supply of amino acids from the medium is required for the initiation of DNA synthesis. Temin's report (63) indicating that amino acids are not required could be explained if it is assumed that his dialyzed serum underwent proteolysis. External supply of amino acids seems to be the major, and perhaps the sole, exogenous substrate requirement for the initiation of DNA synthesis. It is unlikely, however, that the rate of uptake of the amino acids determines the difference in growth rates between rapidly growing and density-inhibited cultures as Holley has suggested (64). For one thing we find little difference in the rate of uptake between fast- and slow-growing cells (54), and the differences found by others are small (60,61). For another, decreasing the external concentration of amino acids by a factor of 5 has little effect on growth rate (11), and adding high concentrations of amino acids to density-inhibited cultures does not accelerate their growth (11).

It has also been proposed that the supply of glucose is a critical determinant for cell multiplication *in vivo* on the basis of wound-healing experiments in the mouse ear (65). More recent results, however, show that external glucose is not required to initiate a full round of DNA synthesis after wounding the mouse ear and that *excess* glucose, rather than stimulating mitosis, actually impedes it (66). This is not to say that the rate of glucose catabolism and energy production are not critical parameters for cell multiplication, but only that the rate of supply of glucose from the external medium does not serve as a direct control on DNA synthesis or mitosis. There seems little reason to doubt that enhanced availability of glucose would be an important factor in sustaining multiplication over an extended period of time.

TRANSMEMBRANE POTENTIAL AND THE MULTIPLICATION OF CELLS

The classic response of animal cells to external stimuli is expressed in excitable tissues: nerve and muscle. There the stimulus causes a depolarization of the membrane, so that the transmembrane potential becomes at first less negative and then finally positive, thus initiating an action potential (67). This explosively rapid event is propagated down the length of the cell and its extensions. Nonexcitable cells also have a negative transmembrane potential. Since they are not known to manifest action potentials, their standard state is referred to as a steady potential rather than a resting potential (68). There has come to light in recent years evidence that this steady potential in nonexcitable tissue may not be so steady after all. This information is slow in coming because the cells are much smaller than excitable cells and therefore difficult to penetrate with microelectrodes without being severely damaged. This seems to be particularly true in tissue culture and the more so with sparse cultures in which the cells are greatly extended and flattened out. However, there is some suggestion that cells stimulated to multiply are partly depolarized. Olague and Rubin (69) reported that sparse, rapidly growing cells have an average transmembrane potential of about -10 mV, whereas cells in dense cultures have an average potential of about -32 mV.

In comparing membrane potential measurements in sparse and dense cells there is always the possibility that the reduced membrane potential of sparse cells is caused by their greater susceptibility to damage on penetration by the microelectrode. Attempts were made in the cited study to minimize this possibility by accepting only results obtained from cells that maintained their initially recorded potential for some time after penetration.

Another approach is to record potentials in dense populations before and after the addition of a growth-stimulating concentration of serum. Such studies show a partial depolarization of the dense cells to the potential found in sparse cells (70). Unlike the case of excitatory cells, however, the potential does not return to its original high negative value for several hours.

If they are confirmed, these findings could offer an important clue to the mechanism of density-dependent inhibition. It has been shown that both cell surface area and cell movement are restricted by contact with other cells. These restrictions might reasonably be expected to decrease the ion permeability of the cell surface and thereby increase the membrane potential. The events could provide the first link in a chain leading to the inhibition of DNA

synthesis. A prolonged increase in membrane potential might be expected to lower the intracellular pH if the distribution of protons were at thermodynamic equilibrium (71,72); that is, the decrease in intracellular pH could be predicted directly from the Nernst equation. Even if the protons were not passively distributed, the electrochemical force tending to drive up their concentration within the cell would increase with an increase in membrane potential.

Depolarization such as described by Hulser and Frank (70) would tend to decrease the internal proton concentration and raise the intracellular pH. There is at least one fly in the ointment of the proposition that depolarization is the initial response of cells to growth-stimulating signals. It should be possible to depolarize cells by raising the concentration of K^+ in the medium. I have varied the concentration of K^+ in the medium of chick fibroblast cultures over the entire range from 0 to 150 mM without stimulating multiplication among the cells. Indeed, very high concentrations of K^+ inhibit multiplication. It is possible, though, that any stimulus to multiplication provided by the depolarization is counterbalanced by other effects of the shift in ion balance. It may also be significant that the potential change caused by serum arises from altered permeability, whereas that caused by raising the external K^+ concentration does not.

GROWTH REGULATION IN EPITHELIAL CELLS

Epithelial cells are more difficult to grow in culture than are fibroblastic cells. In particular they do not stand up well to dispersal into suspensions of individual cells. In order to maintain the characteristic morphology and appearance of many types of normal epithelium, it is necessary to keep them in clumps containing more than 10 or 20 cells. Under such conditions it is not possible to have a uniform population density of cells distributed at random on the dish, as is the case for fibroblasts, and direct biochemical analysis of the entire mass of a culture gives a meaningless average of results from dense and sparse cells. It is, however, possible to apply a variety of microscopic techniques to determine the relationship between population density and growth rate. Measurements have been made of cross-sectional surface area, rate of movement, frequency of mitosis, and the proportion of cells synthesizing DNA during a given time span (15,73,74). The results show that the proportion of epithelial cells in mitosis or synthesizing DNA is proportional to the cross-sectional surface area of the cells and therefore inversely proportional to population density. The rate of cell movement is also decreased with population density, once more raising the question about the relationship between cell movement and multiplication. The results with epithelial cells concur with those obtained with fibroblasts in showing the effects of population density on the rate of cell multiplication.

THE MALIGNANT TRANSFORMATION OF CELLS

Perhaps the definitive property of malignant cells is their insensitivity to normal growth-regulating influences. This of course manifests itself in the animal in a progressively growing tumor. Tissue culture provides the opportunity for direct comparison of normal and malignant cells under the same conditions. It also provides an unparalleled opportunity for initiating the malignant transformation and observing all its sequen-

tial stages. This is done by infecting dispersed cell cultures with a tumor virus. The most efficient viruses for producing rapid and uniform transformation are the RNA-containing sarcoma viruses. The most widely studied of these are various strains of Rous sarcoma virus. This virus can produce the change from normalcy to malignancy within 2 days in a high proportion of cultured chick embryo fibroblasts (75).

A difference in the regulatory properties of the normal and infected cultures can be observed within 24 hours after infection. Within this short period of time the infected cells undergo their first signs of morphological transformation, and their growth rate becomes independent of population density (75,76). In addition, the rate of glucose transport increases to even higher levels than those occurring in fast-growing normal cells (56).

The changes in behavior caused by infection with the sarcoma viruses are reversible. This can be demonstrated by infection with mutants that are temperature sensitive for the capacity to transform (56). Cells transformed by infection with one of these mutants at a permissive temperature revert to normal appearance and behavior when switched to a nonpermissive temperature. The cellular changes in both directions are not accompanied by changes in the rate of virus production, indicating it is not the mere synthesis of virus that causes transformation.

The increased utilization of glucose by the transformed cells results in increased glycolysis and lactic acid production (77). Glucose is rapidly depleted from the medium, and pH is lowered. It is thus difficult to maintain a constant environment for the cells, and they are easily damaged by these self-imposed changes. Unless precautions are taken to avoid these problems, the infected population may be found to be growing at an even slower rate than the uninfected one within a few days after the transformation (78). In addition it is likely that a continuous process of change and selection occurs in populations of transformed cells in culture; this complicates the determination of the changes that are directly related to the loss of growth-regulating properties of the cells. Thus it is crucial to focus on those events that precede and accompany the loss of replicative control if one wishes assurance that these events are part of the underlying mechanism of transformation.

CELL CYCLE AND BIOCHEMICAL CHANGES IN DENSITY-DEPENDENT INHIBITION

The decreased rate of DNA synthesis in density-inhibited cultures is paralleled by a decrease in the proportion of cells synthesizing DNA (78). The cells that synthesize DNA do so at the same rate as cells in a rapidly growing population. This indicates that most of the cells in a density-inhibited population are arrested or delayed in either the G1 or the G2 period of the mitotic cycle. Estimates of the DNA content of such populations show that the $2N$ amount of DNA is present, indicating that most of the cells are in the G1 period (79).

When the cells are stimulated to grow by the addition of serum or by wounding, there is a delay of a few hours before DNA synthesis begins. The length of the delay is of the same order as the length of the G1 period (21,54,63,80). The great majority of cells initiate DNA synthesis without undergoing a prior mitosis, which would have been necessary had the cells of the density-inhibited population been in the G2 period (21,79,80). The weight of evidence therefore indicates that density inhibition is effected either by an arrest or delay during the G1 period or a diversion during the period to G1 a state apart from the regular mitotic cycle. There is some evidence supporting the latter view in specialized tissues of the intact animal (81), but comparable evidence has not been obtained for the less specialized fibroblasts most commonly used in tissue culture studies.

Most of the biochemical differences found between rapidly growing and density-inhibited populations are of a quantitative nature. The transport of glucose and its analogs is faster in the growing population. The rates of glycolysis and respiration are increased, as measured by lactic acid and carbon dioxide production, respectively (77). The relative rates of synthesis of various phospholipids in a density-inhibited population of 3T3 cells changes within a few minutes after stimulation by serum (82).

Perhaps the most marked difference between inhibited and rapidly multiplying populations aside from the rate of DNA synthesis itself, is in the activity of enzymes associated with DNA synthesis. The activities of thymidine kinase, ribonucleotide reductase, and DNA polymerase are very low in a density-inhibited population (83–85). These enzymes undergo a marked increase in activity after application of a growth stimulus. This increase almost coincides in time with the increased rate of DNA synthesis in the cell population. The increase in the activity of these enzymes requires the synthesis of RNA and protein. It can be prevented by the addition of low concentrations of actinomycin D, which reputedly inhibit only ribosomal RNA synthesis (84). This of course suggests that increased numbers of ribosomes are required for the synthesis of these enzymes. It should be noted, however, that it does not rule out the possibility that

RNA species other than ribosomal RNA are required for the synthesis of the enzymes. The requirement for protein synthesis to precede DNA synthesis suggests that the enzymes do not merely preexist in an inactive form, but must be synthesized anew. There are, however, pitfalls in experiments using inhibitors of protein synthesis which make interpretation somewhat treacherous (86).

The proportion of ribosomes existing as polysomes is higher in rapidly growing than in inhibited cultures (87). Within a few hours after addition of serum to an inhibited culture there is an increase in the number of polysomes (88). It should be noted, however, that there is a high rate of protein synthesis even in the inhibited cultures. Indeed, in several cell lines there is little difference between growing and inhibited cultures, and between transformed and normal cultures, in the overall rate of incorporation of radioactive amino acids into protein (89,90). This suggests that the rate of protein turnover is increased in the inhibited cultures, and this has been found to be the case (91).

Changes in the rate of RNA synthesis with altered states of growth have been intensively studied in lymphocytes stimulated to DNA synthesis by phytohemagglutinin. In the resting lymphocytes there is a relatively low rate of RNA synthesis and no detectable DNA synthesis (92). The rate of RNA synthesis in general, and of ribosomal RNA synthesis in particular, increases sharply within a few hours after the addition of phytohemagglutinin. An increase in the activity of the ribosomal RNA polymerase occurs without delay after stimulation, whereas the activity of the messenger RNA polymerase occurs after a short delay. Neither increase requires protein synthesis, indicating that the RNA polymerases exist in the resting lymphocyte in an inactive form.

Studies on RNA synthesis in density-inhibited cultures of fibroblasts have given equivocal results. It was first reported that the addition of serum to density-inhibited cultures of 3T3 cells was immediately followed by a sharp spike of increased RNA synthesis followed by a decline (93). Later it was found that the uptake rates of the radioactive precursors used for the study of RNA synthesis were increased after serum stimulation, and this increased uptake created the false impression of increased RNA synthesis (57). However, other differences in RNA synthesis have since been found between growing and inhibited cultures. There is evidence that the rate of ribosomal RNA synthesis is more rapid in fast-growing cells than in inhibited ones, and some suggestion that the processing and turnover of RNA differ with growth rate (94,95). No clear idea has emerged of how these findings are related to the regulation of DNA synthesis and to cell multiplication.

COMPARISON OF *IN VIVO* AND *IN VITRO* SITUATIONS

Growth regulation *in vivo* involves cells that are in contact with one another over their entire surface. This holds true even for the rapidly multiplying cells of the epidermis near the edge of a wound. Cells in sparsely seeded cultures are in the extraordinary situation of being isolated from one another. If this abnormal situation does not cause excessive damage to them, the cells multiply at their maximal rate. They continue to do so until they form a confluent sheet, at which time their freedom of movement and surface area are reduced and their rate of DNA synthesis slows down. It is important to note, however, that even in the state of confluency in culture cells do not approach the closeness of packing and extent of contact that characterize cells near the edge of a wound *in vivo.* Superficial analysis of the *in vitro* situation has led many to believe that the very act of contact is itself growth inhibitory and an important physiological regulatory mechanism. The term "contact inhibition" reflects this point of view. Even if there were some specific effect of cell contact, it would not appear to be applicable to growth-regulating situations *in vivo,* where loss of contact does not occur (30).

A modified view is that contact between cells on a planar surface restricts the surface area and movement of the cells because they have difficulty in moving over one another (8). The restrictions on surface area and movement become more severe as the cells become more and more crowded, and more of their surface area is engaged with other cells. Under this view, any influence that restricts surface area and movement is inhibitory to growth, whereas under the strict interpretation of contact only another cell can be inhibitory. The facts seem to favor the view that contact per se is not a specific inhibitory signal, but just one of a variety of effects that restrict cell surface activity. For example, cells from different species and indeed genuses are mutually inhibitory (6,18, 96). It seems unlikely that the same specific signals are so widely shared.

If the modified view of the effect of contact is correct, it should be possible to restrict surface area and movement by means other than contact with other cells and achieve the same effect on multiplication. One such means is to prevent attachment and spreading of cells on a solid substratum. It is well known that normal cells are incapable of sustained growth in suspension. Indeed, this is the basis for discriminating between normal and transformed cells by the agar suspension technique (97). The suspended cells assume a spherical, or nearly spherical, shape, which minimizes surface area. And without a solid substratum, there is no opportunity for movement. Under the strict contact view, there is no *a priori* reason that cells in suspension should not multiply. Under the modified view, though, the failure of normal cells to grow in suspension follows simply from their altered surface dimensions and activities.

It is in the application to *in vivo* regenerative situations that the surface restriction view seems to have its greatest advantage over the specific contact signal view. As already noted, cells adjacent to wounds do not lose contact with one another (30). They do, however, move over the surface of the wound and over each other. In doing so their surface membranes must be in constant motion, and the shape of the cells continuously changing. It seems likely that surface area would increase. Therefore the idea that surface restriction is growth regulatory would be applicable both *in vitro* and *in vivo.*

There are other explanations of growth regulation that are not at all dependent on direct cell-to-cell interaction. One such explanation is the chalone hypothesis of Bullough (98). This hypothesis states that tissue-specific inhibitors of growth are continually released by cells. These inhibitors are called chalones, and are reputed to be proteins. Under this hypothesis the growth rate of cells in a tissue is inversely related to concentration of chalone. When a full population of cells at maximum density is present, the chalone concentration is high and multiplication is held to a low level. When a portion of the cell population is damaged or removed, the local concentration of chalone is reduced and the rate of cell multiplication is increased. This situation continues until the original contiguous population is restored. Many attempts have been made to demonstrate chalones in tissue culture situations. As already noted, even those positive reports of the existence of inhibitors in culture refer to low-molecular-weight inhibitors (23), and no evidence of specificity has been obtained. If chalones do indeed exist, they are extraordinarily difficult to demonstrate *in vitro* (99), where the most precise and sensitive techniques for their detection are presumably available.

Another explanation that does not invoke cell contact is the wound hormone hypothesis. This hypothesis holds that damaged tissue releases growth-stimulating substances (100). It seems to have derived support in the early days of tissue culture from the observation that embryo extract was essential for the growth of tissue explants in plasma clots (3). With the advent of dispersed cell culture, embryo extract proved unnecessary for growth. It might still be maintained that embryo extract is required for the growth of cells in explants that retain the three-dimensional relationship with one another that is characteristic of their tissue of origin. However, the wound hormone theory does not explain how the growth stimulus is maintained after the original wounding until, and only until, the original mass and form of the tissue are restored. Thus, while there is a suggestion that growth-stimulating material from damaged tissue might play a role in helping to initiate multiplication, there is no support for the idea that its role is a major, or a continuing, one.

CONCLUSIONS

The regulation of animal cell multiplication seems to be a consequence of changes initiated at the cell surface. The synthesis of DNA and the division of cells occur hours after application of a stimulus. Indeed, the stimulus can be withdrawn several hours before DNA synthesis or mitosis has begun in most cells of a population without affecting the onset of either process (62,80). This indicates that the synthesis of DNA and mitosis are the end products of a chain of events that are more directly affected by the stimulus. A variety of processes are altered by stimuli to multiplication, including cell movement (101), glucose transport (54), glycolysis (77), respiration (77), pentose phosphate shunt (102), synthesis of phospholipids (82), and processing of RNA (94,95) and protein (91). Of these, the ones most likely to have wide-ranging effects are those concerned with glucose catabolism since they provide not only the main source of energy for the cell but also the building blocks for nucleic acid and protein synthesis. The pathways of glucose catabolism also respond rapidly to environmental alteration (103–105). One such environmental alteration that is known to affect both DNA synthesis and glucose catabolism is pH (11,13,35). Increases in pH, which stimulate DNA synthesis, stimulate the activity of phosphofructokinase, the prime regulatory enzyme of glycolysis (104), both as isolated enzyme (106) and in the intact cell (107–109). Glycolysis is the chief pathway of glucose utilization by cultured cells (110) and it appears to be more critical to DNA synthesis than is the respiratory pathway (111). The activity of several of the enzymes of glycolysis require zinc (53), which is also required to initiate DNA synthesis (48). Since the availability of divalent cations within cells increases with pH (112) and internal pH should respond to membrane alterations (72), one can make a coherent working hypothesis for the regulation of cell multiplication through alterations in cellular pH. Such changes would be expected to influence the rate of glycolysis and thereby perhaps the rate of DNA synthesis. This is little more than a speculative notion, but it points up several areas of experimentation. Cell membrane potential and internal pH should be determined under a variety of growth situations. The speed and extent of changes in glucose catabolism should also be determined after stimulation of growth-inhibited cells. And the precise nature of the relation between glucose catabolism and DNA synthesis needs to be clarified.

REFERENCES

1. Goss, R., *Adaptive Growth*, Academic Press, New York, 1964.

2. McMinn, R., *Tissue Repair*, Academic Press, New York, 1969.

3. Fischer, A., *The Biology of Tissue Cells*, Glydendalske Boghandel Nordisk Forlag, Copenhagen, 1945.

4. Dulbecco, R., *Proc. Natl. Acad. Sci. U.S.* **38**, 747 (1952).

5. Laws, J., and Strickland, J., *Nature* **178**, 309 (1956).

6. Abercrombie, M., and Heaysman, J., *Exp. Cell Res.* **5**, 111 (1953).

7. Rubin, H., *Cancer Res.* **21**, 1244 (1961).

8. Stoker, M., and Rubin, H., *Nature* **215**, 171 (1967).

9. Rubin, H., *Exp. Cell Res.* **41**, 138 (1966).

10. Rubin, H., and Rein, A., in *Wistar Institute Symposium Monograph No. 7*, Defendi, V., and Stoker, M., eds., Wistar Institute Press, Philadelphia, 1967, p. 51.

11. Rubin, H., in *Growth Control in Cell Cultures*, Ciba Foundation Symposium, Wolstenholme, G., and Knight, J., eds., Ciba Foundation Symposium, Churchill & Livingstone, Edinburgh, 1971, p. 127.

12. Rein, A., and Rubin, H., *Exp. Cell Res.* **49**, 666 (1968).

13. Rubin, H., *J. Cell Biol.* **51**, 686 (1971).

14. Todaro, G., and Green, H., *J. Cell Biol.* **17**, 299 (1963).

15. Castor, L., *J. Cell Physiol.* **72**, 161 (1968).

16. Dulbecco, R., *Nature* **227**, 802 (1970).

17. Abercrombie, M., Heaysman, J., and Karthauser, H., *Exp. Cell Res.* **13**, 276 (1957).

18. Abercrombie, M., and Ambrose, E., *Exp. Cell Res.* **15**, 332 (1958).

19. Griffiths, J., *J. Cell Sci.* **8**, 43 (1971).

20. Kruse, P., and Miedema, E., *J. Cell Biol.* **27**, 273 (1965).

21. Gurney, T., *Proc. Natl. Acad. Sci. U.S.* **62**, 906 (1969).

22. Dulbecco, R., and Stoker, M., *Proc. Natl. Acad. Sci. U.S.* **66**, 204 (1970).

23. Yeh, J., and Fisher, H., *J. Cell Biol.* **40**, 382 (1969).

24. Rubin, H., *Exp. Cell Res.* **41**, 149 (1966).

25. Balk, S., *Proc. Natl. Acad. Sci. U.S.* **68**, 271 (1971).

26. Holley, R., and Kiernan, J., *Proc. Natl. Acad. Sci. U.S.* **60**, 300 (1967).

27. Holley, R., and Kiernan, J., in *Growth Control in Cell Cultures*, Ciba Foundation Symposium, Wolstenholme, G., and Knight, J., eds., Churchill & Livingstone, Edinburgh, 1971, p. 3.

28. Altman, P., *Blood and Other Body Fluids*, Federation of American Society of Experimental Biology, Washington, D.C., 1961.

29. Schultze, H., and Heremans, J., *Molecular Biology of Human Proteins*, Vol. I, Elsevier, Amsterdam, 1966.

30. Winter, G., in *Wound Healing*, Montagna, W., and Billingham, R., eds., Pergamon Press, London, 1964, p. 113.

31. Sefton, B., and Rubin, H., *Nature* **227**, 843 (1970).

32. Burger, M., *Nature* **227**, 170 (1970).

33. Vasiliev, J., *J. Cell Physiol.* **75**, 305 (1970).

34. Rubin, H., and Koide, T., *J. Cell Physiol.*, in press.

35. Ceccarini, C., and Eagle, H., *Proc. Natl. Sci. U.S.* **68**, 229 (1971).

36. Johnson, G., and Pastan, I., *J. Natl. Cancer Inst.* **48**, 1377 (1972).

37. Whitfield, J., MacManus, J., and Rixon, R., *J. Cell Physiol.* **75**, 213 (1970).

38. Otten, J., et al., *J. Biol. Chem.* **247**, 1632 (1972).

39. Sheppard, J., *Proc. Natl. Acad. Sci. U.S.* **68**, 1316 (1971).

40. Willingham, M., Johnson, G., and Pastan, I., *Biophys. Biochem. Res. Commun.* **48**, 743 (1972).

41. Froehlich, J., and Rachmeler, J., *J. Cell Biol.* **55**, 19 (1972).

42. Smets, L., *Nature New Biol.* **239**, 123 (1972).

43. Hauschka, P., Rubin, R., and Everhart, L., *J. Cell Biol.* **55,** 108a (1972).

44. Burger, M., *Current Topics in Cell Reg.* **3,** 135 (1971).

45. Inbar, M., and Sachs, L., *Proc. Natl. Acad. Sci. U.S.* **63,** 1418 (1969).

46. Clive, M., and Livingston, D., *Nature New Biol.* **232,** 154 (1971).

47. Ozanne, B., and Sambrook, J., *Nature New Biol.* **232,** 156 (1971).

48. Rubin, H., *Proc. Natl. Acad. Sci. U.S.* **69,** 712 (1972).

49. Rubin, H., and Koide, T., *J. Cell Biol.,* **56,** 777 (1973).

50. Brutlag, D., Shekmann, R., and Kornberg, A., *Proc. Natl. Acad. Sci. U.S.* **68,** 2826 (1971).

51. Scrutton, M., Wu, C., and Goldthwait, D., *Proc. Natl. Acad. Sci. U.S.* **68,** 2497 (1971).

52. Chang, L., and Bollum, F., *Proc. Natl. Acad. Sci. U.S.* **65,** 1041 (1970).

53. Vallee, B., and Wacker, W., *The Proteins,* Vol. V, Academic Press, New York, 1970.

54. Sefton, B., and Rubin, H., *Proc. Natl. Acad. Sci. U.S.* **68,** 3154 (1970).

55. Hatanaka, M., Huebner, R., and Gilden, R., *J. Natl. Cancer Inst.* **43,** 1091 (1969).

56. Martin, G., et al., *Proc. Natl. Acad. Sci. U.S.* **68,** 2739 (1971).

57. Cunningham, D., and Pardee, A., *Proc. Natl. Acad. Sci. U.S.* **64,** 1049 (1969).

58. Weber, M., and Rubin, H., *J. Cell Physiol.* **77,** 157 (1971).

59. Rubin, H., in *2nd International Symposium on Tumor Viruses at Royaumont, France,* Edition du C.N.R.S., Paris.

60. Foster, D., and Pardee, A., *J. Biol. Chem.* **244,** 2675 (1969).

61. Isselbacher, K., *Proc. Natl. Acad. Sci. U.S.* **69,** 585 (1972).

62. Griffiths, J. B., *J. Cell Sci.* **6,** 739 (1970).

63. Temin, H., *J. Cell Physiol.* **78,** 161 (1971).

64. Holley, R., *Proc. Natl. Acad. Sci. U.S.* **69,** 2840 (1972).

65. Bullough, W., *Biol. Rev.* **27,** 133 (1952).

66. Gelfant, S., *Ann. N.Y. Acad. Sci.* **90,** 536 (1960).

67. Katz, B., *Nerve, Muscle and Synapse,* McGraw-Hill, New York, 1966.

68. Woodbury, W. W., in *Physiology and Biophysics,* 19th ed., Ruch, T., and Patton, H., eds., Saunders, Philadelphia.

69. Olague, P., and Rubin, H., in *Growth Control in Cell Cultures,* Ciba Foundation Symposium, Wolstenholme, G., and Knight, J., eds., Churchill & Livingstone, Edinburgh, 1971, p. 257.

70. Hulser, D., and Frank, W., *Z. Naturforschung* **26,** 1045 (1971).

71. Conway, E. J., *Physiol. Rev.* **37,** 84 (1957).

72. Carter, N., et al., *J. Clin. Inv.* **46,** 920 (1967).

73. Zetterberg, A., and Auer, G., *Exp. Cell Res.* **62,** 262 (1970).

74. Castor, L., *J. Cell Physiol.* **75,** 57 (1969).

75. Rubin, H., and Colby, C., *Proc. Natl. Acad. Sci. U.S.* **60,** 482 (1968).

76. Colby, C., and Rubin, H., *J. Natl. Cancer Inst.* **43,** 437 (1969).

77. Bissell, M., Hatie, C., and Rubin, H., *J. Natl. Cancer Inst.* **49,** 555 (1972).

78. Rubin, H., *Proc. Natl. Acad. Sci. U.S.* **67,** 1256 (1970).

79. Nilausen, K., and Green, H., *Exp. Cell Res.* **40,** 166 (1965).

80. Shodell, M., and Rubin, H., *In Vitro* **6,** 66 (1970).

81. Epifanova, O., and Tersikh, V., *Cell Tissue Kinetics* **2,** 75 (1969).

82. Cunningham, D., in *Growth Control in Cultures,* Ciba Foundation Symposium, Wolstenholme, G., and Knight, J., eds., Churchill & Livingstone, Edinburgh, 1971, p. 207.

83. Lieberman, I., et al., *J. Biol. Chem.* **238,** 3955 (1963).

84. Lieberman, I., and Ove, P., *J. Biol. Chem.* **237,** 1634 (1962).

85. Nordenskjold, B., et al., *J. Biol. Chem.* **245,** 5360 (1970).

86. Cox, R., et al., *J. Mol. Biol.* **58,** 197 (1971).

87. Stanners, C., and Becker, H., *J. Cell Physiol.* **77,** 31 (1970).

88. Gurney, T., personal communication.

89. Eagle, H., Piez, K., and Levy, M., *J. Biol. Chem.* **236,** 2039 (1961).

90. Hare, J., *Cancer Res.* **27,** 2357 (1968).

91. Hershko, A., et al., *Nature New Biol.* **232,** 206 (1971).

92. Pogo, B., *J. Cell Biol.* **53,** 635 (1972).

93. Todaro, G., Lazar, G., and Green, H., *J. Cell Comp. Physiol.* **66,** 325 (1965).

94. Emerson, C., *Nature New Biol.* **232,** 101 (1971).

95. Weber, M., *Nature New Biol.* **235,** 58 (1972).

96. Eagle, H., and Levine, E., *Nature* **213** (1967).

97. MacPherson, I., and Montagnier, L., *Virology* **23,** 291 (1964).

98. Bullough, W. S., *Biol. Rev.* **37,** 307 (1962).

99. Smets, L. A., *Cell Tissue Kinetics* **4,** 233 (1971).

100. Abercrombie, M., *Symp. Soc. Exp. Biol.* **11,** 235 (1957).

101. Baker, J., and Humphreys, T., *Proc. Natl. Acad. Sci. U.S.* **68,** 2164 (1971).

102. Warshaw, J., and Rosenthal, M., *J. Cell Biol.* **52,** 283 (1972).

103. Chance, B., Estabrook, R., and Williamson, J., *Control of Energy Metabolism,* Academic Press, New York, 1965.

104. Scrutton, M., and Utter, M., *Ann. Rev. Biochem.,* **37,** 249 (1968).

105. Koobs, D., *Science* **178,** 127 (1972).

106. Trivedi, B., and Danforth, W., *J. Biol. Chem.* **241,** 4110 (1966).

107. Minakame, S., et al., *Biochem. Biophys. Res. Commun.* **17,** 748 (1964).

108. Halperin, M., et al., *J. Biol. Chem.* **244,** 384 (1969).

109. Ui, M., *Biochim. Biophys. Acta* **124,** 310 (1966).

110. Paul, J., in *Cells and Tissues in Culture,* Willmer, E., ed., Academic Press, New York, 1965.

111. Polgar, P., Foster, J., and Cooperband, S., *Exp. Cell Res.* **49,** 231 (1968).

112. Nakamaru, Y., and Schwartz, A., *J. Gen. Physiol.* **59,** 22 (1972).

CHAPTER EIGHT

Contact Inhibition of Cell Locomotion

ALBERT HARRIS

Department of Zoology
University of North Carolina
Chapel Hill, North Carolina

INTRODUCTION

Cell Locomotion

The cells of most vertebrate tissues are capable of a peculiar type of "crawling" locomotion, which is most easily observed and studied when the cells are placed in culture on glass or other solid substrata. Although this crawling locomotion is often spoken of as being "amoeboid," it is in fact quite different from the movement of *Amoeba proteus;* amoeboid cytoplasmic flow occurs only in certain white blood cells, and even these lack a true fountain zone (1–3).

Tissue cells, as exemplified by the most commonly studied "fibroblastic" cells, can extend themselves and move only by attaching to a solid substratum such as glass, plastic, or plasma clot (4). Such cells adhere to the substratum at their margins and somehow exert a tractional force that tends to pull the cell margins outward. When equal amounts of traction are exerted by opposite margins of the cell, spreading results, but if one part of the margin dominates the rest, the whole cell is moved bodily in the direction of this part of the margin (5).

Monolayering

When large numbers of such cells are allowed to attach and spread on a solid substratum, they tend to distribute themselves more or less evenly in a thin layer (6,7). A most commonly observed instance of such spreading is that of outgrowth from an explanted piece of solid tissue, such as embryonic heart ventricle. Cells from such explants migrate out radially, covering the surrounding substratum with a circular expanding sheet of cells (2).

This tendency of cultured cells to occupy space evenly and to spread out in thin layers ("monolayering") was first studied in detail by Abercrombie and Heaysman (8,9). These workers sought to explain this colonial behavior of cells in terms of the locomotor activities of individual cells and, more especially, the interactions of these cells with one another. In a series of meticulous statistical analyses of cell movements they were able to demonstrate that cell locomotion is in some way inhibited by contact with neighboring cells, with the result that cells tend not overlap one another and populations of cells spread out evenly. Abercrombie and Heaysman called this directional inhibition of locomotion "contact inhibition" (9,10).

Contact Inhibition of Movement and Growth

Other workers have found that cell multiplication (growth and division) also becomes inhibited in dense cultures of many types of cultured cells (11–13). Unfortunately this inhibition of growth has come to be referred to by the same term as had been applied to the inhibition of movement, that is, "contact inhibition." This would not have been too confusing if the two phenomena had turned out to be closely or causally linked, as many once suspected. However, even though cells of many types either display both types of contact inhibition or have

neither, enough exceptions to this rule have been found to show that the two phenomena are at least separable and may possibly be entirely different (14).

Nevertheless, the continued use of the identical term "contact inhibition" to describe both phenomena has caused much confusion, even though Abercrombie and his collaborators have never maintained that there was any direct relationship between the two types of contact inhibition. At present it is likely that most nonspecialists interpret the term "contact inhibition" to refer specifically to contact inhibition of multiplication. In the remainder of this chapter the term "contact inhibition" will refer specifically to the inhibition of cell locomotion only. It is unfortunate how often growth and locomotion have been confused with one another, as witness the use of the term "outgrowth" to mean migration.

THE ORIGINAL OBSERVATIONS ON CONTACT INHIBITION

The original papers in which Abercrombie and Heaysman first described contact inhibition have been very widely cited in subsequent studies of cell behavior. Nevertheless, enough misconceptions about contact inhibition have survived to suggest that these original papers may be referred to in bibliographies somewhat more often than they actually are read. Be this as it may, it is certainly true that the original papers on contact inhibition contain a wealth of quantitative information and still deserve to be considered in detail.

Inverse Correlation Between Contact and Speed

The first of these papers (8) relates the speed of moving chick heart fibroblasts to the number of contacts made with neighboring cells. Cell movements were recorded by time-lapse cinemicrography, and the resultant films were analyzed quantitatively. In this way the movements of a total of 246 cells were followed over periods of 4–8 hours. The speeds of individual cells were found to vary greatly and often to fluctuate in an apparently erratic manner. Cell velocities ranged from 0 to as much as 250 μ/hour and often changed greatly within a few minutes.

Seeking to determine the causes of this variability, Abercrombie and Heaysman categorized the cells observed according to the number of other cells with which they were in contact (their "contact number"). Individual cell speeds and contact numbers were measured over 13.5 minute periods of observation, the duration of these periods corresponding to 1 foot lengths of film. When the speeds of cells were compared with their contact numbers, a strong and statistically significant inverse correlation was found (Fig. 1). On the average, cells moved more slowly, the greater the number of other cells with which they were in contact. Although this correlation did not explain all of the variation in cell speed, it was highly significant ($P < .001$ according to the t test). The extent of this correlation did, however, vary somewhat between the six cultures analyzed.

Cells that gained or lost a contact with another cell during one of the 13.5 minute periods were assigned a fractional contact number equal to the average number of cells contacted. For example, a cell that began a period with a contact number of 4 but then became detached from one of its neighbors would be given a contact number of 3.5 for the period in which the detachment occurred. Thus the pronounced zig-zag pattern of the curve plotted in Fig. 1a reflects the fact that cells making or breaking contacts

move somewhat more rapidly than other cells. A further analysis of the data by Abercrombie and Heaysman showed that the increase in the speed of cells losing contacts was somewhat greater than that of cells gaining contacts. By direct observation of the films they found that cells breaking contacts often spring apart suddenly, whereas those forming new contacts frequently expand their area of apposition, thus drawing together. In either case, a slight temporary acceleration results from the change in contact number.

It is also important to notice that cells in contact with as many as five or six of their neighbors (and thus effectively surrounded) continued to move at an average speed approximately half that of uncontacted cells. This continued movement of contacted cells has sometimes been neglected in more recent discussions of the phenomenon, which are apt to give the impression that intercellular contact paralyzes all motion.

Possible Misinterpretations

One difficulty with this type of analysis is that actual cell speed can be underestimated if cells change direction during the time intervals. As a result, more frequent changes in direction might be mistaken for reduced speed. To avoid this possible misinterpretation, Abercrombie and Heaysman redetermined cell speeds over a shorter time interval (half that mentioned above). When these speeds were compared with contact number, the same inverse correlation was found. Likewise, when cell movements were followed continuously, by projection of the film and tracing of individual paths, these movements were still correlated with contact number in essentially the same way as had been determined over the longer intervals. It is clear,

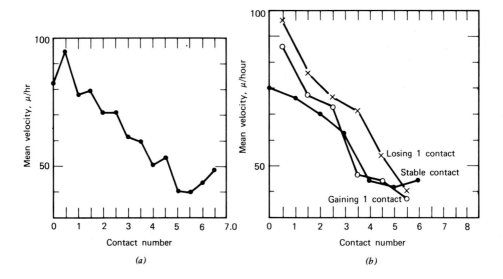

Fig. 1. Cell velocity as a function of contact number: (*a*) mean velocities, all readings, plotted according to contact number; (*b*) mean velocities plotted for readings where the contact number remained stable during the observation of velocity (●), where a contact was gained during observation (○), and where a contact was lost (X).

therefore, that an actual decrease in speed must have been responsible.

Another possibility considered by Abercrombie and Heaysman was that the apparent relationship between contact number and speed might have been coincidental. For example, cells might move more slowly as the cultures became older while at the same time cells became more densely packed due to multiplication. A spurious, or at least noncausal, correlation could be the result. To test this possibility, the data from the first 80 minutes of observation of each culture were compared with those from the last 80 minutes in an analysis of covariance. Although this did reveal a slight decline in cell speed with age, regardless of contact number, this decline was not significant. In contrast, the correlation between contact and velocity remained highly significant, independently of the culture's age.

Still another possible explanation that had to be considered was that cell move-ment might be inhibited by nearby cells whether or not they were actually in contact. For example, these cells might exhaust the medium or coat the substratum or in some other way inhibit their neighbors' movement at a distance. As one would expect, contact number was strongly correlated with the cell density in the immediate area. However, when Abercrombie and Heaysman limited their analysis to cells having the same number of near neighbors, they found the same inverse correlation between speed and contact number, regardless of the local population density. In a similar way, it was shown that the decrease in cell speed is not a consequence of distance from the explant.

Colliding Cells

To determine with more certainty whether intercellular contacts actually *cause* cells to move more slowly, Abercrombie and Heaysman studied the behavior of cells in the area between adjacent explants. As cells migrated outward from the two explants, the cell density between them increased and cells collided. When cell speeds were compared over this period of collision and increased contact number, a significant drop in cell velocity was observed. This indicates that cell speed actually decreases *because* of the formation of contacts (rather than the reverse, for example).

In their second paper on the social behavior of cultured fibroblasts (9) Abercrombie and Heaysman were concerned primarily with the tendency of these cells to form a thin layer and, in general, not to overlap one another extensively. Again using explants placed next to one another ("confronted cultures"), they counted cells in the "collision zone" between explants as cells migrated into this area from both sides. The cell density in this area was found to increase gradually but eventually to reach a plateau level as the cells became confluent and closely packed. This plateau density was significantly less than twice the cell density at an equal distance from the explant on the side away from the collision zone (158%, to be exact). This difference shows that cell migration into this area actually becomes inhibited, presumably due to the crowding and the increased intercellular contact. There was, in fact, no detectable reduction of migration into this area until the two outgrowths actually made contact. Once contact was made, however, the advancing cells slowed to about a quarter of their previous speed.

When the directions of cell movements in the collision zone were analyzed, a similar change in behavior was found. Prior to the junction of the two outgrowths, movement was found to be predominantly outward, with some lateral movement and very little inward movement toward the explants. After junction of the outgrowths, however, not only was the speed of movement decreased, but the remaining movement was nearly equal in all directions, though there was a slight remaining excess of movement in the outward direction. Once again this mutual inhibition of cell movement tended to produce an evenly spread distribution of cells with a minimum extent of overlap.

Extent of Overlapping Between Cells

Unfortunately it was not practical for Abercrombie and Heaysman to measure the amount of cytoplasmic overlap directly, simply because cell margins are too thin and irregular to be distinguished clearly. The nuclei, however, can be clearly seen, and so they chose to count overlaps between nuclei as a measure of overlapping between cells. As a quantitative estimate of the extent to which overlapping was inhibited, they compared the number of nuclear overlaps actually observed with the number to be expected if overlapping were entirely random. This number of expected overlaps can be calculated from the area of individual nuclei and the number of nuclei within a given area. The formula giving the number of expected overlaps is

$$\frac{n(n-1)}{2} \cdot \frac{\pi(2r)^2}{A}$$

where n is the number of nuclei, r is their radius, and A is the total area. The first term corresponds to the number of different possible combinations between n nuclei, and the numerator of the second term equals the area around the center of each nucleus within which the center of another nucleus would have to lie in order for an overlap to be counted.

When Abercrombie and Heaysman counted the actual numbers of nuclear overlaps in many different areas of several cultures of chick heart fibroblasts, they found an average of about one-third of the number to be expected if cells were arranged randomly. In only a few cases was the number observed as much as one-half the "expected" number, and the difference from the randomly expected number was statistically significant in every one of 24 samples to at least the level of $P < .001$.

All of these various instances of mutual inhibition of movement among fibroblasts seem to be attributable to essentially the same phenomenon. To refer to "this directional prohibition of movement" Abercrombie and Heaysman coined the phrase "contact inhibition." Abercrombie has subsequently restated the definition of contact inhibition as "the directional restriction of displacement on contact" (10).

Consequences of Contact Inhibition

It seems likely that this contact inhibition is also the explanation of certain other characteristic features of fibroblast outgrowth. For example, outgrowth zones tend rapidly to assume a rounded or circular shape, even when the explant itself is highly irregular in shape (square, elongate, etc.) (6). Contact inhibition would tend to bring about such circularity by causing cells to migrate most rapidly into the least densely populated areas of the substratum. This variation in the rate of outward migration would tend to compensate for the irregularities in the explant margin, preferentially filling in areas of low population and, in general, promoting an even dispersal of cells. Contact inhibition also helps to explain the outward migration itself. The directional inhibition of movement

by contact will necessarily favor movement in the direction of fewest contacts and will therefore cause cells to disperse outward from an area of high population, such as the explant center. Experienced tissue culturists may feel some surprise that a commonly observed feature of cell behavior, such as outgrowth, should require explanation. However, a little reflection shows that such directional migration is by no means to be expected on purely *a priori* grounds. Clearly it is not the least important aspect of the work of Abercrombie and Heaysman that they were able to see the need to explain certain phenomena that had come to be taken for granted.

POSSIBLE RELATIONSHIP TO CANCER

Abercrombie and Heaysman next turned their attention to the study of contact inhibition between cells of different types, work in which they collaborated with H. M. Karthauser (15). The particular cells studied were embryonic chick heart fibroblasts, neonatal mouse skeletal muscle fibroblasts, and mouse sarcoma cells of the lines Sarcoma-37 and Sarcoma-180. In this study explants of the various cell types were 0.5–1 mm apart on a thin plasma clot over glass. As outgrowth proceeded, the interactions between the outward migrating cells were measured quantitatively by comparing the cell populations between the explants with those that had extended laterally. Cell counts were made within 100 μ wide strips extending from the explants in these two directions. These "between" and "side" populations were compared with respect to total number of cells, distance of outgrowth along the strip, overall population density, and population densities at the proximal and at the distal ends of the strips. The extent of mutual interaction between the confronted cells could then be expressed quantitatively by the ratios of the "between" and "side" measurements ("between/side ratios"). A ratio differing from unity indicates a distortion of the patterns of outgrowth, and ratios less than unity correspond to inhibition of outward migration toward the adjacent outgrowth. These measurements were made both before and after the apposed outgrowths actually made contact.

One complicating factor, which should be borne in mind, is the use in this study of a plasma clot substratum beneath the cells. Moving cells stretch and orient the clot fibers around themselves, and other cells tend to move preferentially along the axis of this orientation (contact guidance). As a result, outgrowing cells from one explant within a clot will tend to migrate toward an adjacent explant and vice versa. This mutual orientation produced at a distance through the medium of a plasma clot has been called by P. Weiss (16) "the two-center effect," and this effect may have been partly responsible for some of the interaction observed by Abercrombie and his collaborators in this study.

Prior to contact between outgrowths, the extent of outward migration was in most cases greater in the area between explants than in the lateral areas (i.e., the between/side ratios were greater than unity). This was presumably due to the two-center effect. However, after the junction of outgrowths, the outward migration of the fibroblasts was inhibited by contact either with confronting fibroblasts or sarcoma cells. The between/side ratios then diminished to values significantly less than unity, presumably due to contact inhibition. In marked contrast to this, the sarcoma cells were able to continue their outward migration relatively unimpeded by contact with fibroblasts in their path, so

that the between/side ratios of sarcoma cells remained approximately unity.

Overlapping and Invasiveness

This comparative insusceptibility of sarcoma cells to contact inhibition was also measured quantitatively by counts of nuclear overlaps. While chick and mouse fibroblast nuclei overlapped one another only about one-third or one-fourth as frequently as would be expected if their distribution were random, sarcoma cell nuclei (of both types) overlapped fibroblast nuclei much more frequently, the number of these overlaps being only a few percent less than the randomly expected number.

The overall conclusions of this paper were thus extremely exciting. Although normal fibroblasts, even those from different tissues of unrelated species, show contact inhibition toward each other, sarcoma cells seem "immune" to this inhibition. Sarcoma cells are, however, able to inhibit fibroblast outgrowth somewhat. It seems quite likely that this relatively unrestricted movement of sarcoma cells in culture could be closely analogous to the greater invasiveness of these cells in the body. This apparent analogy between the behavior of cells *in vitro* and *in vivo* seems to afford an excellent opportunity of studying directly special features of cancerous cells that may be related to their malignancy. In general, it seems fair to say that the widespread enthusiasm for the concept of contact inhibition may have sprung largely from the relative lack of susceptibility to contact inhibition shown by these and other (17–20) cancerous cells. To choose one important example, a very important area of current research involves the use of morphologically different plaques of cultured cells as an assay of the activity of oncogenic viruses in culture (21). This use of altered colonial morphology and, in particular, overlapping as an indicator of neoplastic transformation depends in large part on the supposition that reduced susceptibility to contact inhibition is equivalent to increased invasiveness.

DIRECT OBSERVATION OF CONTACT INHIBITION

In the three studies discussed so far the behavioral characteristics of individual cells were inferred from large numbers of statistical measurements of cell positions. However, in a subsequent study Abercrombie and Ambrose (22) used interference microscopy and time-lapse cinemicrography to analyze the details of individual encounters between cells. They found that normal fibroblasts (again from explants of chick heart and neonatal mouse muscle) react to contact with one another by ceasing forward movement in the direction of contact. Locomotion was, however, able to continue along those parts of the cell margin that were not in contact with another fibroblast. The movement of sarcoma cells, on the other hand, was found to be relatively uninhibited by contact either with fibroblasts or with other sarcoma cells. Abercrombie and Ambrose made one further observation that may relate directly to the mechanism of contact inhibition. They found that the ruffling movements of the cell surface, which are characteristic of the leading margin of moving cells, are inhibited along those parts of the cell margin that come in contact with other fibroblasts. This ruffling was not inhibited in contacting sarcoma cells. Because these ruffling movements of the cell periphery have been suspected of being a visible manifestation of the cellular mechanism of locomotion, this observation of the con-

tact inhibition of ruffling would seem to suggest that intercellular contact somehow "turns off" the cell's locomotory activity.

Contact Paralysis

In the words of Abercrombie and Ambrose, "the adhesion and cessation of activity of a leading ruffled membrane appear to be the visible expression of the contact inhibition previously reported." Because of this probable relationship, the paralysis of ruffling by contact is itself often referred to as "contact inhibition," as if the cessation of ruffling and the slowing of locomotion were entirely equivalent. However, as will be seen, this direct relationship between ruffling and locomotion is subject to some doubt, and so, to avoid confusion, the inhibition of ruffling will be termed contact paralysis, following the usage of Wolpert and collaborators (23,24).

In addition, Abercrombie and Ambrose observed that pinocytosis, which often occurs in association with ruffling, is also inhibited along the contacted margins of fibroblasts. This phenomenon might be called contact inhibition of pinocytosis. It can also be observed that the blebbing activity (surface herniation), which, like ruffling and pinocytosis, frequently occurs along the cell margin, is also inhibited along those parts of the margin in contact with other cells (25). One might call this contact paralysis of blebbing.

The probable significance of these observations is that the cell surface membrane is somehow least stable along the cell margin (resulting in ruffling, pinocytosis, and blebbing) and that contact (at least between normal cells) somehow "stabilizes" this membrane, so that these various types of surface movement are inhibited (cf. ref. 26). The surfaces of neoplastic non-contact-inhibiting cells are apparently not stabilized in this way by intercellular contact, or at least not to the same degree.

POSSIBLE MECHANISMS OF CONTACT INHIBITION

What Processes Are Inhibited?

Almost certainly the greatest single obstacle to the understanding of contact inhibition has been our inadequate knowledge of the physical mechanism of tissue cell locomotion itself. Those studying contact inhibition have been in the unfortunate position of having to explain how the cell's locomotory machinery is inhibited by contact, without actually knowing what this locomotory machinery consists of, and, in particular, without knowing which cellular processes are responsible for generating the propulsive forces (1). Considering this difficulty, it is no surprise that there has been no general agreement regarding the most probable physical explanation of contact inhibition. As one would expect, differences of opinion regarding the mechanism of cell locomotion have been accompanied by corresponding differences in the interpretation of contact inhibition.

The various proposed explanations of contact inhibition can be divided into two main categories. The first of these supposes that cells are in one or another way physically hindered by one another, whereas the second type of explanation proposes that the locomotory mechanism is somehow "turned off" or paralyzed by contact so that traction ceases to be exerted at the point of contact and progress ceases. These two alternative lines of explanation will be discussed separately.

Physical Interference as a Possible Cause of Contact Inhibition

RESTRICTION OF MOVEMENT BY MUTUAL ADHESION. There are several possible ways by which intercellular contact could interfere physically with movement. For example, Gail and Boone (27) have suggested that the strong and long-lasting adhesions formed between two 3T3 cells would tend to prevent either contacted cell from moving as far as it would have alone, since, on the average, their directions of movement would be different and would thus tend to cancel one another out. In support of this possibility, Gail and Boone showed that the average duration of adhesions between untransformed 3T3 cells was about three times as long (152 minutes) as the duration of adhesions between SV-40 virus–transformed 3T3 cells (50.5 minutes). Thus the reduced susceptibility of the transformed cells to contact inhibition might reflect the shorter period of time after each collision (in transformed cells) during which the cells are immobilized by their adhesion. On the other hand, Gail and Boone also showed that SV-40 virus–transformed 3T3 cells undergo many more collisions at high cell density than do untransformed cells (considerably more than three times as many, in fact), so it is hard to see how the threefold longer duration of adhesion of the untransformed cells could explain their inhibition of movement at high cell densities. This unresolved difficulty probably stems from Gail and Boone's measurement of adhesion duration at a low cell density only. Gail and Boone also do not explain how stronger mutual adhesion between untransformed cells would prevent cell–cell overlapping.

CONTACT INHIBITION DUE TO GREATER ADHESION BETWEEN CELL AND SUBSTRATUM THAN BETWEEN CELL AND CELL (DIFFEREN-

TIAL ADHESION). In brief, this theory supposes that contact-inhibited cells are more adherent to the underlying substratum (glass or plastic) than they are to one another, so that in maximizing the extent of their adhesions they form a monolayer. It is argued that overlapping is minimized because it entails a loss of strong cell–substratum adhesions in return for weaker cell–cell adhesions. This explanation of monolayering has been strongly supported by Carter's observation (28,29) that overlapping between cells can be greatly increased by culturing them on cellulose acetate, which is much less adhesive than glass. Abercrombie (10) had also observed such an increase in overlapping when cells were cultured on agar, which is also less adhesive. Carter's explanation of this increased overlapping was that on cellulose acetate the strength of adhesion between cell and substratum is reduced to (or below) the level of cell–cell adhesiveness. As a direct extension of this idea it might be supposed that sarcoma cells (or other relatively non-contact-inhibiting cells) are more or less equally adhesive either to glass or to other cells, so that they move indiscriminately between the two. This same type of explanation of monolayering by differential adhesiveness has recently been treated in detail by Martz and Steinberg (30), who consider that it is a special case of Steinberg's "differential adhesion hypothesis" by which Steinberg (31) has sought to explain the sorting out from one another of cells of different histological types.

HAPTOTAXIS AND "PASSIVE" LOCOMOTION. Carter has also observed that cultured cells move preferentially from less adhesive substrata (such as cellulose acetate) onto more adhesive substrata (such as palladium metal), a phenomenon that he has termed haptotaxis (28,32). Carter has proposed not only that monolayering

may be a special case of such haptotaxis but also that the invasion of normal tissue by cancer cells may also represent an instance of cells moving from less adhesive to more adhesive surroundings, the normal cells being more adhesive than the cancer cells (33). Actually, Carter has gone somewhat further than this and has proposed that tissue cell locomotion may be essentially "passive" in the sense that cells are pulled forward by the work of adhesion between the cell's leading margin and the substratum (34). In other words, he explains haptotaxis on the basis that cell spreading and movement on a solid substratum are essentially similar to the wetting of a solid by a liquid. If this really were the true explanation of cell locomotion, contact inhibition would *necessarily* reflect a greater adhesiveness of cells for the substratum than for each other.

Actually, however, Carter's theory of passive wetting is certainly not the true explanation of cell locomotion. For one thing, his observations on haptotaxis (preferential movement from less adhesive to more adhesive substrata) can be readily explained as the result of competition between several regions of active traction distributed around the cell periphery, each pulling outward in its own direction (5,35). Portions of the cell margin lying on a more adhesive substratum will simply gain a better foothold, so that the cell is pulled in that direction, as a result of active, rather than passive, locomotion. The net movement of such a cell could be compared to a tug of war between two teams, one of which is standing on a slippery surface.

Furthermore, recent observations by Abercrombie et al. (36) and by others (37–40), to be discussed later, have shown that cultured cells exert rearward tractional forces against particles attached to their surfaces, which of course is consistent with active locomotion, but not with passive "wetting" locomotion. In addition, Harris and Dunn (39) have shown that the outermost advancing margin of fibroblasts need not be attached to the substratum, as Carter's theory would require (28,29). Thus the "passive" theory of tissue cell movement seems to have been disproved conclusively. Of course this does not in itself exclude the possibility that monolayering is caused by differential adhesion of the cells to the substratum, as already discussed, since this explanation of monolayering is consistent with either active or passive locomotion.

CRITICISM OF THE DIFFERENTIAL ADHESION MODEL. Several serious arguments can be advanced against this differential adhesion theory of contact inhibition. In the first place the proponents of this theory have simply assumed that contact-inhibited cells are more adhesive to glass than to one another, whereas as a matter of fact there are good reasons to believe that the reverse is true. For example, many, if not most, contact-inhibiting cells form coherent sheets, considerable portions of which may be pulled away from the substratum (41,42). In other words, such cells may be pulled away from the substratum by a pull transmitted to them by intercellular adhesions. It is quite obvious in such cases that the strength of the intercellular adhesions must be considerably greater than the strength of the cell–substratum adhesions. This difference in adhesiveness is in direct conflict with the central assumption of the differential adhesion hypothesis, even allowing for possible discrepancies between the strength of these adhesions and the reversible work of adhesion involved in their formation [see discussion by Martz and Steinberg (30)]. Curiously, it is the less contact-inhibiting types of cell (e.g., sarcoma cells or macrophages) that fail to form

coherent sheets, indicating that in *these* cells (but not in the contact-inhibiting cells), cell–substratum adhesions really are stronger than cell–cell adhesions.

Actually it is not at all clear that living cells really do tend to maximize the area of their adhesions either to one another or to artificial substrata. Intercellular adhesions are known to consist largely of local specialized junctions of various types—tight junctions, desmosomes, and so on (43,44)—and even cell–substratum adhesions are localized to a relatively few points distributed around the cell periphery (42,45,46). Monolayers of some cells—especially epithelial cells, but also some fibroblasts—actually adhere to the substratum primarily or only at the margin of the *sheet,* so that cells in the central portion of the sheet adhere only to one another and are mostly not attached to the substratum at all (46). Likewise, cell monolayers can be formed by masses of fibroblasts stretched between glass fibers, so that most cells of such monolayers are not actually in contact with a substratum at all (47). Differential adhesion can hardly be responsible for these types of monolayering.

Another inadequacy of the differential adhesion theory of contact inhibition is its failure to explain contact paralysis of ruffling. Cells confronted by the margin of a less adhesive substratum continue to be quite capable of vigorous ruffling along this margin, whereas the ruffling would have been abruptly paralyzed by contact with a contact-inhibiting cell (5). In other words, cells do not react to a less adhesive substratum in the same way that they react to the surface of contact-inhibiting cells; thus the causes for failure to overlap must be different in the two cases.

Although it is perfectly true that the frequency of cell overlapping can be greatly increased by culturing cells on less adhesive substrata (29), it has also been found that cells will move onto less adhesive substrata in preference to overlapping one another (5). When cells are sufficiently crowded at the margins of a substratum of lesser adhesiveness, they will move out onto this substratum rather than moving up onto the surfaces of their neighbors. Clearly they would not be expected to behave in this way if their failure to overlap were due to differential adhesiveness alone. The probable reasons for the increased overlapping that occurs on less adhesive substrata will be discussed in the section on retraction clumping.

CONTACT INHIBITION "OF THE SECOND KIND." For the various reasons mentioned it seems clear that contact inhibition cannot be so simply explained as being merely the result of differential adhesiveness. Nevertheless, this concept may be of value in explaining *some* instances of monolayering behavior. For instance, certain cells, such as sarcoma cells and macrophages, tend to form monolayers on glass even though their locomotion continues unabated after they make contact with one another, and they show little or no contact paralysis of ruffling. Judging from the inability of such cells to form cohesive sheets, they would seem to adhere to the substratum more strongly than to one another. When moving in crowded cultures, such cells seem to push bodily against their neighbors, the cell margins frequently displacing one another from the substratum (personal observations). In *these* cases monolayering actually can be explained as the result of stronger adhesion between cell and substratum than between cell and cell. Such behavior could be considered as a special case of contact inhibition which fulfills the strict definition of contact inhibition (cf. ref. 10), but is the result of quite different causes than the contact inhibition

between normal untransformed fibroblasts. Abercrombie (personal communication) has suggested the term "contact inhibition of the second kind" to describe the type of monolayering that is due simply to differential adhesiveness to the substratum. This contact inhibition of the second kind would be characterized by the absence of contact paralysis and failure of cell sheet formation. Such monolayering due to differential adhesion is discussed in a recent paper by Vesely and R. A. Weiss (48), who introduced the term "type 2 contact inhibition" to describe the phenomenon. Since contact inhibition of the second kind would be dependent entirely on the substratum used, it might be considered almost as an artifact of tissue culture. It would certainly be difficult to draw any analogies to behavior *in vivo* without, for example, making assumptions about the relative adhesiveness of cells to one another and to collagen fibers, and so on.

Another possible explanation of some types of monolayering that should be considered is that the cell margins may be very strongly adhesive while the cell's dorsal and ventral (upper and lower) surfaces are nonadhesive (10). This could explain why certain types of cells tend to form coherent sheets and yet do no overlap. Although this possibility has received little experimental study, it is possible that the strict monolayering of many epithelial cells might be brought about in this way. Epithelial cells often show little or no overlapping of their margins, seeming to knit tightly together over a narrow marginal band. The specialized junctions (tight and gap junctions, desmosomes, etc.) with which these margins adhere to other cells certainly suggest some sort of adhesive specialization of the cell margin (43). Fibroblastic cells, however, do not adhere together in such precise sheetlike patterns, and this type of strictly marginal adhesion can probably be considered as a very interesting special case, though one that deserves much further study.

Contact Inhibition by Paralysis of the Locomotory Mechanism

For the various reasons recounted in the preceding sections it appears that the most typical cases of contact inhibition (as between untransformed fibroblasts) cannot be explained as a simple physical restriction on overlapping or movement. Apparently one must postulate instead that the locomotory activities that propel the cell are somehow "turned off" or paralyzed at points of intercellular contact. Unfortunately, as already mentioned, the mechanism by which tissue cells propel themselves is still poorly understood. Although it is absolutely certain that tissue cells do not move by a fountain zone flow of cytoplasm like the large free-living amoebae (1) and equally clear that they are not dragged passively along by the formation of new adhesions to the substratum (5,39), it is not at all certain just how they do move. This presents a problem to those studying contact inhibition since any very specific hypothesis about the "turning off" of locomotion must ultimately be related to some specific theory of the mechanism of cell locomotion itself. Consequently this section will devote considerable space to a discussion of the principal theories of tissue cell locomotion. Unfortunately, however, until locomotion itself is explained in mechanical and biochemical terms, it will be necessary to discuss inhibition without specifying exactly what is being inhibited.

INTERCELLULAR DIFFUSION OF INHIBITORS (ELECTROTONIC COUPLING). In recent years the idea has become popular that contact inhibition might be mediated by a chemical signal of some kind diffusing between cells. Work by Abercrombie and collaborators [(particularly the paper by Abercrombie and Gitlin (49)] has clearly shown that cells do not inhibit one another at a distance, but that actual contact between cells is required. Although this finding seemed at first to exclude entirely the possibility that diffusing substances might produce contact inhibition, the concept of diffusing inhibitors has since been revived by Loewenstein's discovery of low-resistance junctions between adhering cells of various types (50). Such "electrotonic" junctions, which have now been demonstrated to occur between many types of fibroblast, involve the formation of tight, gap, or occluding junctions surrounding zones or areas of high ionic permeability on apposed areas of the cell surface membrane (51,52). Ions and molecules diffuse between such electrotonically coupled cells, even though these ions and molecules are unable to diffuse out into the surrounding medium. In fact, the resistance to current flow between coupled cells is often no greater than that between equally distant sites within the cytoplasm of a single cell. Goshima (53) has shown that electrotonically coupled fibroblasts can transmit the depolarization wave between beating heart muscle cells in culture and has used this as a most elegant means of determining which particular types of fibroblast can form electrotonic junctions. Subak-Sharpe, Bürk, and Pitts (54) have shown that cell metabolites can diffuse between adhering cells, apparently by such electrotonic junctions.

The supposed relationship of such coupling to contact inhibition is based primarily on reports, such as that of Loewenstein and Kanno (55), that certain cancerous, and thus presumably non-contact-inhibiting cells, also show much reduced electrotonic coupling.

Similar correlations between diminished contact inhibition and loss of specialized tight and gap junctions have been reported by Martinez-Polomo, Braislovsky, and Bernhard (56) and by McNutt and Weinstein (52). The basic idea behind this work is that if cells lack the types of junction that are required for coupling, they presumably also lack electrotonic coupling. As a result the inhibitory signal would be unable to diffuse between adjacent cells and contact inhibition would fail, allowing the cells to be invasive and malignant.

Flaxman, Revel, and Hay (57) carried this approach another step forward by observing fibroblasts as they contacted one another and fixing and sectioning these cells at the precise time contact inhibition took place. Unfortunately the practical difficulties of such a procedure kept them from producing very clear-cut results. Despite their almost heroic efforts at sectioning cell contacts, they were only able to find a single definite tight junction. In a more recent ultrastructural study of the contacts formed between colliding chick heart fibroblasts Heaysman and Pegrum (58,59) found that areas of close membrane apposition formed within 20 seconds and plaques of microfilaments accumulated in the cytoplasm underlying these junctions. Although intercellular gaps as narrow as 50–100 Å were observed, no tight or gap junctions were found such as would be required to mediate electrotonic coupling. Other work has cast increasing doubt on the supposed correlation between contact inhibition and electrotonic coupling. For example, many sarcoma cells appear to become coupled, whereas other cells that show contact inhibition may lack any significant number of tight junctions (51). Ironically, L cells, such as were used by Flaxman and associates, were found by Goshima (53) to lack electrotonic coupling with heart myoblasts.

DIFFICULITIES WITH THE MODEL. Despite the popular appeal of this supposed relationship between electrotonic coupling and contact inhibition, the theory does not seem ever to have been given a very precise formulation. Indeed, it is rather difficult to imagine such a model that would be entirely consistent with the known facts of cell interaction. In particular, it must be explained how a diffusing substance could possibly inhibit a cell into which it diffused without already having inhibited the cell of origin. Unless one assumes that each individual cell produces a different inhibitor (to which only it would be immune), it becomes necessary to suppose that the diffusing inhibitor somehow becomes activated as it passes from one cell to its neighbor, while inhibitors passing in the other direction would likewise become activated at the cell–cell boundary. Otherwise any cell that was capable of producing contact inhibition would necessarily already be paralyzed by its own inhibitors. In actual fact, actively moving cells are perfectly capable of producing contact inhibition. Considerations such as these would seem to rule out the possibility that any simple ion like calcium could be responsible, since such ions could not be modified ("activated") at the cell boundary. Clearly this general type of model becomes less and less plausible as it is considered in more detail.

Some more recent evidence seems to exclude altogether such theories of diffusing inhibitors. Trinkaus, Betchaku, and Krulikowski (60) studied the pattern of inhibition (paralysis) of ruffling in colliding chick heart myoblasts and fibroblasts. They found that ruffling is paralyzed *only* at points of intercellular contact and that this paralysis does not gradually extend outward beyond the point of contact as one would expect if a diffusing inhibitor were responsible. In my own work I have found this to be true of other cell types as well (Fig. 2). Although it was already known—for example, from the work of Abercrombie and others (2,22)—that ruffling and locomotion can continue at the opposite end of a contacted cell relatively far from the contact, it is now clear that the paralysis, at least of ruffling, does not extend even as much as a few micrometers beyond the contact, even after several minutes.

Unless the hypothetical diffusing inhibitor were of extremely high molecular weight, it would be expected to diffuse through the contacted cell, so that the paralysis of motion would spread gradually even to the cell's opposite end. That this does not occur is exemplified by the ability of epithelial cell sheets to spread. In such cases the cells at the margin of the sheet exert traction and pull the rest along, while the cells in the interior of the sheet are inhibited by their neighbors. If the contact inhibition of the cells of the sheet were the result of a diffusing substance, the margin, also, would soon be inhibited. This failure of the inhibiting effect to extend any appreciable distance

beyond the contact might possibly be accounted for as the consequence of a very unstable inhibitor that breaks down rapidly as it diffuses. In general, however, it seems more and more likely that contact inhibition results from a mechanical interaction of some kind rather than from the diffusion of an inhibitor.

INHIBITION OF MOTILITY BY CYCLIC AMP. One specific substance that has been suggested as the possible diffusing inhibitory agent is cyclic adenosine 3′,5′-monophosphate (AMP). It has been proposed that contact between adjacent cell membranes might serve to activate the enzyme adenyl cyclase within the membrane, increasing the synthesis of cyclic AMP and raising its concentration within those cells having contacts with their neighbors. This is supported by the observation that the cytoplasmic concentration of cyclic AMP in 3T3 fibroblasts rises threefold or fourfold at confluency, but that such an increase in concentration does not occur in a number of comparable cell lines that lack contact inhibition of locomotion (61). In addition, it is found that the addition of dibutyryl cyclic AMP to the culture medium

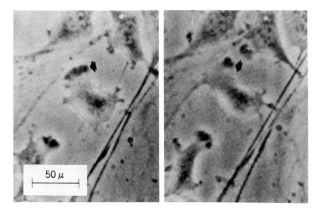

Fig. 2. Contact paralysis of ruffling in a Sarcoma-180 cell resulting from contact with the margin of a 3T3 mouse fibroblast. *Left:* sarcoma cell advances toward upper left, with active ruffling activity along its leading margin (arrow). *Right:* 2 minutes later the sarcoma cell has contacted the fibroblast and ruffling has ceased along the contacted margin, but ruffling continues where the cells are not yet in contact (arrow). (Original photograph from time-lapse film.)

causes a marked inhibition of cell loco-motion. Motility can also be inhibited by the addition of prostaglandin E, which is known to activate membrane adenyl cyclase and thus to raise the intracellular concentration of cyclic AMP.

Taking all these facts together, it would appear very likely that at least part of the inhibition of movement in confluent 3T3 cells (as well as other untransformed cells) may be due to their increased content of cyclic AMP, resulting from the activation of adenyl cyclase at points of contact. Furthermore, it has also been shown that when dibutyryl cyclic AMP is added to certain relatively non-contact-inhibiting cells, these cells undergo a marked change in morphology that supposedly makes them look more like normal untransformed contact-inhibiting fibroblasts (62). Insulin has now been shown to produce comparable effects on cell morphology, probably by stimulating adenyl cyclase activity (63). Since high levels of cyclic AMP have also been found to inhibit cell multiplication, these observations suggest a possible link between contact inhibition of locomotion and contact inhibition of growth.

However, there are some criticisms that may be raised against this explanation of contact inhibition. For one thing, exposure to dibutyryl cyclic AMP does not inhibit ruffling activity as strongly as does intercellular contact (3). Likewise, the inhibition of locomotion seems to be more gradual than that produced by contact. It is also most difficult to explain why contact paralysis occurs only at the point of intercellular contact. If the inhibiting substance were diffusible and of relatively low molecular weight (as is cyclic AMP), it should become rapidly distributed to all parts of the cell, and inhibition should not be confined to the point of contact as is actually observed.

CONTACT INHIBITION BY KILLED CELLS? Some of the questions raised here can be approached indirectly by observing the behavior of moving cells as they encounter killed and fixed cells. S. A. Cairns and R. A. Weiss [unpublished observations cited by Abercrombie (10)] found that fibroblasts could move readily onto the surfaces of glutaraldehyde-fixed fibroblasts. This shows that contact inhibition cannot be due simply to the physical obstruction of the cell margin, since the fixed cells retained their original shape. In addition their observation suggests that whatever property of the cell does produce the inhibition must itself be destroyed or inactivated by fixation. I also have studied the effects of fixed cells on living cells, albeit with slightly different results. Both 3T3 and other fibroblasts were fixed rapidly by perfusion either with formaldehyde (freshly prepared from paraformaldehyde) or with glutaraldehyde, and the cells were thoroughly washed with several changes of saline before new medium and living cells were introduced. At first cells treated in this way seemed to elicit normal contact inhibition, there being relatively little overlapping between the living cells and the fixed ones. However, closer analysis by time-lapse cinemicrography showed that ruffling was not inhibited by contact with these fixed cells. Apparently the tendency not to overlap these fixed cells is attributable merely to contact inhibition "of the second kind" (as defined above), that is, to a simple difference in adhesiveness between the substratum and the fixed cells. In Fig. 3 one can observe the relative lack of spreading among the few cells that have reached the surfaces of the fixed cells. The recent paper by Vesely and R. A. Weiss (48) reports a similar result.

It is probably very significant that no one has ever found contact inhibition of

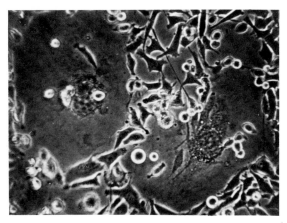

Fig. 3. Failure of live Sarcoma-180 cells (small cells) to overlap aldehyde-fixed 3T3 fibroblasts (the two large flat cells). Since the ruffling activity of the sarcoma cells is not inhibited by contact with the fixed fibroblasts and since the fixed cells were apparently less adhesive than the substratum (note reduced spreading of the few sarcoma cells on the fibroblast surface), this tendency not to overlap was judged to be contact inhibition "of the second kind" (due to reduced adhesiveness) rather than true contact inhibition (due to inhibition of the locomotory mechanism). (Original photograph from time-lapse film.)

a cell by anything other than another living cell. Attempts to produce contact inhibition with nonliving objects should by no means be abandoned, however, since there are few accomplishments that potentially could tell us more about the physical interaction responsible for contact inhibition.

RELATIONSHIP BETWEEN RUFFLING AND LOCOMOTION. It has been widely assumed that there is some direct, probably causal, relationship between the "ruffling" behavior characteristic of the advancing cell margin and the tractional forces that produce cell locomotion. These rearward-moving ruffles certainly give a strong subjective impression that they are somehow pulling the cell margin outward, this impression being especially convincing when these movements are accelerated by time-lapse cinemicrography. This impression of a causal relationship is strengthened even further by the observation that when ruffling ceases either spontaneously or because of contact inhibition, or because of inhibi-

tion by some drug or change in culture conditions, locomotion also comes to a halt (in most cases) along the affected part of the cell margin (22). However, since the physical mechanism of tissue cell locomotion is as yet unknown, it has never been possible to determine directly what if any linkage exists between ruffling and movement. For this same reason it is quite unclear what the relationship may be between contact inhibition of locomotion and contact paralysis of ruffling.

LOCOMOTION BY PERISTALTIC WAVES? From his light-microscopy observations on fibroblast locomotion, Ambrose (64) has developed a very plausible theory explaining traction as a result of ruffling. Although more recent observations make it appear that this theory is almost certainly invalid, nevertheless Ambrose's theory has become sufficiently well known to merit a thorough consideration. His basic idea was that ruffles might represent peristaltic waves moving across the cell surface and that these waves might serve to exert rearward traction against the substratum in much the same fashion as the contractile waves that serve to propel snails, nemertine worms, and certain annelid worms. To explore this possibility, Ambrose developed a special type of microscope designed to make visible the cell's contacts with the substratum (65). This "surface contact microscope" worked on the principle of the critical angle refractometer. Cells moving on a glass surface were illuminated from below with a ray of light directed at an oblique angle to the glass–water interface. The essential principle is that the light beam will not pass from a region of higher refractive index to a region of lower refractive index if the angle of incidence at the boundary is too oblique. If the sine of this angle of incidence were to exceed the ratio of the refractive indices of the two media, total internal reflection would occur at the interface and objects above the surface would not be illuminated. Ambrose reasoned that, because cytoplasm has a slightly higher refractive index than the culture medium, it should be possible to adjust the angle of illumination so that only contact points would be illuminated, since only at these contact points would light be able to enter the cells. Consequently it should be possible to observe contact points either as bright spots of transmitted light if the cells were observed from above or as dark spots if the reflected beam was observed from below.

Using this method, Ambrose reported the existence of many tiny points of contact between fibroblasts and their glass substratum (64). These points appeared to move rapidly in a rearward direction, and Ambrose supposed that they must be propelling the cell forward. If Ambrose's hypothesis were correct and if

tissue cells really were pulled along by such peristaltic contractions, the relationship between the contact paralysis of ruffling and the contact inhibition of movement would be clear. If the supposedly wavelike ruffles and their associated contact points ceased to be propagated rearward whenever cells came into contact, the rearward traction would cease to be exerted and movement would stop. Such explanations of contact inhibition have been widely considered in the literature, and at least one theoretical model has been developed to explain how ruffling movements might be damped out at points of intercellular contact (66).

EVIDENCE AGAINST PERISTALTIC LOCOMOTION.   Attractive though this hypothesis may be, more recent studies have failed to substantiate it. In the first place, the ruffles of the cell margin have been shown to arise by direct upfolding of the cell margin (37,67–69), and although these thickenings do sometimes propagate rearward, most of their apparent motion turns out to consist of backward folding. In fact, these upfolded ruffles are neither undulatory or peristaltic and seem only to reflect contractions in the cell's upper surface.

Furthermore, Ambrose's observations with the surface contact microscope have not been confirmed by other methods designed to locate the points of contact and adhesion between cell and substratum. For example, Curtis (70), using interference reflection microscopy, found that the closest contacts between fibroblasts and glass were localized in a strip near the cell margin. Likewise, it has been found by micromanipulation that cell–substratum adhesions are usually restricted to the cell periphery and that such adhesions are rarely found in the more central areas beneath the main cell body, where Ambrose reported seeing "contact points" (42).

It appears that Ambrose did not consider carefully enough the optical principles underlying surface contact microscopy, particularly as applied to nearly transparent flattened objects such as fibroblasts. Even if a ray of light were to gain entrance to the cell at a contact point, this light ray would still undergo total internal reflection when it reached the cell's upper surface (so long as this surface were also parallel to the substratum). This light ray could escape total internal reflection and be seen by the observer *only* if it were somehow scattered or bent, as, for example, by encountering a mitochondrion, a pinocytotic vesicle, or a ruffle. In such a case one would actually "see" the object scattering the light rather than the contact point itself; that is, the point of light would be imaged at the point of scattering rather than at the point where the light ray originally entered the cell. Such a microscope would actually function in a way analogous to dark-field illumination and thus could not accurately locate contact points (at least not in flat transparent cells such as fibroblasts).

MOVEMENT OF PARTICLES ATTACHED TO THE CELL SURFACE. More recent research on tissue cell locomotion has revealed that various particles that become attached to the cell's outer surface are transported rearward (i.e., in the direction opposite locomotion). This rearward transport of attached particles occurs not only on the cell's upper surface (36,37,40) but also on the lower surface (38,39), and so particle movement would appear to result from the cell's application of rearward traction to small movable objects (i.e., the particles) as well as to large immovable ones (i.e., the substratum). Since Ambrose's peristaltic wave theory

cannot account for the exertion of traction against small objects, it can probably be discarded as being inconsistent with the facts. In addition, the observations on particle transport have given rise to an entirely new theory of tissue cell locomotion, according to which the entire cell surface, including the surface membrane, moves continually in a rearward direction, on both upper and lower surfaces, with membrane being continually reassembled along the leading margin, disassembled somewhere toward the rear, and recycled to the front through the cytoplasm, perhaps in the form of vesicles, perhaps dissolved as "subunits." The rearward membrane flow would exert traction, pulling the cell forward as well as flattening it. The evidence supporting this somewhat revolutionary theory of membrane flow has been described elsewhere (25,38). (see Fig. 4)

DOES CONTACT INHIBIT MEMBRANE ASSEMBLY? Assuming for the moment that tissue cells actually are propelled by membrane flow, one might speculate that contact inhibition occurs when the assembly of membrane at the leading margin is inhibited by contact with another cell. This would not only explain the inhibition of locomotion but would also account for the contact paralysis of ruffling and of blebbing, since both these types of surface movement seem to involve rapid local assembly of surface membrane (25). Cells whose membrane assembly process is for some reason not inhibited at points of intercellular contact would thus be non-contact-inhibiting and probably invasive. Although it is not at all clear why intercellular contact should inhibit membrane assembly, one possibility is that the two apposed surface membranes are somehow physically reinforced or stabilized by their mu-

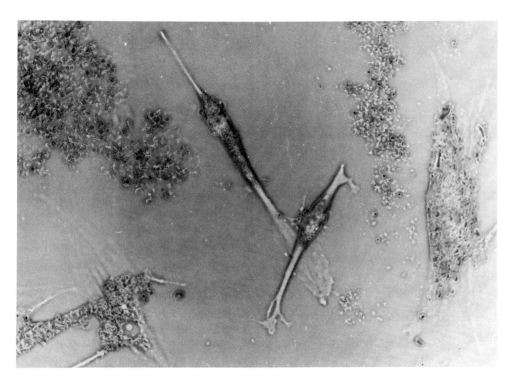

Fig. 4. Fibroblasts from rabbit synovium cultured on glass. Fine glass particles had been scattered evenly over the substratum several hours previously, and the cells have picked many of these up (note the cleared areas) and transported them centripetally to form accumulations surrounding and over-lying the nucleus. This process of particle transport does not involve phagocytosis and is believed to reflect the cells' locomotory traction being applied to movable objects. Note that the two cells in the center overlap, the large one having extended a process beneath the unattached central region of the smaller. Such "underlapping" does not involve the movement of one cell onto the surface of another.

tual adhesion. Unfortunately it is by no means certain, yet, that membrane assembly and flow really do occur in moving fibroblasts, much less that it is these processes that are inhibited during contact inhibition. Better and more unequivocal means of observing membrane flow are urgently needed to test this theory. For example, one might expect a reduction in "capping" (of surface antigens) in confluent contact-inhibited cells. On the basis of current evidence this does seem to be the most probable explanation, however.

Now that ruffles are known to be marginal upfoldings, contact paralysis of ruffling could be readily explained as a result of the contacting cell margins sticking together and thus being physically prevented from folding backward away from each other. It has long been known that intercellular adhesions form at sites of contact inhibition (71), and, as a general rule, contact-inhibiting cells seem to be more adhesive than non-contact-inhibiting cells. On the other hand, the simple physical prevention of ruffle upfolding is not in itself sufficient to prevent locomotion or even appreciably to inhibit it. Cells susceptible to contact inhibition will continue to move even when overlain with layers of agar or plasma clot, which prevent ruffling, but not locomotion (personal communication from M. Abercrombie).

CAUSES OF OVERLAPPING (WITHOUT LOSS OF CONTACT INHIBITION)

If we assume that monolayering among contact-inhibiting cells is not actually the result of greater adhesiveness to the substratum than to other cells, as already discussed, how can one explain the increase in overlapping that occurs when cells are cultured on agar (10), cellulose acetate (28), and the like, instead of on glass or wettable plastic? Ordinarily one tends to assume that if one cell comes to lie on top of another, it must have "crawled" there by the same means as those that propel it across the substratum. Thus it is usually concluded that extensive overlapping between cells necessarily indicates that they have an especially increased ability to use the surfaces of other cells as a substratum. Although this is sometimes the true explanation, there are at least three additional ways in which overlapping can occur without one cell moving directly onto the surface of another. These additional types of overlapping are (1) *retraction clumping,* in which one cell is pulled by contraction onto another; (2) *underlapping,* in which one cell crawls under (rather than over) another; and (3) the secretion of extracellular materials over which cells can move without actually contacting the surface of the cell beneath.

Retraction Clumping

Whenever one or more of a mutually adherent cluster of cells pulls away from the substratum, it is pulled back toward the other cells by its own contractility (and/or elasticity). Quite frequently such retracted cells come to rest on top of their neighbors, so that an overlap results, even though neither cell has actually crawled up onto the other. This process is called retraction clumping and occurs sporadically among cultured cells, especially when the adhesiveness of the substratum is reduced relative to intercellular adhesiveness, as, for example, when cells are grown on cellulose acetate instead of glass (see Fig. 5) (5). Because retraction clumping is basically a phenomenon of deadhesion and contraction (rather than locomotion), the overlaps produced by this means occur in clusters next to more or less vacant areas of the substratum from which the overlapping cells had been pulled away. Such clusters of overlaps next to clear areas are apparent in Carter's illustrations (28) in which he shows the supposedly reduced contact inhibition of L cells grown on cellulose acetate (note especially Carter's Fig. 6). Presumably, therefore, the increased overlapping that Carter observed must have been due at least in large part to increased retraction clumping. Extensive retraction clumping has also been observed among 3T3 fibroblasts and other cells cultured on cellulose acetate and nonwettable polystyrene (5). In such cases new clumps of overlapping cells can be easily produced by sharply rapping the culture chamber so that a few cells pull away from the substratum. These occurrences of cell detachment and retraction can be observed directly since they are complete in only a minute or so.

Because the overlaps produced by retraction clumping are not actually the result of one cell actively crawling onto the surface of another, they should definitely not be considered as a true failure of contact inhibition. In this sense it is misleading to say that reduced adhesive-

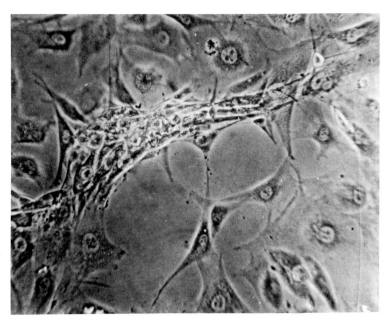

Fig. 5. Retraction clumping in a sheet of 3T3 mouse fibroblasts cultured on a (relatively) nonadhesive) cellulose acetate substratum. Note that the vacant area adjacent to the clumped cells had detached and retracted elastically. Since this type of overlapping does not involve locomotion, it should not be considered a failure of contact inhibition. From Harris (5), reprinted by permission from *Experimental Cell Research*.

ness of the substratum reduces the susceptibility of cells to contact inhibition. In order to justify the conclusion that cells are actually less subject to contact inhibition, one would have to show that they actually have an increased ability to crawl across the surface of other cells.

On the other hand, it is worth noting that when contact-inhibiting cells such as 3T3 fibroblasts are cultured on relatively nonadhesive substrata, they do undergo a striking morphological change, causing them to resemble transformed non-contact-inhibiting cells. When 3T3 cells are cultured on glass or wettable polystyrene, they are highly flattened with broad lamellar extensions. When these cells are cultured on cellulose acetate or nonwettable polystyrene, however, they become much more spindle-shaped, with narrow ruffled lamellae and thick refractile centers around the

nucleus (5). Cells with such a morphology might easily be mistaken for transformed fibroblasts or even sarcoma cells, and it is difficult to believe that this close similarity is simply fortuitous. This alteration in morphology by the substratum is only temporary, however, and 3T3 cells regain their original flattened shape when returned to a glass substratum.

Underlapping

Another important way by which overlaps can occur without cells actually using one another as substrata is underlapping. Underlapping occurs when the advancing margin of one cell encounters the margin of another cell at a point where this second cell is not adhering to the substratum. The advancing cell is thus able to crawl *beneath* the second, so that an apparent overlap is formed even though neither cell has actually crawled onto or over the surface of the other, as one would be apt to assume if observing only the result. Boyde, Grainger, and James (72) have conclusively demonstrated underlapping using scanning electron microscopy and time-lapse cinemicrography, and others have recorded the phenomenon by cinemicrography and micromanipulation (42,45).

Underlapping can only occur because of the curious fact that fibroblasts and many other cells cultured on flat substrata adhere to the substratum only near certain parts of their margins, leaving a large nonadherent area near the cell center. In fact, the usual stellate or spindle-shaped fibroblasts adhere to substrata only near the tips of their several extended lamellae, so that they are actually stretched out between a few small peripheral adhesions. This is what gives them their shape (42,45,73). The reasons for this restriction of adhesions to such a small proportion of the cell surface are not known, though the phenomenon may be related to the localization of intercellular adhesions in the form of tight junctions, desmosomes, and the like, which are observed by electron microscopy. Ordinarily it is the convex extensions of the cell margin (where ruffling occurs) that *do* adhere to the substratum, and the concave indentations along the margin that *do not* adhere. Likewise elongate cells always adhere at the ends but often not at the sides, with the result that other cells are able to extend beneath (underlap) the cell sides. Consequently, where there is extensive underlapping, a pattern of criss-crossing results, with underlapping occurring principally between cells whose long axes lie at a steep angle to one another. (see Fig. 4)

Obviously a given cell's susceptibility to underlapping will vary according to the proportion of its periphery that adheres to the substratum. When most or all of a cell's margin is stuck down firmly to the substratum, it will be difficult or impossible for another cell to crawl beneath it, and a population of such cells will show little underlapping. A good example is provided by 3T3 cells since they adhere to the substratum over a large proportion of their margin and show little underlapping. Conversely, more stellate cells (such as those of the line Sarcoma-180), having only a small fraction of their margin stuck down to the substratum, are highly susceptible to underlapping, and many or most of the overlaps occurring among such cells are attributable to this underlapping (42).

As mentioned in the preceding section, 3T3 cells become stellate when cultured on cellulose acetate instead of glass. This change in shape reflects a smaller fraction of their margin adhering to the substratum and allows an increased amount of underlapping. Thus it appears that at least some of the increased "overlapping" observed on less adhesive substrata may be attributable to underlapping (in addition to retraction clumping).

In fact, some studies have suggested that many or most of the overlaps occurring in cultures of transformed, relatively non-contact-inhibiting cells are actually due to underlapping, the cells locomoting beneath one another instead of over one another. In such cases the apparent loss of contact inhibition would not necessarily involve an increased ability of the cells to use one another's surfaces as substrates, as has usually been assumed. Using time-lapse cinemicrography to observe the formation of overlaps between normal and polyoma-virus-transformed 3T3 fibroblasts, Bell (74) has found that virtually all overlaps between transformed as well as untransformed cells occur by underlapping, and *not* by locomotion onto one another's surfaces, as others had previously concluded. Because the transformed cells have a reduced area of adhesion to the substratum, they have a greater susceptibility to being underlapped by other cells, and because they have narrower lamellar extensions, they are more able to extend between the marginal adhesions of other cells, thus underlapping them. The consequent underlapping seems to be the primary cause of the extensive "criss-crossing" observed among transformed 3T3 cells (and presumably among transformed derivatives of other cell lines as well). As Bell says, "the radical difference in culture pattern of the two cell types at confluency seems due to differences in patterns of cell substratum adhesions and cell morphology." In addition he observed that contact inhibition occurs in transformed as well as untransformed cells, presumably when an advancing cell makes contact with another cell margin at a point where this second cell *is* in contact with the substratum. He concludes that "an end result of criss-crossing therefore cannot be taken as evidence of reduced contact inhibition of movement." Similar observations have recently been reported by Guelstein and associates (75), who compared the relative frequencies of underlapping, overlapping, and immediate halting in collisions between untransformed mouse fibroblasts and between transformed mouse fibroblasts. They found that true overlapping occurred infrequently and that underlapping occurred with approximately the same frequency among untransformed cells as it did among transformed ones. They therefore conclude that the greater degree of criss-crossing observed among transformed cells is not attributed to decreased contact inhibition of movement. However, they do mention that untransformed cells usually

underlapped one another only slightly before retraction occurred, whereas transformed cells often extended completely beneath one another. In other words, the *frequency* of underlapping was approximately equal in the various cells studied, but the *extent* of underlapping was much greater in the transformed cells than in the untransformed ones, presumably accounting for the criss-crossing of the former.

Should Underlapping Be Considered a True Failure of Contact Inhibition?

The observations of Bell and of Guelstein et al. raise the very serious question of whether the criss-crossing of cells due to underlapping should be considered as a true failure of contact inhibition. Ordinarily a lack of contact inhibition is taken to mean that cells have an increased ability to crawl on the surfaces of other cells, although the concept was not so defined by its originators (8–10). Underlapping, however, does not involve such an ability to use other cells as substrata. In fact, DiPasquale and Bell (76) have found that neither Sarcoma-180 cells, polyoma-virus-transformed 3T3 cells, KB (carcinoma) cells, or chick heart fibroblasts were able to spread or move on the surfaces of various epithelial sheets or on one another's surfaces (when allowed to settle out from suspension), even though they were able to form adhesions with one another.

Clearly, the ability of these cells to "overlap" one another cannot be attributed to a capacity to crawl on one another's surfaces. On the other hand, Abercrombie's films (personal observation) quite clearly do show sarcoma cells crawling over (rather than under) fibroblasts, so true overlapping *can* occur at least in some cases. It would seem that each of the previously studied cases of non-contact-inhibiting cells must now be restudied in order to determine which of these cases are attributable to underlapping and which actually involve cells crawling over one another.

Perhaps the central question in these considerations should be, which aspects of cell behavior in culture are most closely and reliably correlated with invasiveness and malignancy *in vivo*? It should certainly not be forgotten that one of the primary reasons for being concerned with contact inhibition has been the belief that neoplastically transformed cells tend to overlap more than normal ones. Unless contact inhibition can be redefined in such a way that this correlation with noninvasiveness is retained, the concept of contact inhibition will certainly lose much of its utility and importance. From this point of view it would perhaps be best to consider underlapping as a failure of contact inhibition only when it is *very extensive,* with large parts of cells including the nuclear region passing freely beneath their neighbors (criss-crossing). On the other hand, cases in which cells extend only short distances beneath one another before their ruffling is inhibited and retraction ensues would be considered as instances of true contact inhibition, even though some underlapping had occurred. Although a reduced area of adhesion to the culture substratum may be a necessary condition for underlapping to occur, it is not a sufficient condition, because contact inhibition, contact paralysis, and retraction of the extended process can all still occur during underlapping. As a consequence, contact inhibition can reduce the *extent* of underlapping regardless of the proportion of the cell margin adhering to the substratum or the *frequency* of underlapping. So if transformed cells are able to extend long distances beneath other cells and remain there, this ability probably does reflect

a decreased susceptibility of the locomotory apparatus to paralysis, and not just a reduction in the area of the cell periphery adhering to the substratum. In this respect the loss of contact inhibition may be related to invasiveness after all, so long as only very extensive underlapping is classed as a failure of contact inhibition.

Effect of Urea on Overlapping

Weston and coworkers have discovered that one can temporarily increase the extent of overlapping between chick heart fibroblasts and 3T3 cells by adding urea to the culture medium (77). For example, the addition of 0.2 M urea to the medium more than triples the nuclear overlap ratio. When returned to normal medium, cells rapidly return to the normal overlap ratio within about 1 hour, but this reversion does not occur if cycloheximide is added to the culture medium, inhibiting the cells' synthesis of new proteins (78). However, if the urea is dialyzed out of the original medium and if the cells are returned to this medium, the cells will regain their previous contact-inhibited state even if cycloheximide is added. This factor that allows a return to contact inhibition is sensitive to heat and trypsin. The apparent explanation of these findings is that urea causes the release from the cell surface of some protein that is required for cells to contact-inhibit one another in the usual way. Cycloheximide prevents the cell from resynthesizing this protein, but the protein can reattach to the cell and restore its contact inhibition once the urea is removed. Urea-treated fibroblasts were found to be much less adhesive to other cells, although their adhesiveness to glass was apparently undiminished (77) (cf. ref. 79). They were also much more readily agglutinated by concanavalin A, which suggests that urea exposes α-mannosyl residues on the cell surface, as Burger and coworkers (80–83) have demonstrated in virus-transformed fibroblasts. Urea treatment was also found to release fibroblasts temporarily from contact inhibition of growth, suggesting some direct link between the two varieties of contact inhibition (78). Unfortunately it has not yet been determined whether transformed cells are made still less contact-inhibiting by urea, whether they release a factor like that released by normal fibroblasts when treated with urea, or whether transformed cells are made more contact-inhibiting by the urea-released factor from normal cells.

It is apparent from Weston's photographs and films of urea-treated fibroblasts that the increased overlap ratio is due primarily to underlapping. Urea-treated cells crawl under one another, showing little ability to form intercellular adhesions. Weston and Roth (77) have proposed a rather complicated theory of contact inhibition to account for these results and those of Carter (28). Briefly, they propose that if cells are more adhesive to the substratum than to each other, they will crawl under each other, whereas if they are more adhesive to each other than to the substratum, they will crawl over one another. On the other hand, cell overlapping will be minimized if the cells' adhesion to one another matches their adhesion to the substratum so that they crawl neither under nor over. As an alternative, one might suggest the possibility that the cellular locomotory apparatus is inhibited at sites of intercellular adhesion and that urea simply diminishes this adhesion. The increase in overlapping that results from reduced substratum adhesiveness can be attributed to retraction clumping (5). One interesting prediction yet to be tested is that urea

should *reduce* the extent of overlapping among cells on cellulose acetate.

Overlapping Due to Secreted Materials

Many cell types, especially fibroblasts, secrete collagen and other extracellular protein materials that may accumulate in cultures in the form of layers, both above and below the cells. By moving across or between such layers, cells may overlap one another without actually ever coming in contact, their surfaces being insulated by the extracellular material between them. As a result cells that would otherwise be contact-inhibited by one another would be enabled to move over one another in multiple layers, the extent of overlapping becoming more pronounced with increasing amounts of extracellular protein. The electron-microscopic sections of Yardley (84) demonstrated that between fibroblasts there are layers of material, probably collagen, that seem to be responsible for their extensive overlapping. Elsdale and Foley (85) studied multilayering in fetal lung fibroblasts and found that these cells secrete a collagenous matrix substrate around and above themselves. Cells moving onto this matrix in turn secreted more matrix, eventually creating a sort of "Dagwood sandwich" of alternate layers of cells and collagen, sometimes up to 30 cell layers thick. When these cells were cultured in a medium containing the enzyme collagenase, this matrix was apparently digested as fast as it was secreted, with the result that multilayering was entirely prevented.

Elsdale and Foley (85) were particularly interested in the formation in their cultures of thickened ridges composed of very many layers of cells. To explain this tendency of localized overlapping, they suggested that only immobilized cells secreted continuous layers of matrix, so that inhibition of movement at some point would tend to promote subsequent multilayering in that area due to the increased deposition of the matrix material. Elsdale and Bard (86) made vertical sections through multilayered cultures of fetal lung fibroblasts. When they studied these sections by electron microscopy, they found that the extracellular fibrous material occupied not more than about 30% of the area between overlapping cell layers. They concluded that this amount of extracellular material would not be adequate to insulate the overlying cell layers from contact and thus prevent their contact inhibition. To explain the observed overlapping they propose that this extracellular material (collagen) may serve to stabilize a special bipolar, highly spindle-shaped cell that they believe is not subject to contact inhibition. Thus instead of simply preventing contact, the extracellular material would alter the cells to a new form immune to contact inhibition. Elsdale and Bard do not make it entirely clear why cells of this altered shape should necessarily lose contact inhibition, however, and it is also unclear whether a 30% coverage of the cell surface would or would not be sufficient to insulate it from contact inhibition. Cultured cells are known to adhere to solid substrata over a relatively small proportion of their surface area, often less than 30%, so perhaps the observed amount of extracellular material is, after all, sufficient to insulate the overlying cells (42).

Quite probably there are still more ways in which overlapping can occur without an actual breakdown of contact inhibition. One must therefore be quite wary of concluding that an increase in overlapping is necessarily due to a reduction or loss of contact inhibition. Instead, this overlapping might reflect (1) an increased secretion of collagen; (2) a reduction in the proportion of the cell margin adhering to the substratum (allowing underlapping); (3) increased cellular contractility or decreased substratum adhesiveness (resulting in retraction clumping). Probably the only way to be certain of the extent of contact inhibition is actually to observe cell movement directly (or by time-lapse cinemicrography) and to determine whether or not cells are capable of active locomotion directly on the surface of other cells. However, even direct observation might fail to detect such complicating factors as the presence of collagen layers between cells, and it could sometimes be difficult to distinguish between contact inhibition of the first kind (actual inhibition of the locomotory mechanism) and contact inhibition of the second kind (failure of intercellular adhesion). These uncertainties will probably remain with us until the mechanism of cell locomotion itself is sufficiently well understood to permit determination of when this mechanism is functioning and when it has been inhibited.

ALIGNMENT

A characteristic feature of many types of fibroblast (fetal lung fibroblasts, cells of the lines BHK and WI-38, etc.) is a tendency to form tracts of mutually aligned elongate cells. Such alignment is believed by many to be a sign that the cells in question are contact-inhibited, and, conversely, loss of alignment is sometimes considered to be equivalent to loss of contact inhibition. For example, in much of Stoker and MacPherson's work on the neoplastic transformation of BHK fibroblasts by oncogenic viruses, loss of cell alignment is taken as a sign of transformation, and such transformed cells are spoken of as being non-contact-inhibiting (21). Actually it is rather doubtful whether the degree of cellular alignment is always a reliable measure of the extent to which cell movement is inhibited by contact. On the other hand, these two phenomena do very often seem to be closely correlated, and, of course, alignment is interesting in its own right.

The work of Elsdale and collaborators (85–87) has been particularly concerned with the extreme degree of parallel orientation that is observed among cultured human fetal lung fibroblasts. After these cells are initially plated out, they begin to form small regions of mutually oriented cells, and these tracts gradually intermingle and fuse, becoming larger and larger. After extended periods of culture in a given dish, all of the cells become aligned parallel with one another. Using time-lapse cinemicrography, Elsdale studied the locomotion of the individual cells and discovered that these mutually aligned fibroblasts move freely along the surfaces of one another, continually shuffling back and forth in parallel paths. However, when contact is made between cells whose long axes are not parallel (the end of one cell running into the side of another), something very interesting is observed. If the axes of the two cells differ by less than about 20 degrees, forward movement is not inhibited and the advancing cell swings in beside the other, becoming parallel and remaining so, thus joining its tract of mutually aligned cells. On the other hand, when the angle between the axes of the colliding cells is greater than about 20 degrees, the advancing cell undergoes contact inhibition (87). Apparently it is as a result of this *directional* contact inhibition that elongate fibroblasts are gradually recruited into parallel tracts. Unfortunately Elsdale has not yet provided a complete analysis of this most intriguing directionality of contact inhibition. One is especially curious as to whether transformed (crisscrossing) cells show a reduced directionality of contact inhibition.

CONTACT RETRACTION

One of the more characteristic features of intercellular collision is an abrupt retraction of the contacting cells, often causing them to spring apart, though they can also spring together. This phenomenon has been termed contact retraction by P. Weiss (88,89) and is probably at least partly responsible for the transient increases in cell speed that Abercrombie and Heaysman (8) observed to occur when contact between fibroblasts is lost. This type of retraction is presumably an important clue to the nature of the cellular reactions involved in contact inhibition, and the phenomenon has usually been interpreted as evidence that contact induces cells to contract—that is, to increase the tensional forces tending to retract their extensions (10). This is not the only possible explanation, however. Spread cells are always stretched out under tension, so that any cell extension will be retracted once it becomes detached from its support (42,90). Because cells will always retract if detached, one cannot be certain that contact retraction is due to increased contractility at the point of contact. It could as well be the result of a loss of substratum adhesion at the point of contact. Of course it is quite possible that the observed detachment and retraction are in fact due to increased contraction, but one would have to have some means of measuring changes in cell tension before this could be determined. One alternative possibility is that cells might tend to detach from the substratum at points of intercellular contact, so that if the adhesion formed between the cells were too weak to support the cells' contractile tension, contact retraction would occur. Cells of many types actually do tend to form sheets adhering to one another with a loss of substratum adhesion along those parts of the cell margin adhering to other cells (42).

CONTACT INHIBITION BETWEEN NERVE CELLS

Radially directed outgrowth is characteristic of cells extending outward from explants, and this phenomenon was one aspect of cell behavior that Abercrombie and Heaysman sought to explain as a consequence of contact inhibition. Such radial outgrowth is especially pronounced in nerve cell explants, around which the outward extending axons effectively "trace out" the paths followed by their actively migrating tips (growth cones) (91). Such outgrowth of nerve fibers in culture will only occur, however, if the cells are embedded within a thick plasma clot, and P. Weiss (92) has presented persuasive evidence that the fibers become oriented along the axes of stretched plasma clot micelles. Weiss showed that these micelles for some reason become oriented radially around tissue explants, an effect he attributed to their dehydration by the metabolic activity of the cells. Extending Harrison's original observations (4) of cellular dependence on contact (stereotaxis), P. Weiss and coworkers showed that cultured cells of most types (including nerves) will orient themselves along the axis of tension in a stretched plasma clot, this being an example of the phenomenon he termed contact guidance. Thus P. Weiss explained the radial outgrowth of nerve fibers as a result of contact guidance along the radially oriented fibers.

However, in a recent reexamination of nerve cell orientation, Dunn (93) has shown that the radial outgrowth of nerve axons is primarily due to contact inhibition between adjacent growth cones (Fig. 6). Using chick lumbar root ganglia, Dunn measured the length of the outgrowing nerve fibers. By deliberately stretching plasma clots by known amounts and then measuring their optical birefringence produced by this stretching, Dunn determined the quantitative relationship between clot birefringence and clot fiber orientation. Using this relationship he could measure the orientation of the clot surrounding explants simply by observing the clot's birefringence. When he compared the degree of orientation of the outgrowing nerve fibers with the calculated degree of orientation of the clot fibers, he found that the nerve fibers were always much more highly oriented in a radial direction than were the fibers of the clot itself.

When Dunn studied the zone of confrontation between two adjacent nerve explants, he found that the population density of outgrowing axons in this area becomes much *lower* than that around other parts of the explant, as if outgrowth were being inhibited by the presence of the cells from the other explant. The axons in this confronted area were also much less radially oriented than those around other parts of the explant margin. This is of course very different from the "two-center effect" reported by P. Weiss, in which the nerve fibers extended between explants were more highly oriented and even *more* dense than around the rest of the periphery of the explant. Of course, such an increase in fiber orientation and population density between explants would be expected if the direction of nerve extension were determined primarily by the orientation of the clot fibers. Dunn explained the contradiction between his observations and those of P. Weiss by showing that

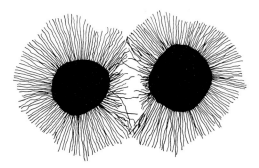

Fig. 6. Contact inhibition of nerve axon out-growth. Confronted cultures of two chick dorsal root ganglia cultured for 27 hours in a thick plasma clot and silver stained. Note that axon outgrowth has been strongly inhibited in the "collision zone" between the two explants. Also note the abrupt dropoff of axon population density at the periphery of the outgrowth. Both phenomena are due to contact inhibition between neuron growth cones. After Dunn (93).

the two-center effect occurs only when the clot is sufficiently liquid to allow nerve fibers in contact with one another to pull together into fascicles as described by Nakai (94). So long as the clot was not liquefied, the two-center effect did not occur, even though the clot fibers were still oriented radially around the explants. Clearly some other orienting factor must be at work, most probably a mutual inhibition between nearby fibers, judging from the reduced growth in the confronted areas.

Dunn was able to demonstrate this inhibition directly, using time-lapse cine-micrography and sequential photography. He showed that the advancing growth cone continually protrudes long, thin filopodia. When these filopodia contact another cell, the whole growth cone rapidly retracts a short distance and is then reextended in a new direction. In other words, contact inhibition occurs between the advancing growth cones and is apparently responsible for generating the predominantly radial pattern of out-

growth around explants of nerve tissue in essentially the same way as contact inhibition produces radial outgrowth of fibroblasts and other cells. Dunn also stresses that the characteristically abrupt termination of the outgrowth zone around nerve explants can be readily explained as a result of contact inhibition (or any other close-range negative interaction) between the advancing growth zones. Apparently this rapid falloff in population density occurs because the radial direction of outgrowth is maintained primarily by interaction with adjacent fibers, so that a growth cone that extends beyond the rest soon begins to wander erratically, and its consequent decline in radial velocity allows the others to catch up with it. This explanation of the phenomenon was supported by Dunn's observation that this demarcation in population density is reduced in cultures with a lower outgrowth density. In addition the degree of radial orientation of the nerve fibers was found to be directly proportional to population density, as would be expected if orientation were the result of contact inhibition.

CONTACT INHIBITION BETWEEN EPITHELIAL CELLS

Contact inhibition among a type of epithelial cell (chick pigmented retina) has been analyzed quantitatively by Middleton (95). Using time-lapse cinemicrography, he observed that such cells are capable of active locomotion resulting in intercellular collisions, so that the characteristic monolayering of such cells is attributable to contact inhibition. When Middleton determined the frequency of overlapping among such cells and compared this with the expected rate of overlapping if cells were distributed randomly, he found that the overlap index was only about one-fourth that observed by Abercrombie and Heaysman. This is apparently indicative of a much higher degree of contact inhibition among epithelial cells than among fibroblasts. Middleton also found that the rate of overlapping increased significantly with increasing population density and that the mitotic index decreased with increased density.

CELL TYPES LACKING CONTACT INHIBITION

It is perhaps surprising that there have not been more studies of the contact inhibition (or the lack of it) shown by the various types of white blood cells, such as macrophages, leukocytes, and lymphocytes. These white blood cells differ from the cells of solid tissues not only in that they circulate in the blood and lymph but also in their ability to migrate with relative freedom over, and even through, solid tissues. In view of their migratory ability, it is probably to be expected that these various white blood cells should be relatively immune to contact inhibition either by contact with one another or with fibroblasts. In fact, this is exactly what has been observed.

Oldfield's very painstaking study on this subject (96) was patterned on the earlier papers by Abercrombie and Heaysman. She studied leukocytes and macrophages from chicken blood (buffy coat), analyzing not only the interaction of these cells with one another but also their interaction with chick heart fibroblasts. She found that polymorphonuclear leukocytes overlap one another very little, having overlap ratios of 0.01 or lower even in dense cultures. These leukocytes seemed under some circumstances to repel each other at a distance. For example, when buffy coat explants were placed near each other on a plasma

clot surface, the length of the outgrowth between the explants was significantly less than the outgrowth length at the side. A peculiar "no-man's-land" formed between such explants, consisting of a cell-free zone into which cells failed to migrate. This apparent mutual repulsion occurred only when the clot substratum was "dry," that is, without overlying liquid medium. Both leukocytes and monocytes (macrophages) were observed to disperse away from one another when confined in a thin space beneath a coverslip. These instances of mutual repulsion between cells are indicative of negative chemotaxis, as if each cell secreted some substance that would accumulate in the thin film of medium around it, somehow repelling other cells. Twitty and Niu (97,98) reported some evidence of negative chemotaxis between melanophores and even fibroblasts from newt embryos and attributed the speckled arrangement of melanophores in certain species to such mutual repulsion. However, the no-man's-land effect has never been observed among fibroblasts, melanocytes, or other solid-tissue cells. Incidentally, it is a somewhat puzzling question of how a cell can generate a gradient of a negative chemotactic substance while simultaneously responding to the gradients of its neighbors. Of course, leukocytes (and possibly macrophages) are known to show positive chemotaxis, as, for example, migrating toward the exudates of certain bacteria (99). Positive chemotaxis has never been demonstrated in any other differentiated cell types, however.

Oldfield further reported that both macrophages and polymorphonuclear leukocytes (from chick buffy coat) were capable of overlapping chick heart fibroblasts more or less at random, as if there were *no* contact inhibition. This is somewhat surprising since both macrophages and leukocytes tend *not* to overlap one another even when cultured on glass in relatively dense populations. My own unpublished observations of "monolayers" of mouse macrophages and of rabbit and human leukocytes lead me to conclude that the failure of these cells to overlap one another is due entirely to their greater adhesiveness to the substratum than to one another (i.e., contact inhibition of the second kind). Time-lapse films of such cells show no inhibition of ruffling at points of contact between cells and indeed little or no mutual adhesion. Although Ramsey and I (100) were able to detect a small but statistically significant diminution in the speed of human polymorphs on contact with one another, this was barely appreciable and did not involve paralysis of the surface activity associated with their locomotion. I have also found that macrophage motility is not inhibited to the least degree by contact with fibroblasts, not even 3T3 fibroblasts, which are capable of eliciting contact inhibition from Sarcoma-180 cells. In my experience virtually all of the overlapping between macrophages and fibroblasts is actually underlapping, with the macrophages crawling beneath the fibroblasts. In mixed cultures macrophages actually tend to *accumulate* beneath fibroblasts in large numbers, as if intercellular contact actually promoted, rather than inhibited, macrophage locomotion (Fig. 7). Likewise, Oldfield reported that there was a significant excess in overlaps of fibroblasts with polymorphonuclear leukocytes and monocytes.

Fig. 7. Nine mouse peritoneal macrophages underlapping a single large binucleate 3T3 mouse fibroblast. This tendency of macrophages to accumulate beneath fibroblasts in culture seems to be the exact opposite of the contact inhibition shown by the fibroblasts for each other. This peculiarity may be related to the normal invasiveness of macrophages. (Tracing from a single frame of time-lapse film.)

of growth, whereas cells (such as neoplastically transformed cells) that have a reduced susceptibility to contact inhibition of movement are also much less subject to contact (or density-dependent) inhibition of growth. It is primarily because of this correlation in the presence or absence of these two types of inhibition that the same term "contact inhibition" has come to be interchangeably and ambiguously applied to both. These two types of inhibition are operationally quite distinct, however, and one should never assume that they are equivalent, or even that some causal connection necessarily exists between the two. In fact, a few instances have been found in which certain cells show a reduction in one of these types of contact inhibition, but not in the other.

CORRELATION BETWEEN CONTACT INHIBITION OF LOCOMOTION AND CONTACT INHIBITION OF GROWTH

As a general rule, cell strains or lines that are subject to contact inhibition of movement also show contact inhibition

For example, Macieira-Coelho (14) found that cells of a human fibroblast line attained a higher plateau density after transformation with Rous sarcoma virus (reduced contact inhibition of growth), but, surprisingly, the nuclear overlap ratio in the virus-transformed cells was about half that of the original untransformed line. Thus contact inhibition of movement was apparently enhanced while contact inhibition of growth diminished, showing the separability of the two phenomena. Unfortunately the details of cell-to-cell interaction, such as the inhibition of ruffling or of translocation, were not analyzed in this case.

Studying mixed cultures of mouse and chick fibroblasts, Njeuma (101) observed that the growth of chick fibroblasts could be inhibited by either mouse or chick fibroblasts, but that mouse fibroblast growth was not inhibited by contact with chick fibroblasts. On the other hand, Abercrombie, Heaysman, and Karthauser (15) found reciprocal contact inhibition of movement between mouse and chick fibroblasts. Thus a chick fibroblast can apparently produce contact inhibition of movement in a mouse fibroblast, but *not* contact inhibition of growth. Once again this demonstrates the separability of the two phenomena.

Actually, it remains to be seen whether some such apparent exceptions may not be explicable as the results of retraction clumping and underlapping or contact inhibition "of the second kind" (see p. 158). Despite such exceptions, the general rule is clear and strongly suggests the existence of some link, whether direct or indirect, between the ability (or inability) of certain cells to move while in contact with other cells and their reduced growth in confluent cultures. The correlation is simply too prevalent to be accidental.

In fact this correlation even extends to experimental situations. For example, as already mentioned, when urea is added to the culture medium of either chick heart fibroblasts or 3T3 fibroblasts, these cells show increased overlapping and movement, but are also released from contact inhibition of growth (78). Likewise, when their culture medium is replaced, cultured fibroblasts undergo a wave of increased movement, followed by a wave of mitoses. The phenomenon of "wound healing" in culture provides another example of the coordinate release of cells from contact inhibition both of movement and of growth (102). For example, when part of a confluent monolayer of contact inhibited fibroblasts is scraped so as to make a "wound," that is to say, an area of vacant substratum, not only do the cells bounding the vacant area show increased movement, causing them to migrate into the gap, but they also show a temporary increase in mitotic rate (103,104). In future, it might be advisable to look for other ways to stimulate cell motility within confluent monolayers, to determine whether such a stimulation of movement is always accompanied by an increase in the rates of growth and mitosis. It has been found that a surprising variety of treatments can temporarily release fibroblasts from contact inhibition of growth (105). It would be quite interesting to know whether these treatments also stimulate motility.

GROWTH CONTROL IN RELATION TO CONTACT INHIBITION OF MOVEMENT

One possible relationship between the inhibition of movement and the inhibition of division is suggested by the observations of Montagnier (106) and others that untransformed cells will not grow and divide in culture unless supplied with a suitable stable substratum on which they can spread. This "anchorage dependence" of growth is absent in transformed cells. The ability of such cells to continue growth while suspended in agar is highly correlated with malignancy and loss of contact inhibition of movement and of growth. Maroudas (107) has recently determined just how large a substratum is required to allow the growth and division of untransformed fibroblasts. Since spreading, which is a form of locomotion, is required for the growth of untransformed cells, possibly there is some linkage between the locomotory mechanism and some aspect of growth control such as the initiation of DNA synthesis, such that growth cannot proceed while the locomotory machinery is turned off. This would explain why growth cannot occur without cell spreading on a substratum and would likewise account for the observation that growth is reduced when locomotion is inhibited by intercellular contact and that such treatments as wounding or exposure to urea, which stimulate movement, also stimulate growth. Alternatively, it is possible that the connection between locomotion and growth is much less direct. For example, contact inhibition of movement might serve only to immobilize confluent cells so some other, unrelated, inhibitor of growth would be able to produce its effect.

The periodic replacement of culture medium at intervals of several days (as is usual in tissue culture work) has most seriously complicated the interpretation of much of the work done on the control of growth and locomotion. Between such feedings, the medium is gradually depleted of nutrients, waste products and other materials accumulate in the medium, and, as a rule, the pH declines gradually. In itself this gradual exhaustion of the medium depresses both movement and growth, and this depression can be confused with the inhibiting effect of cell–cell contact unless careful controls are made. The periodic addition of new medium stimulates both movement and division, further obscuring the phenomena of contact inhibition. These effects have even led some to suggest that contact inhibition might be only an artifact of medium exhaustion.

In several recent studies these problems have been circumvented by the use of continuous perfusion to maintain con-

stant conditions of nutrition and pH, and to avoid the accumulation of any cell-secreted inhibitors. Castor (108) has determined the rates of motility and division as functions of population density in three strains of 3T3 fibroblasts differing in their serum requirements and susceptibility to contact inhibition of growth. He found that the average rate of cell locomotion declined up to sevenfold with increasing population density (number of cells per area) and that this inhibition occurred at reduced population densities when lower concentrations of serum were used in the medium. Castor also showed that at any given level of population density there was a direct correlation between the average rate of mitosis in the population and the average rate of cell translocation. Higher cell densities inhibited both division and locomotion, but the proportionality factor between these two variables was altered slightly by changing the serum concentration.

In a very similar study, also using continuous perfusion and time-lapse cinemicrography to study 3T3 cells, Martz (109) confirmed for this cell type Abercrombie and Heaysman's original observation of an inverse correlation between speed and contact number. Although both cell speed and mitotic rate were found to be reduced at higher population density, when Martz compared the rates of nuclear translocation and the intermitotic periods of individual fibroblasts, he found no significant correlation. Thus, although growth and movement are correlated on the population level, they are not on the cell level, which strongly suggests that there is not in fact any direct connection between contact inhibition of locomotion and of growth. However, these data do not exclude some sort of delayed effect of speed on growth or possibly the existence of a brief critical period during the cell cycle when locomotion could alter generation time. It may be noted parenthetically that Martz and Steinberg (110) also found no correlation between individual cell contacts and growth rate in these same cells.

POSSIBLE ROLE OF CONTACT INHIBITION *IN VIVO*

Cell locomotion, especially of mesenchymal cells, is quite widespread during early embryonic development, but although a few cell types (the white blood cells) retain their motility, most tissue cells become quiescent in the older embryo and the adult. Nevertheless, these quiescent cells still retain a latent ability to resume active locomotion when placed in tissue culture, and it seems probable that the function of this latent motility is to allow these cells to move into gaps left by wounding or other tissue damage. Explantation of tissue into culture can be considered as an extreme form of wounding, and outgrowth at the free margin of an explant tends to fill the surrounding areas rather like filling a wound. Presumably the phenomena of contact inhibition (both of locomotion and of growth) correspond in some way to the means by which motility and growth are normally repressed in the body, either in the absence of a wound or following completion of wound healing. In other words, contact inhibition may be the body's means for minimizing cell movement within intact solid tissues while retaining the ability of cells to move into gaps wherever they have a contact-free margin. The relative lack of contact inhibition shown by malignant cells, macrophages, and other invasive cells would reflect an unrestrained activity of the locomotory capacity present in latent form within nearly all tissue cells. Unfortunately this presumed homology be-

tween the inhibition of locomotion in crowded cell cultures and that occurring in solid tissues is hypothetical and will remain so until the precise mechanism of cell locomotion is understood, together with the means by which this locomotion is repressed.

Actually little or no direct information seems to be available concerning the amount of locomotion of fibroblastic cells within the body. Weston and Abercrombie (111) found very little intermingling of (tritium-labeled and unlabeled) heart or liver cells within fused tissue fragments. It should not be forgotten that substantial cell movement does continue among contact-inhibited fibroblasts in culture, even when these cells are totally surrounded by others. For example, both Abercrombie et al. (8) and Martz (109) found that fibroblasts having contact with six neighbors moved at an average rate fully one-half that of those fibroblasts having no contacts. Likewise, both these reports show a great variability of cell speed independent of cell contact. In fact, Martz has calculated that only about 9% of the variance in cell speed can be accounted for by the observed correlation between contact and speed. Clearly, the interactions involved in controlling cell motility must be much more complicated than one might have wished.

Although contact inhibition may play an important role in controlling the morphogenetic cell movements of embryonic development, one can only speculate about this at present. Certainly there is extensive cell locomotion during development, and to the extent that it occurs even among cells surrounded by their neighbors, this locomotion presumably must reflect either a total loss or a reduced degree of contact inhibition. Trelstead, Hay, and Revel (112) have hypothesized that the lateral migration of mesoblast cells during avian gastrulation may be directed by mutual contact inhibition tending to keep these cells moving predominantly toward their free (lateral) edge. Presumably these cells would have lacked contact inhibition during their earlier ingression through the primitive groove, however. As suggested previously, the eventual immobilization of most tissue cells once their morphogenetic rearrangement is complete can be considered to be an acquisition of contact inhibition. Similarly contact inhibition may be involved in the arrangement of epithelial cells into continuous sheets, precisely one cell thick. After all, this is a case of the formation of a cell monolayer.

PROSPECTS AND PRIORITIES FOR FUTURE RESEARCH

In conclusion, I would like to look back over the observations and experiments described in this chapter and to point out what seem to me the primary inadequacies in the present state of our knowledge and the probable new directions that should be taken to extend our understanding of contact inhibition. First of all, one should always seek to clarify what is meant when cells are described as being contact-inhibited (even apart from distinguishing between contact inhibition of movement and contact inhibition of growth). Do their nuclei overlap at a less than random frequency? Is their rate of movement inversely correlated with contact number? Is their ruffling activity paralyzed on contact? Are their cells only aligned in parallel rows? Of course, it is quite true that all of these properties are frequently shared by cells of a given type and all may be lost together in transformed cells. Nevertheless, as Martz and Steinberg (30) have

recently stressed in another review, these various indications of contact inhibition are not necessarily inseparable. For example, certain cells, such as macrophages, may show little overlapping with one another but lack contact paralysis of ruffling. Unless the criteria of contact inhibition are specified in each instance, it will be most difficult to compare various reported observations of contact inhibition and by extrapolating from one case to another eventually to arrive at an explanation of the phenomenon. Ideally, each of these major criteria of contact inhibition should be determined in every instance studied.

To approach this same basic problem from a different point of view, one should seek to determine whether a given case of monolayering results from the paralysis of movement on contact (true contact inhibition "of the first kind") or whether monolayering is simply the consequence of stronger adhesion to the substratum (contact inhibition "of the second kind"). Clearly these are two distinct phenomena, and only confusion will result if observations made on one type of monolayering are applied to the interpretation of the other type. The same is true of those various cases in which extensive overlapping occurs between cells. While one might choose to consider *any* case of overlapping as a failure of contact inhibition (in a very strict sense), overlapping need not be due to the greater ability of cells to crawl onto one another's surfaces, though that is what has usually been concluded. In many (perhaps most) cases, overlaps result from cells crawling beneath one another (underlapping) (74,76). Overlaps can also be formed by retraction clumping, which is basically a detachment phenomenon having little to do with locomotion. Overlapping can also result from the secretion of insulating layers of collagen or other extracellular fibers. These various types of overlapping should never be confused with one another, and especially in view of the observation of DiPasquale and Bell (76), each of the various previously studied instances of overlapping should probably now be restudied to determine which type of overlapping is involved in each case.

In order to approach the question of what happens to cells on contact that inhibits their motility, more attention should be paid to the ultrastructural, electrical, and biochemical events accompanying inhibition. The very rapid formation of specialized junctions and adhesion plaques at points of contact is especially interesting in this regard (57, 58).

Other tantalizing subjects for study are the cases of apparent nonreciprocal contact inhibition, where one of the contacted cells is inhibited and the other is not. Such instances promise to reveal much of the interaction involved (15, 113). It should be remembered that malignant cells can invade normal ones, and the most direct analogy to this in culture is the contact between transformed and untransformed cells. Such heterologous contacts therefore provide a better system for studying malignancy than do homologous contacts between like cells.

As the mechanism of locomotion itself becomes better understood, the cellular processes responsible for movement should be compared in contact-inhibited versus noninhibited cells. If tissue cells really are propelled by rearward membrane flow, as current evidence suggests, significant differences in membrane fluidity are to be anticipated in contact-inhibited cell types as opposed to transformed non-contact-inhibiting cells (114). Likewise, considerable changes in membrane behavior should be looked for in contact-inhibited cells at the time they become confluent.

ACKNOWLEDGEMENTS

I am indebted to the following persons for the help and advice they have given me during the preparation of this review: Michael Abercrombie, J. P. Trinkaus, Graham Dunn, Kurt Johnson, and Abby Rich. Thanks to them many errors have been removed, the remaining errors being my own. Without the secretarial assistance of my wife, Dr. Elizabeth H. Harris, the manuscript would never have been completed. Portions of the paper were prepared during the tenure of a postdoctoral fellowship from the Damon Runyon Memorial Fund for Cancer Research, Inc., and it would never have been written without this support.

REFERENCES

1. Abercrombie, M., *Exp. Cell Res. Suppl.* **8**, 188 (1961).
2. Abercrombie, M., in *Cells and Tissues in Culture*, Vol. 1, Willmer, E. N., ed., Academic Press, New York, 1965, pp. 177–202.
3. Ramsey, W. S., *Exp. Cell Res.* **72**, 489 (1972).
4. Harrison, R. G., *J. Exp. Zool.* **17**, 521 (1914).
5. Harris, A. K., *Exp. Cell Res.* **77**, 285 (1973).
6. Ephrussi, B., *Arch. Anat. Micros.* **29**, 95 (1933).
7. Loeb, L., *Am. J. Physiol.* **56**, 140 (1921).
8. Abercrombie, M., and Heaysman, J. E. M., *Exp. Cell Res.* **5**, 111 (1952).
9. Abercrombie, M., and Heaysman, J. E. M., *Exp. Cell Res.* **6**, 293 (1954).
10. Abercrombie, M., *In Vitro* **6**, 128 (1970).
11. Willmer, E. N., *J. Exp. Biol.* **10**, 323 (1933).
12. Todaro, G. J., Lazar, G. K., and Green, H., *J. Cell. Comp. Physiol.* **66**, 325 (1965).
13. Stoker, M. G. P., and Rubin, H., *Nature* **215**, 171 (1967).
14. Macieira-Coelho, A., *Exp. Cell Res.* **47**, 193 (1967).
15. Abercrombie, M., Heaysman, J. E. M., and Karthauser, H. M., *Exp. Cell Res.* **13**, 276 (1957).

16. Weiss, P., *J. Exp. Zool.* **68**, 398 (1934).

17. Abercrombie, M., and Heaysman, J. E. M., *Nature* **174**, 697 (1954).

18. Abercrombie, M., *Cold Spring Harbor Symp. Quant. Biol.* **27**, 427 (1962).

19. Abercrombie, M., in *Mechanisms of Invasion in Cancer*, Denoix, P., ed., Springer-Verlag, Berlin, 1967, pp. 140–144.

20. Abercrombie, M., *Natl. Cancer Inst. Monogr.* **26**, 249 (1967).

21. Stoker, M., *Virology* **24**, 165 (1964).

22. Abercrombie, M., and Ambrose, E. J., *Exp. Cell Res.* **15**, 332 (1958).

23. Gustavson, T., and Wolpert, L., *Biol. Rev.* **42**, 442 (1967).

24. Wolpert, L., and Gingell, D., *Symp. Soc. Exp. Biol.* **22**, 169 (1968).

25. Harris, A. K., in *Tissue Cell Locomotion*, Ciba Foundation Symposium, 1973, pp. 3–20.

26. Rubin, R. W., and Everhart, L. P., *J. Cell Biol.* **57**, 837 (1973).

27. Gail, M. H., and Boone, C. W., *Exp. Cell Res.* **47**, 193 (1967).

28. Carter, S. B., *Nature* **203**, 1183 (1965).

29. Carter, S. B., *Nature* **213**, 256 (1967).

30. Martz, E., and Steinberg, M. S., *J. Cell Physiol.* **81**, 25 (1973).

31. Steinberg, M. S., *J. Exp. Zool.* **173**, 395 (1970).

32. Carter, S. B., *Exp. Cell Res.* **48**, 189 (1967).

33. Carter, S. B., *Nature* **220**, 970 (1968).

34. Carter, S. B., *Nature* **225**, 858 (1970).

35. Wolpert, L., MacPherson, I., and Todd, I., *Nature* **223**, 512 (1969).

36. Abercrombie, M., Heaysman, J. E. M., and Pegrum, S. M., *Exp. Cell Res.* **62**, 389 (1970).

37. Ingram, V. M., *Nature* **222**, 641 (1969).

38. Harris, A., *Acta Protozool.* **11**, 145 (1972).

39. Harris, A., and Dunn, G. A., *Exp. Cell Res.* **73**, 519 (1972).

40. Bray, D., *Proc. Natl. Acad. Sci. U.S.* **65**, 905 (1970).

41. Coman, D. R., *Cancer Res.* **21**, 1436 (1961).

42. Harris, A., *Dev. Biol.* **35**, 97 (1973)

43. Farquhar, M. G., and Palade, G. E., *J. Cell Biol.* **17**, 275 (1963).

44. Devis, R., and James, D. W., *Nature* **194**, 695 (1962).

45. Chambers, R., and Fell, H. B., *Proc. Roy. Soc. (London)* **109**, 380 (1931).

46. Vaughan, R. B., and Trinkaus, J. P., *J. Cell Sci.* **1**, 407 (1966).

47. Curtis, A. S. G., and Varde, M., *J. Natl. Cancer Inst.* **33**, 15 (1964).

48. Vesely, P., and Weiss, R. A., *Int. J. Cancer* **11**, 64 (1973).

49. Abercrombie, M., and Gitlin, G., *Proc. Roy. Soc. (London)* **B162**, 289 (1965).

50. Loewenstein, W. R., *Dev. Biol. Suppl.* **2**, 151 (1968).

51. Furshpan, E. J., and Potter, D. D., *Curr. Topics Dev. Biol.* **3**, 95 (1968).

52. McNutt, N. S., and Weinstein, R. S., *Science* **165**, 597 (1967).

53. Goshima, K., *Exp. Cell Res.* **58**, 420 (1969).

54. Subak-Sharpe, H., Burk, R. R., and Pitts, J. D., *J. Cell Sci.* **4**, 353 (1969).

55. Loewenstein, W. R., and Kanno, Y., *J. Cell Biol.* **33**, 225 (1967).

56. Martinez-Polomo, A., Braislovsky, C., and Bernhard, W., *Cancer Res.* **29**, 925 (1967).

57. Flaxman, B. A., Revel, J. P., and Hay, E. D., *Exp. Cell Res.* **58**, 438 (1970).

58. Heaysman, J. E. M., and Pegrum, S. M., *Exp. Cell Res.* **78**, 71 (1973).

59. Heaysman, J. E. M., and Pegrum, S. M., *Exp. Cell Res.* **78**, 479 (1973).

60. Trinkaus, J. P., Betchaku, T., and Krulikowski, L. S., *Exp. Cell Res.* **64**, 241 (1971).

61. Johnson, G. S., Morgan, W. D., and Pastan, I., *Nature* **235**, 54 (1972).

62. Hsie, A. W., and Puck, T. T., *Proc. Natl. Acad. Sci. U.S.* **68**, 358 (1971).

63. Jimenez de Asua, L., Surian, E. S., Flawia, M. M., and Torres, H. N., *Proc. Natl. Acad. Sci. U.S.* **70**, 1388 (1973).

64. Ambrose, E. J., *Exp. Cell Res. Suppl.* **8**, 54 (1961).

65. Ambrose, E. J., *Nature* **198**, 1194 (1956).

66. Curtis, A. S. G., *J. Natl. Cancer Inst.* **26**, 253 (1961).

67. Harris, A. K., *J. Cell Biol.* **43**, 165a (1969).

68. Abercrombie, M., Heaysman, J. E. M., and Pegrum, S. M., *Exp. Cell Res.* **59**, 393 (1970).

69. Abercrombie, M., Heaysman, J. E. M., and Pegrum, S. M., *Exp. Cell Res.* **60**, 437 (1970).

70. Curtis, A. S. G., *J. Cell Biol.* **19**, 197 (1964).

71. Kredel, F. E., *Johns Hopkins Hosp. Bull.* **40**, 216 (1927).

72. Boyde, A., Grainger, F., and James, D. W., *Z. Zellforsch.* **94**, 46 (1969).

73. Algard, F. T., *J. Exp. Zool.* **123**, 499 (1953).

74. Bell, P. B., Jr., *J. Cell Biol.* **55**, 16a (1972).

75. Guelstein, V. I., Ivanova, O. Yu, Margolis, L. B., and Vasiliev, Yu. M., *Proc. Natl. Acad. Sci. U.S.* **70**, 2011 (1973).

76. DiPasquale, A., and Bell, P. B., *J. Cell Biol.* **55**, 60a (1972).

77. Weston, J. A., and Roth, S. A., in *Cellular Recognition*, Smith, R. T., and Good, R. A., eds., Appleton-Century Crofts, New York, 1969, pp. 29–37.

78. Weston, J. A., and Hendricks, K. L., *Proc. Natl. Acad. Sci. U.S.* **69**, 3727 (1972).

79. Coman, D. R., *Cancer Res.* **4**, 625 (1944).

80. Burger, M. M., in *Growth Control in Cell Culture*, Ciba Foundation Symposium, 1971, pp. 45–69.

81. Burger, M. M., and Noonan, K. D., *Nature* **228**, 512 (1970).

82. Burger, M. M., *Nature* **219**, 499 (1968).

83. Burger, M. M., *Proc. Natl. Acad. Sci. U.S.* **62**, 994 (1969).

84. Yardley, J. H., in *Conference on the Biology of Connective Tissue Cells*, Arthritis and Rheumatism Foundation, New York, 1962, p. 179.

85. Elsdale, T., and Foley, R., *J. Cell Biol.* **41**, 298 (1969).

86. Elsdale, T., and Bard, J., *Nature* **236**, 152 (1972).

87. Elsdale, T. R., *Exp. Cell Res.* **51**, 439 (1968).

88. Weiss, P., *Int. Rev. Cytol.* **7**, 391 (1958).

89. Weiss, P., *Growth, Suppl.* **5**, 163 (1941).

90. James, D. W., and Taylor, J. F., *Exp. Cell Res.* **54**, 107 (1969).

91. Harrison, R. G., *Anat. Rec.* **1**, 116 (1907).

92. Weiss, P. J., *J. Exp. Zool.* **100**, 353 (1945).

93. Dunn, G. A., *J. Comp. Neurol.* **143**, 491 (1971).

94. Nakai, J., *Z. Zellforsch.* **52**, 427 (1960).

95. Middleton, C. A., *Exp. Cell Res.* **70**, 91 (1972).

96. Oldfield, F. E., *Exp. Cell Res.* **30**, 125 (1963).

97. Twitty, V. C., and Niu, M. C., *J. Exp. Zool.* **108**, 398 (1948).

98. Twitty, V. C., and Niu, M. C., *J. Exp. Zool.* **125**, 541 (1954).

99. Harris, H. J., *J. Pathol. Bacteriol.* **66**, 135 (1953).

100. Ramsey, W. S., and Harris, A. K., *Exp. Cell Res.*, **82**, 262 1973.

101. Njeuma, D. L., *Exp. Cell Res.* **66**, 244 (1971).

102. Vasiliev, J. M., Gelfand, I. M., Domnina, L. V., and Rappoport, R. I., *Exp. Cell Res.* **54**, 83 (1969).

103. Todaro, G., Matsuya, Y., Bloom, S., Robbins, A., and Green, H., *Wistar Symp. Monogr.* **7**, 82 (1967).

104. Kolodney, G. M., and Gross, P. R., *Exp. Cell Res.* **57**, 423 (1969).

105. Vasiliev, J. M., Gelfand, I. M., Guelstein, V. I., and Fetisova, E. K., *J. Cell Physiol.* **75**, 305 (1970).

106. Montagnier, L., in *Growth Control in Cell Cultures*, Ciba Foundation Symposium, pp. 33–44.

107. Maroudas, N. G., *Exp. Cell Res.* **74**, 337 (1972).

108. Castor, L. N., *Exp. Cell Res.* **68**, 17 (1971).

109. Martz, E., *J. Cell Physiol.* **81**, 39 (1973).

110. Martz, E., and Steinberg, M. S., *J. Cell Physiol.* **79**, 189 (1973).

111. Weston, J. A., and Abercrombie, M., *J. Exp. Zool.* **164**, 317 (1967).

112. Trelstead, R. L., Hay, E. D., and Revel, J. P., *Dev. Biol.* **16**, 78 (1967).

113. Abercrombie, M., Lamont, D. M., and Stephenson, F. M., *Proc. Roy. Soc. (London)* **B170**, 349 (1968).

114. Edidin, M., and Weiss, A., *Proc. Natl. Acad. Sci. U.S.* **69**, 2456 (1972).

CHAPTER NINE

Cellular Interactions in the Immune Response

A. BASTEN

Immunology Unit
Department of Bacteriology
University of Sydney
Sydney, Australia

J. F. A. P. MILLER

Walter and Eliza Hall Institute
of Medical Research
Melbourne, Australia

Activated thymus cell, ATC
Antibody-forming cell, AFC
Antibody-forming-cell precursor, AFCP
Bovine serum albumin, BSA
Third component of complement, C3
2,4-Dinitrophenylacetic acid, DNP
Donkey red blood cells, DRC
Flagellin, Fla
Fowl immunoglobulin G, FγG
Immunoglobulin, Ig
Immunoglobulin G, IgG
Immunoglobulin M, IgM
Immunoglobulin on T cells, IgT
Keyhole limpet hemocyanin, KLH
Mouse-bone-marrow-derived lymphocyte
    antigen, MBLA
4-Hydroxy-3-iodo-5-nitrophenylacetic
    acid, NIP
Neonatal thymectomy, Ntx
Ovalbumin, OVA
Polymerized flagellin, POL
Sheep red blood cells, SRC
Adult thymectomized, irradiated, mar-
    row protected, TxBM

The immune response consists of the complex series of events initiated by the interaction of immunocompetent cells and antigen. It is thought to have originated as an evolutionary by-product of the need on the part of multicellular organisms to develop a cell–cell recognition system for differentiation and morphogenesis (1). Possession of such a mechanism confers on the host certain selective advantages: the capacity to distinguish and eliminate or suppress both foreign (extrinsic) material and, in addition, intrinsic alien patterns arising by somatic mutation or an equivalent process (2). The immune system therefore plays a key role in resistance to pathogenic organisms, "surveillance" against malignant cells, and control of autoaggressive processes (3). To do so effectively a fine balance is required between antigen and effector mechanisms such as antibody or sensitized lymphocytes. This

balance depends in turn on interactions between various types of immunocompetent cells and their products. The purpose of this chapter is to outline recent advances in the field of cell interactions in immunity with particular reference to their biological and practical implications.

CELLS INVOLVED IN
THE IMMUNE RESPONSE

The cells taking part in immune responses can be broadly classified into two groups: *lymphocytes* and various *accessory cells*. The latter include macrophages; fixed cells of the reticuloendothelial system, particularly the dendritic reticulum cells; and various types of blood leukocytes (monocytes as well as eosinophilic and neutrophilic granulocytes). Lymphocytes are responsible for the induction phase of the response that is specific, whereas accessory cells act nonspecifically either in antigen processing (4) or, at a later stage, as amplifiers of ongoing effector mechanisms (see section entitled "Cellular Interactions"). Specificity in this context is determined by the presence on the lymphocyte surface of immunoglobulin (Ig) molecules, the combining sites of which act as recognition units or receptors for antigen (5–7). In contrast, accessory cells lack a comparable recognition system and are not involved in active antibody synthesis. Macrophages do, however, carry surface receptors for Ig, particularly IgG, the third component of complement (C3), and an α_1-globulin (8), which render them capable of participating in a variety of interactions with lymphocytes and their products (see "Cellular Interactions").

In the last decade the central role of lymphocytes in adaptive immunity has been established (9–11). They are the cells that interact with antigen and initiate the immune response (12). Their origin, like that of other leukocytes, is from a pool of hematopoietic stem cells that, in embryonic life, are present in the yolk sac and fetal liver, and postnatally are found in bone marrow (13). The stem cells migrate via the bloodstream to selected sites where their subsequent differentiation is dictated by the inductive influences prevalent in the particular microenvironment (14). In the case of lymphocytes differentiation processes occur in what are now termed primary (central) lymphoid organs. These are the thymus in all vertebrates (10) and the bursa of Fabricius in birds (15) or some equivalent organ (as the fetal liver) in other vertebrates. Antigen plays no role at this point in lymphocyte maturation, the number of cells emerging being dependent only on the size of the stem cell pool. Cells migrating from primary lymphoid organs settle in peripheral lymphoid tissues or "secondary" lymphoid organs, such as spleen and lymph nodes. At this stage they are sensitive to antigen, which induces their further differentiation and proliferation to effector cells capable of carrying out immune responses. Cells that differentiate in the thymus have been termed T lymphocytes or T cells and play a major role in cell-mediated immunity. In contrast, cells derived from the bursa or its mammalian equivalent, referred to as B lymphocytes or B cells, are involved in humoral immunity and produce the various immunoglobulins secreted into the serum (16). Although T and B cells are morphologically indistinguishable by light microscopy, they have a number of characteristic features that are of particular importance in analyses of cellular interactions in immune responses.

Surface Properties

Recent work, particularly in the mouse, has revealed the existence of a distinct pattern of surface markers for each type of lymphocyte (Table 1). For example, B cells have a high density of Ig determinants on their surface, whereas little or no exposed Ig can be detected by similar methods on T cells (17,18). The B

Table 1. Surface Markers on T and B Lymphocytes in the Mouse

Markers	B Lymphocytes	T Lymphocytes
Immunoglobulin determinants	High density (well exposed)	Low density (poorly exposed)
θ Alloantigen[a]	Absent	Present
Plasma cell alloantigen	Present on plasma cells only	Absent
MBLA (heterospecific antigen)	Present on B cells, plasma cells (and marrow precursors)	Absent
Receptors for the Fc piece of antibody molecules	Present on all B cells (not on plasma cells)	Absent
Receptors for C3	Present on a subpopulation of B cells	Absent
Histoincompatibility[b] (H_2) antigens	High density	Low density

[a] Present also in brain and skin.

[b] A. Basten, unpublished observations.

cells lack the θ antigen found on thymus cells and on the great majority of T cells in peripheral lymphoid tissue (17,19). A search for an alloantigen on B cells, analogous to θ on T cells or to PC_1 on plasma cells (20), has not been successful. Mouse-bone-marrow-derived lymphocyte antigen (MBLA), although present on all B cells (21,22), has in addition been found on the surface of antibody-forming cells (AFC) and hematopoietic cells (up to 60% of nucleated bone marrow cells) (23). Mature B lymphocytes (but not T cells or plasma cells) do, however, possess a receptor for the Fc piece of antibody molecules. It has been detected by means of soluble immune complexes (19,24) or by aggregated antibody molecules alone (25). A subpopulation of B lymphocytes has in addition been shown to carry a receptor for C3 (26).

These markers can be exploited to study the distribution of the two classes of cells in lymphoid tissue (19,27) and to separate one class from the other (11). Thus anti-θ serum and complement is cytotoxic for T cells (17) and anti-light-chain Ig serum and complement for B cells (28). Furthermore, Degalan beads (29) or Sepharose beads (30) coated with anti-Ig reagents will selectively retain B cells, as will an antigen-coated column when the cells are preincubated with the corresponding antibody (31). The production of lymphocyte populations purified *in vitro* and of animals depleted or enriched for either T or B cells has been described in detail elsewhere (11).

Physiological Properties

The role of lymphocytes in the immune response depends not only on their surface properties but also on their lifespan, homing properties, and capacity to recirculate. After initial differentiation within primary lymphoid organs (i.e.,

thymus and bursa), they leave via the bloodstream and settle in peripheral lymphoid tissues where full immunological competence is acquired (10,15). The T lymphocytes become part of a complex of long-lived small lymphocytes that recirculate continuously from blood to lymphoid tissue and back to blood. Their lifespan in the mouse has recently been estimated to be approximately 16 weeks (32). The sites of recirculation of T cells form discrete areas within secondary lymphoid tissues, "thymus-dependent areas"—notably the paracortical areas of lymph nodes, periarteriolar lymphocyte sheaths of the spleen, and diffuse interfollicular tissue of Peyer's patches (33). The B cells, on the other hand, migrate to different "thymus-independent areas," which include the lymphoid follicles and, following antigenic stimulation, the germinal centers. Contrary to earlier reports (34), the lifespan of B cells is longer than a few days (6 weeks in the mouse). Not surprisingly, evidence has been found of the recirculation of B cells, although it is slow in tempo (35,36). This is particularly important in the light of recent experiments indicating that B cells are essential for the expression of immunological memory (37,38).

Functions

The B lymphocytes are the precursors of AFC (plasma cells). Their prime function is therefore related to the humoral group of immune responses. In addition, these cells may well play a number of subsidiary roles—for example, in antigen concentration (39), antibody-dependent killing of target cells (40), and antibody-mediated suppression of the immune response (41). The T lymphocytes, on the other hand, are responsible for initiating cell-mediated immune responses, the prototype of which is delayed hypersensitivity. In addition, they are essential to allow optimal antibody production to many antigens (11, 42,43). They do not, however, themselves form antibody in the sense of a classical Ig molecule, which is secreted into the serum. Instead they exert their effect via various soluble factors, which may react with other T cells, B cells, or nonlymphoid cells such as macrophages. These interactions, to be discussed in greater detail in the section entitled "Cellular Interactions," are of considerable importance in many situations, including graft rejection, tumor immunity, and resistance to certain infections, particularly fungal, viral, and parasitic.

ANTIGEN RECOGNITION

Recognition of antigen has been most satisfactorily explained in terms of clonal selection (44). According to Burnet's hypothesis, lymphocytes are heterogeneous in the sense that each cell is genetically endowed with the capacity to synthesize a single type of antibody molecule or receptor for antigen (i.e., it possesses specificity). The theory does not contravene the dogma of molecular biology since antigen serves only as an activator of a potentiality already present in the cell genome. Implicit in any satisfactory explanation of antigen recognition is the capacity to discriminate between self and nonself. A failure to do so would of course result in autodestruction, or "horror autotoxicus," to use Ehrlich's original phrase (45). At the cellular level this implies that antigen–lymphocyte interactions ought to result either in immune activation leading to a response (appropriate for foreign materials) or, alternatively, in a state of nonresponsiveness or "tolerance" (appropriate for self components) (46). The demonstration by

several groups (47–49) that a given antigen can act both as a "tolerogen" and an "immunogen," depending on its mode of presentation, provides strong support for selectionist theories. Signal discrimination between antigen and receptor is of relevance to certain types of cell interaction and will be referred to again in the section entitled "Cellular Interactions."

Antigen recognition would thus depend on the existence of specific recognition units or receptors on the surface of immunocompetent lymphocytes. Their presence on both T and B lymphocytes has now been conclusively established by a wide variety of techniques. For example, T cells as well as B cells can be selectively eliminated by exposure to a radioactively labeled antigen in vitro—the "suicide effect" (7,50,51). Furthermore, immunological memory (37,38) and immunological tolerance (52,53) have been shown to be associated with each cell line. To account for the specificity of antigen recognition, Mitchison (54) proposed that the receptors on lymphocytes are accurate samples of the antibodies that the cell or its progeny would produce after interaction with the antigen. Subsequent studies have amply confirmed this proposal, particularly in the case of B cells (50,55,56), where the receptor appears to be a monomeric IgM molecule. The nature of the T-cell receptor, however, remains controversial and is worthy of particular comment. Three experimental approaches have been used to examine this problem. In the first, suspensions of pure T cells or of cells containing a known number of T cells have been exposed to fluorescein-conjugated or radio-iodinated anti-Ig reagents (57). Some groups (58–60) have claimed that it is possible with such techniques to demonstrate Ig determinants on T cells, although at a much lower density than on B cells. Others

(17,56,61) have been unable to show any Ig on T cells. The second approach has involved preincubation of cell suspensions with anti-Ig reagents, followed either by exposure to labeled antigen *in vitro* or by testing *in vivo* the capacity of such cells to mount a T-cell-dependent response. Conflicting results have again been obtained with this technique, some investigators (50,62,63) claiming success, others not (64). The limitations of indirect methods of this kind have been well summarized in a recent Brook Lodge Symposium (65). Since then a third, and more direct, approach to the identification of the T-cell receptor has been developed, which overcomes the problem of inaccessibility of determinants. Cell membrane molecules are isolated either by solubilization of lymphocytes or by active metabolic release and are analyzed directly for Ig content. Even with such a direct approach, however, contradictory results have been obtained. Marchalonis and associates claimed that the receptor on T cells, like that on B cells, is a monomeric IgM molecule (66,67), whereas another group (56), utilizing a similar but not identical technique, could identify Ig only on B cells. Marchalonis's group has gone on to demonstrate specificity in T-cell Ig (IgT) in both humoral (68) and cell-mediated responses (69). The controversy about the nature of the T-cell receptor is likely to persist until individual workers standardize both their techniques and their reagents, particularly anti-Ig sera, which can vary in content and potency, and may contain a variety of antibodies, some of which may be directed, for example, against cell membranes (64). If the receptor on T cells is not composed of identifiable Ig, the questions then arise as to its nature and the apparent specificity of T cells. The receptor could possibly represent a more primitive form of recognition unit (64),

but if that is so, no data are yet available to support this contention.

The molecular mechanisms involved in triggering lymphocytes in the immune response are not yet fully understood and will remain unclear until such problems as the nature of the T-cell receptor are resolved. Recently, however, models have been put forward that are at least amenable to experimentation (46,70). Furthermore, some information is available, particularly for B cells, on the consequences of antigen recognition at the cellular level. This derives from *in vitro* studies of the interaction between lymphocytes and either labeled antigen (71) or labeled anti-Ig (72,73). The initial step is the binding of label to the cell surface, which occurs readily at both 4 and 37°C. It is independent of molecular configuration in the sense that in the case of anti-Ig reagents both monomeric Fab′ fragments and F(ab′)$_2$ fragments can attach. Subsequently activation of the cells appears to depend on the aggregation of antigen–receptor complexes at one pole with "cap" formation. This phenomenon is highly temperature dependent (37°C) and will occur only if multivalent ligands [including F(ab′)$_2$] are used (71, 74). Capping is a function of an actively metabolizing cell since agents like sodium azide and cytochalasin B inhibit the process. The final event in cell triggering appears to be internalization of antigen with resynthesis of receptors (72, 75). In contrast, if cells are exposed to multivalent antigens in tolerogenic form, cap formation fails to take place even when the experiment is carried out at 37°C over a prolonged period of time. Taken together, the data suggest that antigen activation of (B) lymphocytes depends on the crosslinking of receptors by antigen, followed by modulation of the cell membrane with increase in receptor density. When, however, the dose of antigen is increased, interlinking of receptors may exceed a critical point beyond which aggregation and cap formation cannot occur. A "frozen" state of this kind results in tolerance, not in immunity (75). Lymphocytes can therefore be activated or become unresponsive depending on the mode of antigen presentation. The importance of this concept will become apparent when the mechanisms of cell interactions are discussed.

The subcellular events responsible for lymphocyte behavior remain unknown. For example, a role for cyclic adenosine 3′,5′-monophosphate (AMP), following antigen activation (76,77), has been postulated, but to date no direct evidence to support or refute this is available.

CELLULAR INTERACTIONS

Collaboration between T Cells and B Cells in Antibody Production

In a preceding section immune responses were broadly divided into those that are cell-mediated (T cell) and those that are humoral (B cell), that is, antibody dependent. Some years ago, however, it was found that methods of T-cell depletion, such as neonatal thymectomy (Ntx), apparently affected both types of immunity (10). Thus, not only was skin graft survival prolonged in Ntx animals but the production of antibodies, particularly the IgG phase, to certain particulate (e.g., heterologous erythrocytes) and soluble antigens (serum proteins) was markedly diminished. These antigens have been referred to as thymus-dependent or T-dependent antigens. In contrast, the antibody response to a number of bacterial and synthetic antigens was unaffected by thymectomy. The latter, which have been termed "thymus-independent (T-independent) antigens, include pneumococcal polysaccharide,

brucella, polymerized flagellin,* lipopolysaccharides, and levan [reviewed by Basten and Howard (78)]. They differ from T-dependent antigens in four particular ways:

1. They possess a molecular configuration with a large number of repeating identical determinants.
2. They are slowly metabolized and therefore persist in the host tissues for prolonged periods of time (79).
3. They evoke an antibody response that often consists exclusively of IgM.
4. At least some of them (e.g., lipopolysaccharide and levan) are activators of C3 (80).

A requirement for T cells in antibody production to T-dependent antigens was at first difficult to reconcile with the concept of a dual immune system. It is now known, however, that T cells do not themselves synthesize antibody in the classical sense of an Ig molecule secreted into the serum but facilitate its production by B cells (43). In other words, whether T-independent or T-dependent antigens are used, antibody synthesis is a property confined exclusively to B cells. Evidence of the collaboration between T and B lymphocytes in T-dependent antibody production was obtained in two systems: thymus–marrow synergism and the immune response to haptenic determinants.

THYMUS–MARROW SYNERGISM. The first suggestion for a need for more than one cell type in antibody formation stems from the experiments of Claman and colleagues in mice (81,82). In these, thymus cells, bone marrow cells, or both were injected into irradiated or thymectomized irradiated recipients of the same strain, together with sheep red blood cells (SRC). The mice that had been

---

* T-Independent *in vitro,* but not *in vivo.*

given both thymus and marrow cells produced a far greater hemolysin response than could be accounted for by summing the activities of either cell population alone. Synergism of this kind has since been confirmed and in addition observed with both other heterologous erythrocytes (43,83,84) and serum proteins (11, 42,52). When the cooperating cells were allogenic, or histoincompatible (i.e., derived from different donor strains), no response was obtained (85). Hence it was not possible in this system to establish by immunogenetic techniques the origin of the AFC. Davies and colleagues (83, 84) adopted an alternative approach to the problem by using T-depleted mice obtained by thymectomy in adult life followed by irradiation and protection with bone marrow (TxBM) (86). Immunological competence was restored by thymus grafts from donors that had slight immunogenetic differences from those used to provide the marrow cells. Thirty days after grafting the mice were immunized with SRC, and their spleen cells were transferred to irradiated hosts, previously immunized either against the thymus or the marrow donor. The recipients sensitized against the thymus donor produced antibody to SRC, whereas those preimmunized against the marrow donor failed to do so. These findings suggested, but did not conclusively demonstrate, that the precursors of AFC (AFCP) were derived from bone marrow—that is, that T cells did not secrete antibody (for review see ref. 12).

Definitive proof of the bone marrow origin of AFCP was obtained by Miller and colleagues (43,87). They, like Chaperon and Claman (85), were unable to obtain an antibody response with mixtures of H_2-incompatible cells, which prevented direct identification of AFCP in such a system. An alternative model was therefore created. Mice that were T-cell depleted (Ntx or TxBM) were restored to immunological competence, not with a thymus graft but by infusions of cells derived from the thymus or thoracic duct lymph (containing mainly T cells). Since the thymectomized mice themselves lacked T cells, it proved possible to achieve nearly normal reconstitution with cells from semi-allogeneic as well as from syngeneic donors. Thus the identity of AFC arising in these chimeras after challenge with SRC could be directly established. The AFC proved to be derived not from the inoculum of thymus or thoracic duct cells but from cells already present in the host, originally derived from the bone marrow (i.e., B cells). Furthermore, these cells failed to respond optimally in the absence of cells derived from thymus or thoracic duct lymph, which indicates that, although T cells did not secrete antibody, they facilitated its production by B cells. The T cells thus function as "helper" cells.

The recent development of sophisticated tissue culture systems for mouse lymphocytes has made it possible to study the interaction between T and B cells *in vitro* (88,89). Although caution is essential in the interpretation of *in vitro* experiments, this approach does provide a convenient way of dissecting the cellular events in antibody production. Two models have proved particularly informative. In the first, normal spleen cells were separated into various fractions by techniques that grade cells according to size, adhesiveness, or density. By combining certain fractions *in vitro*, a synergistic response to SRC was demonstrated (90–92) that apparently required at least three cell types: an adherent radioresistant accessory cell, probably a macrophage; a nonadherent radiosensitive AFCP (i.e., a B cell); and a second nonadherent lymphocyte identified as a T cell. In addition, an *in vitro* approach of this kind, employing purified populations of lymphoid cells, permitted analysis of the possible mechanisms by which T and B cells interact. These will be discussed in detail in the subsection beginning on page 196.

Evidence of lymphocyte collaboration has now been obtained in species other than mice, including rats (93,94), rabbits (95,96), pigs (97), and chickens (98). The phenomenon appears to be of general importance in immune responses.

PRODUCTION OF ANTIBODIES TO HAPTENIC DETERMINANTS. The term "hapten" is used to denote a simple organic compound that on its own fails to elicit an immune response but will do so when coupled to an immunogenic substance or "carrier," for example, heterologous protein (99). Antihapten responses are characterized by what is termed carrier specificity. Thus when an animal is primed to a particular hapten conjugated to one carrier and then challenged with the same hapten conjugated to a different carrier, the response is much lower than that obtained by challenge with the hapten coupled to the original carrier. Furthermore, induction of tolerance to the carrier inhibits the production of antibody to a hapten conjugated to that carrier (100,101). Carrier specificity has been demonstrated in situations involving cell-mediated (102,103) and humoral immunity (54,104). It has also been reported in the immune response to complex antigens, such as lactic dehydrogenase (105), DNA (106), and glucagon (107), individual molecules of which presumably bear determinants capable of behaving as either hapten or carrier for each other.

Carrier specificity was originally explained in terms of the "local environment" hypothesis, according to which the hapten-sensitive lymphocyte recognizes both the hapten and, in addition, determinants on the carrier molecule ad-

jacent to the hapten (108,109). Two lines of investigation in particular rendered this explanation unlikely. First, insertion of an inert spacer molecule between hapten and carrier (e.g., tetraalanine or tetraproline) failed to interfere with carrier specificity; that is, it caused no reduction in the magnitude of the carrier contribution in a secondary hapten response (54,110). Second, no evidence of a contribution from the carrier to the energy of binding of hapten by antibody could be detected using inhibition techniques (111). In other words, the recognition unit (receptor) on the surface of the hapten-sensitive cell could not be an accurate sample of the antibody that that cell would produce on stimulation by hapten–protein conjugates. This was difficult to reconcile with the "minimum theory of antigen recognition" or accurate sample hypothesis (referred to on p. 188), unless it was assumed that two or more cells were involved, one with a receptor for only the hapten, and others with receptors directed toward determinants on the carrier molecules not necessarily adjacent to the hapten.

Several lines of evidence have now established that two distinct and separate classes of lymphocytes are involved in the induction of antihapten antibody responses (112,113). For example, cells from mice primed to a hapten, 4-hydroxy-3-iodo-5-nitrophenylacetic acid (NIP) conjugated onto the carrier ovalbumin (OVA) produced after transfer to heavily irradiated hosts and challenge with NIP-OVA a good antihapten response. By contrast a poor response was obtained if NIP conjugated onto a heterologous carrier, bovine serum albumin (BSA), was used for challenge. This response could, however, be significantly enhanced by adding to the system cells from mice primed to the heterologous carrier, BSA (Fig. 1). Administration of anti-BSA (anticarrier serum antibody) did not

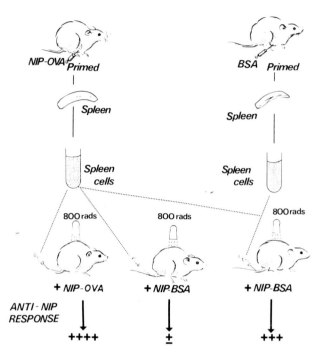

Fig. 1. Diagram illustrating the carrier effect in antihapten antibody production (see text).

cause enhancement. The BSA-primed cells must therefore have cooperated by increasing the efficiency of the NIP-OVA-primed cells in some way other than by providing anti-BSA antibody (114).

Collaboration between carrier-sensitive and hapten-sensitive cells has recently been confirmed in tissue culture (115,116). With this system it was possible to demonstrate two important points: first, thymus cells activated in heavily irradiated hosts to the carrier protein (117) could substitute for carrier-primed cells; second, the fact that semiallogeneic T and B cells cooperated *in vitro* (although not *in vivo*) permitted the use of specific anti-H_2 sera to determine the origin of the antihapten AFC. The AFC was shown to be derived from the hapten-sensitive cell population, not from the carrier-primed population. The carrier–hapten response is thus another example of synergism between T and B cells. Similar conclusions

may be drawn from the *in vivo* experiments of Raff (118), who demonstrated that antihapten antibody production was abolished by pretreatment of the carrier-primed, but not of the hapten-sensitive, cell population with anti-θ serum and complement.

MECHANISMS OF INTERACTION BETWEEN T AND B CELLS. The possible mechanisms whereby T and B lymphocytes interact in antibody formation are as follows:

1. Transfer of subcellular information.
2. Antigen focusing.
3. Elaboration of a soluble mediator:
   a. Specific factor (IgT).
   b. Non-antigen- non-Ig-specific factor (125).
   c. Nonspecific factor.

Two lines of work in particular have made it unnecessary to consider the transfer of subcellular information as a likely explanation. The first of these, which we have already discussed, was the demonstration of specificity in both cell lines. The second was obtained in experiments showing that the B cell, as the direct precursor of AFC, was already committed to the production of antibody of one class and of one allotype (reviewed in ref. 119). In other words, if B cells are committed both with respect to antigen recognition and antibody synthesis, they already possess the genetic potential to secrete material without any help from T cells.

The discovery that the hapten–carrier system is in fact another example of interaction between T and B cells led Mitchison (114) to put forward the "antigen-focusing hypothesis" as an alternative explanation. The T cells were envisaged as picking up antigen by its carrier determinants and focusing the inducing haptenic determinants, also on the same molecule, onto the corresponding receptors on B cells. An antigen bridge (Fig. 2) would thus be formed between receptors on T cells and B cells, both of which must be in physical contact. This hypothesis implies a relatively passive role for T cells in collaborative responses. Positive evidence for such a concept has,

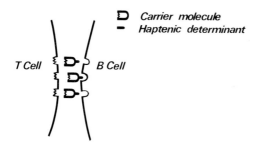

Fig. 2. Diagram illustrating the "antigen-focusing" hypothesis of Mitchison (114). The T cell picks up antigen by its carrier determinant and focuses the inducing determinant, also on the same molecule, onto the appropriate B-cell receptor. The formation of antigen bridges triggers the B cell to synthesize antibody.

however, been difficult to acquire. Thus, if T cells acted simply as passive carriers of antigen, their function should be mimicked by inert antigen-coated particles distributed *in vivo* in the same way as T cells. In fact neither antigen-coated B cells nor antigen-coated tolerant or mitomycin C–treated T cells could substitute for normal T cells in an adoptive transfer system (120). Furthermore, the capacity of normal or primed T cells to collaborate with normal or primed B cells was abrogated both *in vivo* (120) and *in vitro* (121) by preincubating the T cells with mitomycin C at a dose sufficient to inhibit T-cell proliferation completely. A similar conclusion was reached by Segal, Globerson, and Feldman (116), with vinblastine-treated cells *in vitro*. In contrast mitomycin C did not interfere with the ability of activated thymus cells to collaborate in an *in vitro* IgM response (121). The activated thymus cells used here had recently been activated *in vivo* (117), in contrast to primed T cells, which were obtained several weeks after antigen had been given. They were, in other words, a source of T cells that had recently undergone some degree of differentiation. Anderson, Sprent, and Miller (122) recently extended these observations to an *in vivo* IgG response. Activated thymus cells prevented from dividing by preirradiation, rather than by mitomycin, were injected together with B cells and antigen into heavily irradiated hosts. The recipients of untreated T cells produced an excellent IgG response, but those given irradiated cells failed to do so. The implications of these results are clear: for collaboration with B cells to occur effectively, T cells must differentiate and divide. The effect they exert in facilitating antibody production must therefore involve an active, not a passive, process.

The findings just described do not of

course exclude the possibility that actual contact between T and B cells still occurs and may be essential for the induction of antibody responses. For example, the obligatory expansion in clones of reactive T cells could well compensate for the low order of frequency of specific-antigen-sensitive T and B cells [∼ 10^{-5}–10^{-4} (123,124)]. Additional evidence from *in vitro* studies does, however, suggest an alternative mechanism (121). Thus the helper activity of activated thymus cells, although unaffected by mitomycin C, was completely abrogated by pretreatment with actinomycin D and antimycin A. Since DNA-dependent RNA synthesis is inhibited by actinomycin D but not by antimycin A at the concentrations used, and since both agents interfere with protein synthesis, it was concluded that cooperation depends on active synthesis by T cells of some protein component. Release of such a T-cell product obviates the necessity for actual contact between rare antigen-sensitive cells and is consistent with the third possible mechanism for collaboration listed on page 196.

The nature of a soluble factor of this kind has given rise to much speculation (11). It could, for example, be a special class of Ig (IgT) that concentrates antigen either onto the B cell itself or onto a third-party cell, such as the macrophage or reticulum cell. Alternatively, T cells might elaborate a specific factor that is neither antibody nor antigen but is induced by antigen, carries information in the form of the tertiary structure of the antigenic determinant, and activates B cells—a factor perhaps similar to, or identical with, that described by Valentine and Lawrence (125). The possession, by both T and B cells, of a high degree of immunological specificity (see section entitled "Antigen Recognition") does not, however, preclude actual mediation of the cooperative effect by a non-specific factor or factors. Indeed, activated thymus cells have already been shown to release several pharmacological agents that display a multitude of biological effects (126). By analogy it is conceivable that T cells may on interaction with antigen produce a chemical mediator, nonspecific with respect to the antigenic determinants concerned, but facilitating the responsiveness of the appropriate B-cell population. Whatever the immunological capabilities of such a factor, any satisfactory model for its mode of action ought to incorporate at least the following two points:

1. *A mechanism whereby B cells in the presence of antigens, including univalent haptens, are triggered.* The inability of some antigens (e.g., univalent haptens) to stimulate B cells has already been discussed and implies that, for B-cell activation to occur, a second signal, other than antigen binding, is mandatory. The logical source of that signal in the case of T-dependent antigens (hapten–carrier conjugates) is the T cell. Indeed recent studies with clones of hapten-reactive cells have suggested this to be so (127).

2. *The provision of a final common pathway via which T-dependent and T-independent antigens can stimulate B cells.* It follows from requirement 1 that if T cells provide the second signal for B-cell triggering in T-dependent responses, then T-independent antigens ought to possess some intrinsic property permitting them to switch on B cells directly.

Two groups have put forward models that satisfy these minimum requirements. The findings of the first group (128–132) favor the existence of a specific soluble factor that mediates the collaborative effect by way of the antigen-concentrating mechanism proposed by Miller and associates (11). The second group (80) has devised a model favoring the release, by activated thymus cells, of a nonspecific factor. The factor they suggest is a proactivator of C3 and triggers B cells through their C3 receptors. There is evidence in the literature in support of both specific and nonspecific mediators; in fact they may not be mutually exclusive.

The work on a specific mediator of cell collaboration stemmed originally from an *in vitro* study of the mechanism of thymus dependency. As already mentioned, not all antigens (e.g., POL in vitro) require T cells in order to induce an optimal antibody response. Those that do not—namely, the T-independent antigens—have a number of physicochemical features in common that have already been referred to briefly (78). In general, they are slowly metabolized (79), have a rigid structure, and possess a large number of repeating identical determinants, all of which enhance their capacity to interact multivalently with Ig receptors on the surface of appropriate B cells (78). In this context it should be remembered that an antibody that binds antigen bivalently has an advantage in terms of affinity by a factor of 10^4–10^5 over an antibody binding by one site only (133). The B cell with its multitude of similar receptors resembles a polyvalent antibody particle, and, if therefore it can interact with a polyvalent antigen molecule at more than one site, a substantial cooperative effect would be predicted in terms of energy of interaction. Under these circumstances the probability of B-cell stimulation is likely to be a function of the epitope density of the determinant in the microenvironment of the B-cell receptors, as has indeed been demonstrated by Feldmann (134) and Klinman (135) *in vitro* and by Del Guercio and Leuchars (136) *in vivo*. The second signal for B-cell activation would therefore appear to result from the formation on the cell surface of an appropriate antigenic matrix. In the case of T-independent antigens the antigen itself provides the matrix by virtue of its molecular structure. No additional "help" is thus required.

The corollary to these findings is that the function of T cells in antibody production is to endow immunogens, lacking such a matrix, with suitable physicochemical properties, such as multivalency and slow degradability, that enable them to trigger B cells satisfactorily. One might thus expect the requirement for T cells to be related to the mode of antigen presentation. This prediction has recently been verified by a number of approaches. In one study, for example, the haptenic determinant 2,4-dinitrophenylacetic acid (DNP) was coupled to two different "carrier" molecules, one of which, donkey red cells (DRC), elicits a T-dependent response and the other, polymerized flagellin (POL) a T-independent response *in vitro* (137). When DNP was presented in these two different forms to purified B cells, an antihapten response was obtained to DNP-POL, but not to DNP-DRC. T-Independent antihapten responses have also been obtained *in vivo* with DNP coupled to pneumococcal polysaccharide (138) and to levan (136). Thus the requirement for T cells in antihapten antibody production is not determinant specific but simply depends on the nature of the carrier molecule, which in turn determines how the hapten is presented to the B cell. One theoretical way for T cells to provide an immunogenic matrix would be by release of antigen complexed with receptor structures (IgT) as suggested by Miller and associates (11). Any "factor" of this kind would by definition carry specificity since T cells are known to react with antigen specifically (see p. 191).

Evidence in support of such a specific mediator of cell collaboration has recently been obtained in an *in vitro* system by Feldmann and Basten (128–131). This consisted of a culture chamber with two distinct compartments separated from each other by a cell-tight filter. Usually T cells were placed above and B cells below the filter. Thymus cells from mice of strain CBA were activated in irradiated hosts to keyhole limpet hemocyanin (ATC_{KLH}) and then incubated in single or double chambers with syngeneic spleen cells primed to DNP coupled to a heterologous carrier flagellin (DNP-Fla). In the presence of DNP-KLH (Fig. 3) an anti-DNP response was obtained irrespective of whether the cells were in physical contact or not; this indicates that the interaction between T and B cells does indeed depend on the release by T cells of a soluble product. The specificity of this product for both cell lines is demonstrated by the experiments illustrated in Fig. 4. In these thymus cells activated to a non-cross-reacting antigen, fowl immunoglobulin G (FγG), collaborated much less effectively than did specifically activated T cells. Furthermore, lack of augmentation of the response to a second T-dependent antigen, DRC, present in the lower chamber excluded a significant nonspecific effect on the B-cell component. Collaboration failed to occur if activated thymus cells were treated with anti-θ serum and complement to eliminate T cells. It was, however, unaffected by the incubation of activated thymus cells with a polyvalent anti-mouse Ig reagent and complement under conditions known to kill B cells (see p. 188). It thus appears that the factor released in the upper compartment was derived from T cells. (It cannot, however, be formally excluded that a small number of B cells, contaminating the spleens of the donors of activated thymus cells, may have been the source of this factor). A response was observed

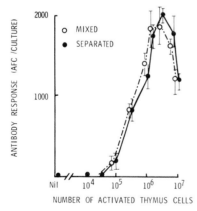

Fig. 3. Demonstration of collaboration between T and B cells across a cell-tight (Nucleopore) membrane. Each point on the graph represents the arithmetic mean of three to six cultures ± S.E. In the case of double culture flasks varying numbers of ATC_{KLH} were placed in the upper chamber and 4×10^7 DNP-Fla-primed spleen cells were placed in the lower chamber as a source of B cells; DNP-KLH (1 μg/ml) was added to both chambers.

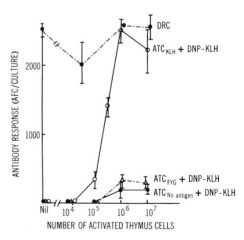

Fig. 4. Demonstration of the specificity of the collaboration between T and B cells across a cell-tight (Nucleopore) membrane. The experimental protocol was similar to that described in the caption of Fig. 3 except that a second antigen, DRC, was added to the cells in the lower chamber.

when Nucleopore membranes (pore size 0.1 μ) were used to separate the T and B cell populations, but not when dialysis membranes were substituted.

These experiments did not, however, define the precise site nor the mode of action of the factor, since the hapten-primed spleen cells in the lower compartment contained T cells and macrophages as well as B cells. A significant role for T cells as targets was excluded by the demonstration that treatment of cells in the lower chamber with anti-θ serum, whilst abrogating the response to the control antigen, DRC, had no effect on the DNP response. Thus the factor presumably triggered B cells directly or by prior interaction with macrophages. A role for macrophages in the collaborative step *per se* was a particularly attractive possibility for several reasons: first, this type of accessory cell is common compared with antigen-specific lymphocytes, thereby enhancing the chance of B-cell triggering; second, its surface provides an ideal site for the formation of an immunogenic matrix in the microenvironment of the B-cell receptors; third, some evidence, albeit indirect, has favored involvement of macrophages in B-cell activation. For example, B cells, not T cells, have been found to adhere to macrophages *in vitro* (139–141). Furthermore, Feldmann (142) has recently shown that antibody production *in vitro* to T-dependent, but not to T-independent, antigens requires the presence of macrophages.

The possibility that the soluble factor acted via an accessory cell, like the macrophage, was examined in a transfer system. Both ATC$_{KLH}$ and DNP-KLH were placed in the upper compartment of a double culture flask and purified macrophages (143) in the lower. After 1–3 days' incubation the macrophages were harvested, washed, and transferred to a single culture vessel containing DNP-Fla-primed spleen cells or purified lymphocytes without further addition of antigen. A highly significant anti-DNP response was obtained 4 days later. This had specificity at both the level of T and B cells.

The capacity of macrophages to trigger B cells in the second chamber was abrogated by trypsinization, by pretreatment with monospecific anti-Ig sera directed against mouse κ or μ chains, and by culturing in the presence of anti-DNP or anti-KLH antiserum. Taken together these findings support the contention that cell collaboration depends on the release by activated thymus cells' of their Ig receptors (which contain μ and κ chains) complexed with antigen. The IgT–antigen complex would be cytophilic for macrophages on which it would form an immunogenic matrix with a surface structure resembling that of a T-independent antigen. The ultimate step in the triggering of antibody production would then occur by the interaction of B cells with the sensitized macrophages. This hypothesis thus satisfies the two minimum requirements stipulated earlier in that it provides a second signal for B cells and, in addition, a single mechanism for B-cell triggering by T-dependent and T-independent antigens (Fig. 5). Its veracity has recently been strengthened by evidence of the release by T cells (presumably free of many contaminating B cells) of an Ig–antigen complex (144).

The concept of a specific T-cell moiety with macrophage cytophilia is consistent both with the "carrier antibody" concept of Bretscher and Cohn (46) and the collaborative models proposed by Lachmann (145) and Gutman and Weismann (146). Furthermore, two other groups using different experimental approaches have recently been able to demonstrate the existence of a T-cell factor with similar characteristics (147–149). The latter in

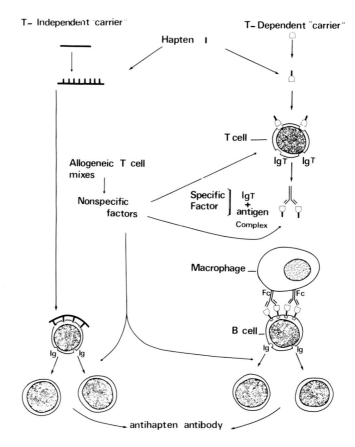

Fig. 5. Diagram illustrating a specific mechanism of collaboration between T and B cells. B-Cell triggering depends on producing a sufficiently high epitope density in the microenvironment of the B-cell receptors In the case of T-dependent conjugates this is achieved by the release of IgT–antigen complexes that adhere to the surface of macrophages, thereby forming a high-density matrix. In the case of T-independent conjugates concentration of determinants is achieved either by virtue of the repeating determinant structure of the carrier molecule or by its slow degradability.

addition found evidence [as did Basten and Feldmann (131) in their model] of a nonspecific as well as a specific moiety. Thus the *in vitro* response to sheep erythrocytes, for example, required four components: B cells, adherent cells (macrophages), a specific T cell, and a nonspecific factor produced by T cells. Although the nature, functions, and relative contributions of the two factors have not been elucidated, these studies do at least suggest that they are not mutually exclusive and may well be produced simultaneously.

Evidence of the existence of *nonspecific soluble mediators* has been obtained by several investigators (150–154). The most effective way of stimulating their production is by means of the "allogeneic effect": a graft versus host reaction *in vivo* or an allogeneic mixture of T and B cells *in vitro*. It was observed both in mice and in guinea pigs (for review, see ref. 154) that the induction of a transient graft versus host reaction [produced by injecting foreign (i.e., "allogeneic") T cells able to react against histoincompatible host cells] obviated the requirement for carrier-specific T cells in the induction of secondary antihapten antibody responses. Furthermore, it enhanced the development of hapten-specific memory in the host B-lymphocyte population. A general proliferative response of T cells, among which exist normally occurring carrier-specific cells, during the graft versus host reaction is unlikely to account for the observed effects on B cells. This is because antihapten antibody responses were elicited, during a graft versus host reaction, after challenge with hapten conjugated to carriers for which no functional T cells existed in the particular strains of animals used. These were "nonresponder" strains genetically incapable of mounting a T-cell-mediated immune response to the particular carrier chosen. *In vivo* the allogeneic effect could be achieved only if the host B cells were the targets of the reaction of the foreign T lymphocytes. This implies that a nonspecific T-cell mediator influencing B-cell responsiveness to antigen can be active only over a very short range.

In contrast to the *in vivo* system, the allogeneic effect could be produced *in vitro* whether or not B cells were the targets of reactivity of T cells (151). This difference may be accounted for by the unique spatial relationships between cells in suspension cultures, which does not exist *in vivo*. On the other hand, the mechanisms of enhancement of B-cell activity *in vitro* may differ from that operating *in vivo*. Whatever the case may be, it is clear that *in vitro*, as *in vivo*, the T cell, not the B cell, is the source of a soluble mediator that tends to act optimally 2–3 days after initiation of the response, that is, in the postinductive phase (152,153). Similar conclusions were reached by Basten and Feldmann using the double-chamber system (131). The capacity of the factor in their model to enhance antibody production to T-

dependent and T-independent antigens to a comparable degree confirmed its nonspecificity. It could therefore be a nonspecific B-cell mitogen that plays a role in the amplification of the antibody response.

More recently a nonspecific mediator of the collaborative effect has been found in syngeneic systems. For example, supernatants from cultures treated with concanavalin A can stimulate antibody production nonspecifically (155). Furthermore, T cells or their products activated to one protein antigen (e.g., KLH) have the ability to reconstitute the response of T-cell-depleted spleen cell cultures to other non-cross-reacting antigens, provided the original antigen (i.e., KLH) is present in the culture (156). The implication of these findings is that B-cell triggering requires two distinct signals, one from the interaction of its specific Ig receptors with antigen and the other from the diffusible factor released from nearby T cells reacting to antigen (12). Various types of mechanism have been invoked to explain the influence of nonspecific T-cell mediators on B-cell responsiveness to antigen. Among these, two will be singled out for special mention.

Activated thymus cells elaborate factors that influence the mobility of phagocytes (monocytes, macrophages, and granulocytes) (157) and the degradation of antigen by phagocytes (158). Phagocytic cells may thus remove antigen from the microenvironment of activated thymus cells and of any antigen-binding B cells within their proximity. T-independent antigens (which may not activate T cells) tend to persist, fail to be degraded, and generally elicit an IgM, not an IgG, response. T-dependent antigens, which in T-cell-depleted mice elicit predominantly an IgM response and little or no IgG, persist longer in the absence of activated thymus cells. It has thus been postulated that activated thymus cells mobilize phagocytic cells that remove antigen from B cells and thus lower the tolerogenicity of this antigen, particularly for Bγ cells, which have a higher antigen-binding capacity than do Bμ cells (Fig. 6) (for review see ref. 159).

The second hypothesis offered to explain the mechanism of T-cell regulation of B-cell responsiveness by means of a nonspecific diffusible mediator is that of Dukor and Hartmann (80). In addition, their hypothesis provides a final common pathway for T-dependent and T-independent antigens. They propose that B cells are switched on via their C3 receptors following activation of C3 (Fig.

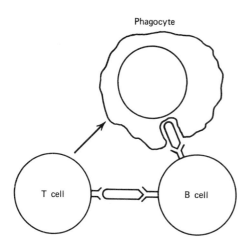

Phagocyte

T cell   B cell

Fig. 6. Diagram illustrating a scheme of collaboration between T and B cells proposed in 1972 by Mitchell and associates (138). When activated by an appropriately presented antigen (e.g., bound on Ig receptors of B cells) T cells produce factors that attract and activate phagocytes, which in turn remove tolerogenic amounts of antigen from the milieu of the B cells. Cells binding large amounts of antigen (particularly those with high-affinity receptors, perhaps most B γ cells) will thereby be spared from tolerance induction and go on to synthesize antibody. Reproduced from the *European Journal of Immunology* with permission from the authors and from the editors.

7). Two experiments in particular favor a role for complement in B-cell triggering. First, *in vivo* depletion of C3 by cobra venom factor was found to inhibit T-dependent antibody production (160); second, when the C3 receptor was blocked *in vitro*, complement-sufficient but not C3-deficient serum was able to enhance the reactivity of B cells to an otherwise T-cell-specific mitogen (161). The question therefore arose whether C3 could be activated during a collaborative response. This indeed seems to be likely. Activated thymus cells undergoing blast transformation secrete lysosomal enzymes (162, 163) that have the capacity to generate active split products of C3 by cleavage (164,165). If the enzymes are released along with other pharmacological factors (see p. 190 and ref. 126), T cells could well trigger B cells. Because of the rapid loss of binding capacity of activated C3 in the fluid phase, only those B cells in the immediate vicinity of activated thymus cells could bind sufficient C3 onto their C3 receptors and hence be triggered. Furthermore, the capacity of phagocytic cells to release lysosomes (166) implies that macrophages (possibly those activated by pharmacological agents released from activated thymus cells) could act as intermediaries or as substitutes for T cells. These observations do not explain how T-independent antigens stimulate B cells since the response to these occurs independently of macrophages and of T cells (142). Earlier in this section it was proposed that T-independency is related to the capacity of antigen to bind simultaneously to multiple receptor sites on B cells (78). Dukor and Hartmann (80), however, have put forward an alternative mechanism for B-cell triggering by this group of antigens. In addition to displaying repeating antigenic determinants they have physicochemical characteristics similar to those

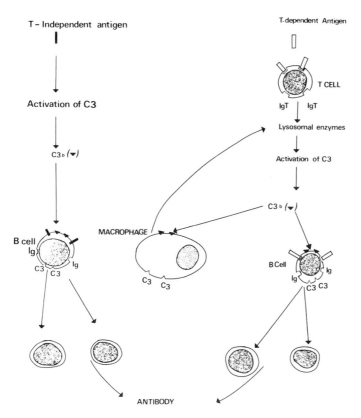

Fig. 7. Diagram illustrating a nonspecific mechanism of collaboration between T and B cells. B-cell triggering depends on interaction between activated C3 (C3b) and the C3 receptor on B cells, as proposed by Dukor and Hartmann (80). In the case of T-dependent antigens T cells provide a source of lysosomal enzymes that activate C3. With antigens like lipopolysaccharide and polymerized flagellin, T cells are not required since these substances can activate C3 by the bypass mechanism. The C3 receptors are indicated by ∨; the Ig and Fc receptors are indicated by ⊔ and ∪, respectively.

of substances that are known to activate C3 via a bypass mechanism (167,168). Indeed levan, bacterial lipopolysaccharides, and possibly polymerized flagellin have now been shown to cleave C3 in this way (80). Should other T-independent antigens do so, binding of activated C3 to the C3 receptors would provide not only a means of initiating the second signal for B cells but in addition a single mechanism of B-cell activation by T-dependent and T-independent antigens (Fig. 7). Since only a subpopulation of B cells carry the C3 receptor, the hypothesis requires that at least the precursors of IgM-producing cells (Bμ) have

this receptor. Dukor and colleagues have evidence supporting this, although preliminary data from Parish (169) indicate that a substantial number of Bγ, rather than Bμ, cells in rat thoracic duct lymph bear C3 receptors.

It is not possible at this stage to determine which T-cell factor—specific IgT, nonspecific diffusible mediator, or both—plays a role in the cooperation between T and B cells under physiological conditions *in vivo*. Neither IgT nor a soluble mediator influencing B-cell responsiveness to antigen has yet been isolated and characterized. However, it is clear that T cells do produce

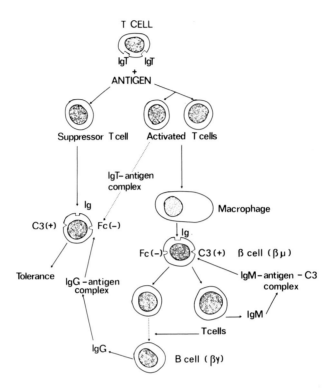

Fig. 8. Diagram illustrating the regulatory factors involved in collaboration between T and B cells. Activated T cells and IgM antibody exert a positive effect on the immune response, whereas suppressor T cells and IgG antibody have a negative effect. Symbols for receptors are the same as those used in Fig. 7A. plus (+) sign indicates that the receptor has a positive role to play in regulation of the response, and a minus (−) sign indicates a negative role.

pharmacological factors active in delayed hypersensitivity reactions (157), but have not unequivocally been shown to synthesize and export Ig molecules. Future studies will clearly need to concentrate on identifying and characterizing products elaborated by purified suspensions of activated thymus cells, freed from any contaminating B cells, and on establishing suitable models for analysis at the level of the whole animal.

REGULATION OF THE IMMUNE RESPONSE. The study of interactions between T and B cells has led to a rapid expansion in our understanding of the regulatory mechanisms underlying the immune response. It has, for example, become apparent that T cells may have the capac-

ity to suppress as well as to enhance antibody production (170,171). Evidence of the existence of "suppressor" T cells stems from a variety of experiments (172–175). The most direct comes from recent work on the induction of tolerance to protein antigens *in vivo* (176), 177). Spleen cells from tolerant mice were transferred into lethally irradiated hosts together with T cells and antigen. No response was observed unless the original tolerant cell population was pretreated with anti-θ serum and complement (to eliminate T cells); only then did the antibody response rise to control levels. In other words, the tolerant state resulted from the presence of suppressor T cells that directly or indirectly inhibited antibody production by B cells.

It should be stressed that neither specific T cells nor specific B cells could have been deleted in this situation. Hence T cells must exert a negative as well as a positive effect on the immune response.

This phenomenon may well have important biological implications, for example, in the development of autoimmune disease. Thus one may postulate that there are no T helper cells but only T suppressor cells for self proteins and nonimmunogenic substances. Activation of these cells in a normal situation would provide suppressive signals and induce tolerance, not immunity, in the corresponding B-cell population. If, however, a new "carrier" molecule is introduced or the suppressor mechanism is bypassed in some way, autoantibody production will occur since the B cells, though unresponsive, may not have been deleted. The mechanism whereby T cells exert such an effect may be explicable in terms of the IgT–antigen concentration model described earlier in this section (Fig. 5). Indeed, preliminary evidence suggests that the IgT–antigen complexes are tolerogenic if they interact directly with the B-cell surface rather than by adherence to macrophages (178). These cells might therefore be rendered unresponsive if macrophages do not "see" the T-cell product (the case with self proteins) or if their surface becomes saturated by overproduction of complexes [as in high zone tolerance or antigenic competion (179)].

The B cells likewise play a vital role in the regulation of antibody production both by secretion of the various classes of Ig (180,181) and by way of the C3 and Fc receptors on their surface. Immunoglobulin G antibody usually exerts a suppressive effect in ongoing responses, a process dependent on the presence of the Fc piece of the molecule (41). The interaction of this class of antibody with antigen results in a complex with high avidity for the Fc receptors on B cells (24) and macrophages (8). A high concentration of antigenic determinants would therefore be created in the vicinity of the Ig receptors on B cells; if dense enough, it could possibly tolerize rather than immunize appropriate AFCP (134). This effect, sometimes termed antibody-mediated tolerance, has in fact been demonstrated *in vitro,* but not yet *in vivo* (182). Passive administration of IgM antibody, on the other hand, exerts a positive feedback effect on the immune response (183). The reason for this seems to be that IgM–antigen complexes, unlike those formed with IgG, are particularly efficient activators of the complement sequence (184) and therefore interact readily with the C3 receptors on B cells. Newly secreted antibody formed early in the immune response would be expected to combine locally with free antigen, complement, and C3 receptors, to trigger appropriate B cells, thereby escalating antibody formation (80). After the switch to IgG synthesis under the influence of T cells (185), C3 activation by IgM–antigen complexes would cease to operate. Thus T cells and B cells exert a critical role in the control of the immune response (Fig. 8). The T helper cells and IgM antibody act early as amplifiers when the supply of antibody may be limited. This could be particularly important, for example, in the control of infections. Subsequent production of unlimited antibody would be controlled by suppressor T cells and IgG, both of which directly or indirectly can render B cells unresponsive.

A further dimension in the regulation of the immune response has been introduced by recent work on an antireceptor site or "aliotypic" antibody, which is thought to interfere with antigen recognition by immunocompetent lymphocytes, probably T cells. Ramseier and Lindenmann (186) have recently reviewed the evidence for this concept in cell-mediated immunity, and Cosenza and Kohler (187) claim to have found a similar type of antibody in humoral immunity. It is, however, as yet too early to determine the precise significance of this work.

Collaboration between Two Classes of T Cells

Lymphocyte interactions in cell-mediated immunity have been less clearly defined than in humoral immunity (11). In graft versus host reactions or in delayed hypersensitivity responses some evidence is available in favor of a synergism between T cells and cells present in bone marrow (188,189). The data as they stand do not, however, establish unequivocally whether the bone marrow cell concerned is a B cell or a cell from a different lineage. There are two reasons in particular for this: first, bone marrow from certain strains of animals may contain significant numbers of T cells and macrophage precursors as well as B cells (94,190); second, the most effective immunogenic target cell for graft versus host reactions is derived from bone marrow (191), which confers a selective advantage on cells from this source in the graft versus host assay. More recently studies of graft versus host reactions have implicated an interaction between two differentiated classes of T cells in this type of response. One class (T_1) comprises the precursors of cells that perform as effectors, and the other (T_2) serves as an amplifier to augment the activity of the former (192,193). Whether T_1 and T_2 cells have intrinsic differences or merely represent different stages in the maturation of T cells in the postthymic phase of their life cycle remains to be determined as does the biological significance of this concept. Sprent and Miller (194) in their studies

of graft versus host reactions were unable to demonstrate synergy between lymphocyte populations enriched for T_1 and T_2 cells.

Collaboration between Two Classes of B Cells

The existence of T-dependent and T-independent antibody responses has raised the possibility of heterogeneity among B cells. Indeed Playfair and Purves (195) recently presented some preliminary evidence in favor of such a subdivision (B_1 and B_2). It was claimed that B_1 cells occurred preferentially in bone marrow rather than in spleen and could respond directly to determinants on a multivalent antigen, whereas B_2 cells were more prevalent in spleen and could only be triggered in the presence of T helper cells. The models for cell collaboration proposed earlier do tend to render an arbitrary subdivision of this kind unnecessary. Thus there is at present no obvious reason why a haptenic determinant coupled to a T-dependent or to a T-independent "carrier" might not interact equally effectively with a single type of B cell.

This conclusion does not, however, exclude the possibility of collaboration between different B cells. Indeed some evidence points in this direction, particularly the work of Andersson and Blomgren (196) and Del Guercio and Leuchars (136). The latter, for example, found that in mice rendered tolerant to levan, DNP-levan failed to induce a significant rise in antihapten antibody levels, whereas DNP conjugated to a heterologous carrier did so. In other words, a B cell recognizing levan appeared to be essential for the expression of the DNP response. The secondary responses elicited by large doses of hapten-heterologous carrier conjugates (110) may be another example of interaction between different B cells (70). On the other hand, the higher efficiency of T, rather than B, helper cells as a concentrating device in the great majority of immune responses, particularly *in vivo* (78), does raise the question of the biological significance of this kind of interaction.

Collaboration between T Cells and Accessory Cells

One example of interaction between T cells and accessory cells like the macrophage has already been mentioned in the discussion of mechanisms of collaboration in antibody production. A similar ancillary role for macrophages has now been proposed in the *in vitro* generation of killer cells against histoincompatible target cells (197), and in the T-cell response to mitogens such as phytohemagglutinin and concanavalin A (198). The macrophage in each case is not playing a major role as an effector cell but is essential for the activation of specific T or B cells. In some cell-mediated responses macrophages themselves behave as effector cells either alone or in conjunction with lymphocytes. The classical example of this is the delayed hypersensitivity reaction. It has been shown that T cells specifically activated by antigen release various pharmacological factors (126,199), including the blastogenic, chemotactic, migration-inhibiting, and macrophage-activating factors, which trap and then stimulate ancillary cells, particularly monocytes or macrophages, in a nonspecific manner. The resultant inflammatory response is largely due to such cells (200). The biological significance of macrophages activated in this manner has been well documented in a number of situations, including antimicrobial immunity [particularly im-

munity to viruses (201), fungi (8) and intracellular bacteria such as listeria (202,203)], transplantation immunity (204), and tumor immunity (8,205).

Activated thymus cells can in addition stimulate and recruit, possibly by the release of soluble factors, other accessory cells, such as eosinophil leukocytes (206, 207) and basophil leukocytes (208,209), which are of particular importance in hypersensitivity states and parasitic infestations.

Interaction between Allogeneic Lymphocytes and Hematopoietic Cells

Lafferty and colleagues (for review see ref. 210) have performed thorough investigations of graft versus host reactions *in vivo* and mixed lymphocyte responses *in vitro*. Their studies led them to the conclusion that immune reactivity can be divided into two major groups:

1. Antigen responsiveness, which depends on the ability of immunocompetent lymphocytes to recognize antigen, interact with it, proliferate, differentiate, and produce the effector cells that mediate humoral and cellular immunity (see p. 188).
2. "Allogeneic stimulation," which occurs when an immunocompetent cell interacts with an allogeneic hematopoietic cell or one of its more differentiated derivatives (e.g., another lymphocyte).

Allogeneic stimulation can have two opposite effects: inactivation of "primitive" stem cells and proliferation of "committed" stem cells or their more mature progeny. The degree of reactivity in allogeneic stimulation decreases rapidly as the phylogenetic separation between the two species involved is increased. Thus no such stimulation is obtained in many xenogeneic combinations (e.g., when cells from sheep and chickens are mixed). The interaction between the stimulating immunocompetent cell and the responding hematopoietic cells probably involves surface contact. Lafferty and associates, however, have reasons to believe that such contact is *per se* insufficient to activate the process of stimulation. They postulate that some specific "recognition" factor (possibly RNA) is transferred from the stimulating cell to the responding cell. This factor is produced only by the immunocompetent cells and could conceivably be under the control of the recently described MLR genetic locus, which governs activation in mixed lymphocyte reactions and maps within the major histocompatibility complex in both mouse and man (211).

The implications of allogeneic stimulation are important. Such an effect could be used to augment antigen responsiveness. Thus the strength of an antigen (e.g., a weak tumor-specific antigen) may be greatly enhanced by the presence of a concomitant allogeneic stimulus. In fact, as mentioned on page 196, the requirement for carrier specificity in secondary antihapten antibody response was obviated by providing an allogeneic stimulus. Furthermore, it has been shown that allogeneic stimulation can break a state of preexisting tolerance (reviewed in ref. 212).

SIGNIFICANCE OF CELLULAR INTERACTIONS IN THE IMMUNE RESPONSE

The immune system is thought to have originated as an evolutionary by-product of the need to develop a cell–cell recognition system (see p. 188). In practical terms this means that multicellular species have acquired the capacity to differentiate self from nonself and to resist invasion by extrinsic influences, particularly pathogenic organisms like parasites, bacteria, and viruses. Primitive species rely exclusively on "constitutive" defense mechanisms that consist of accessory cells such as macrophages and are therefore nonspecific. More sophisticated organisms like mammals have developed a specific "adaptive" system that is a function of immunologically competent lymphocytes (12). Interaction between antigen and lymphocytes confers on the host two distinct advantages. First, it enhances the efficiency of host defenses by virtue of its specificity; second, it permits diversification of the resulting immune response by activating a wider range of effector mechanisms via the cellular interactions described in a preceding section. The key cell responsible for the diversity of reactions is the T lymphocyte. By virtue of its capacity to recirculate it is particularly well equipped for the task of seeking antigen and influencing the response of other lymphocytes and phagocytic cells to that antigen. Support for such a role derives from the recent demonstration of rapid recruitment of specific-antigen-reactive cells from the recirculating pool during both cell-mediated (213,214) and humoral responses (213). The ensuing interactions within lymphoid tissue between activated thymus cells and effector cells like macrophages, B cells, and some granulocytes (see p. 193) result in amplification of the immune response. This is manifest by the simultaneous triggering of a wide range of host defense systems, constitutive as well as adaptive, including the inflammatory response, kinin release, complement and coagulation cascades, and phagocytosis. Although the majority of these reactions are desirable from the teleological point of view, their multiplicity makes analysis and manipulation of the immune response a complex problem.

The T–B cell system is the best characterized of the cellular interactions (11, 43,110,131). Despite this the existence of relatively normal Ig levels in both athymic mice (215) and immunologically deficient humans (216) has raised the question of the biological significance of T-cell influence in antibody production. There is, however, compelling evidence favoring a vital role for T cells in determining the amount, class (185,217, 218), and affinity (219,220) of antibody formed by B cells. Indeed the switch from IgM to IgG production in the immune response seems to be under T-cell control (11,185,218). Furthermore, it would appear that cell collaboration has wide practical implications in the manipulation of the immune response. The concept of antigenic strength, for example, can be reappraised in terms of this system. If T cells in normal animals do indeed fulfill a function in modulating antigen concentration, as suggested by the recruitment studies (213), their activity is likely to enhance the immunogenicity of weak antigens or prevent the tolerogenicity of antigen bound in large concentrations to B cells, in particular Bγ cells. Selective stimulation of T cells has in fact been achieved in several ways, including presentation of antigen in an appropriate form (191), or by the appropriate route (221), and by the use of adjuvants like the polynucleotides (222). Furthermore, chemical modification of antigens has resulted not only in augmentation of cell-mediated immunity but also in the suppression of antibody formation (223,224). This approach has obvious relevance to cancer immunotherapy for two reasons: first, any method capable of enhancing the immunogenicity of tumor-specific antigens or stimulating the T-cell system might lead to a more effective cellular response against the tumor; second, some classes of antibody (e.g., IgG in the mouse) protect, rather than destroy, malignant cells (225). Thus a reduction in a "blocking" or "enhancing" antibody of this kind would be particularly beneficial. The use of antigen coupled to a nonimmunogenic carrier for tolerance induction in B cells may likewise find application here (53).

Both T and B cells differ in their threshold of sensitivity to tolerization, T cells being rendered unresponsive more readily (11). This difference can be exploited to suppress the cell-mediated, rather than the humoral response, to antigen and may be beneficial in transplantation immunity if the circumstances are such as to favor the production of blocking antibody, which in this situation is highly desirable.

The increased sensitivity of T cells to tolerization affords, in addition, an explanation of the termination of tolerance by cross-reacting antigen (226). In this situation T cells recognize the new determinants on the antigen and concentrate the hitherto tolerated determinant onto B cells, thereby switching on antibody formation. This may well be one of the more important mechanisms for the induction of autoimmunity. Conversely the requirement for recognition of two distinct sites on the immunogenic molecule in the initiation of antibody production provides an effective bastion against autoimmunization.

REFERENCES

1. Bodmer, W. F., *Nature* **237,** 139 (1972).
2. Burnet, F. M., *Nature* **232,** 230 (1971).
3. Miller, J. F. A. P., *Rec. Aust. Acad. Sci.* **2,** 82 (1971).
4. Shortman, K., Diener, E., Russell, P. J., and Armstrong, W. D., *J. Exb. Med.* **131.** 461 (1970).

5. Sell, S., *Fed. Proc.* **26,** 528 (1967).
6. Wigzell, H., and Andersson, B., *J. Exp. Med.* **129,** 23 (1969).
7. Ada, G. L., and Byrt, P., *Nature* **222,** 1291 (1969).
8. Nelson, D. S., *CRC Crit. Rev. Microbiol.* **1,** 353 (1972).
9. Gowans, J. L., and McGregor, D. D., *Prog. Allergy* **9,** 1 (1965).
10. Miller, J. F. A. P., and Osoba, D., *Physiol. Rev.* **47,** 437 (1967).
11. Miller, J. F. A. P., Basten, A., Sprent, J., and Cheers, C., *Cell. Immunol.* **2,** 469 (1971).
12. Miller, J. F. A. P., *Int. Rev. Cytol.* **33,** 77 (1972).
13. Moore, M. A. S., and Owen, J. J. T., *Lancet* **2,** 658 (1967).
14. Metcalf, D., and Moore, M. A. S., *Haemopoietic Cells, Their Origin, Migration and Differentiation*, North-Holland, Amsterdam, 1971.
15. Warner, N. L., *Folia Biol.* **13,** 1 (1967).
16. Roitt, I., Greaves, M. F., Torrigiani, G., Brostoff, J., and Playfair, J. H. L., *Lancet* **2,** 367 (1969).
17. Raff, M. C., *Immunology* **19,** 637 (1970).
18. Rabellino, E., Colon, S., Grey, H. M., and Unanue, E. R., *J. Exp. Med.* **133,** 156 (1971).
19. Basten, A., Miller, J. F. A. P., Sprent, J., and Pye, J., *J. Exp. Med.* **135,** 610 (1972).
20. Takahashi, T., Old, L. J., Boyse, E. A., *J. Exp. Med.* **131,** 1325 (1970).
21. Raff, M. C., Nase, M., and Mitchison, N. A., *Nature New Biol.* **229,** 50 (1971).
22. Niederhuber, J. E., Britton, S., and Berqvist, R., *J. Immunol.* **109,** 366 (1972).
23. Niederhuber, J. E., *Nature New Biol.* **233,** 86 (1971).
24. Basten, A., Mandel, T., and Warner, N. L., *J. Exp. Med.* **135,** 627 (1972).
25. Dickler, H. B., and Kunkel, H. G., *J. Exp. Med.* **136,** 191 (1972).
26. Dukor, P., Bianco, C., Nussensweig, V., *Eur. J. Immunol.* **1,** 491 (1971).
27. Raff, M. C., and Owen, J. J. T., *Eur. J. Immunol.* **1,** 27 (1971).
28. Miller, J. F. A. P., Sprent, J., Basten, A., and Warner, N. L., *Nature New Biol.* **237,** 18 (1972).
29. Wigzell, H., Sundkvist, K. G., and Yoshida, T. O., *Scand. J. Immunol.* **1,** 75 (1972).
30. Schlossman, S. F., and Hudson, L., *J. Immunol.* **110,** 313 (1973).
31. Basten, A., Sprent, J., and Miller, J. F. A. P., *Nature New Biol.* **178,** 235 (1972).
32. Sprent, J., and Basten, A., *Cell. Immunol.,* **7,** 40 (1973).
33. Parrott, D. M. V., and de Sousa, M. A. B., *Clin. Exp. Immunol.* **8,** 663 (1971).
34. Everett, N. B., and Tyler, R. W., *Int. Rev. Cytol.* **22,** 205 (1967).
35. Sprent, J., *Cell. Immunol.* **7,** 10 (1973).
36. Howard, J. C., *J. Exp. Med.* **135,** 185 (1972).
37. Miller, J. F. A. P., and Sprent, J., *J. Exp. Med.* **134,** 66 (1971).
38. Mitchell, G. F., Chan, E. L., Noble, M. S., Weissman, I. L., Mishell, R. I., and Herzenberg, L. A., *J. Exp. Med.* **135,** 165 (1972).
39. Nossal, G. J. V., Ada, G. L., Austin, C. M., and Pye, J., *Immunology* **9,** 349 (1965).
40. Perlmann, P., and Perlmann, H., *Cell. Immunol.* **1,** 300 (1970).
41. Sinclair, N. R. St. C., Lees, R. K., Chan, P. L., and Khan, R. H., *Immunology* **19,** 163 (1970).
42. Taylor, R. B., *Transplant. Rev.* **1,** 114 (1969).
43. Miller, J. F. A. P., and Mitchell, G. F., *Transplant. Rev.* **1,** 3 (1969).
44. Burnet, F. M., *The Clonal Selection Theory of Acquired Immunity*, Cambridge University Press, Cambridge, 1959.
45. Ehrlich, P., *Proc. Roy. Soc. (London)* 424 (1900).
46. Bretscher, P. A., and Cohn, M., *Science* **169,** 1042 (1970).
47. Billingham, R. E., Brent, L., and Medawar, P. B., *Nature* **172,** 603 (1953).
48. Mitchison, N. A., *Proc. Roy. Soc. (London)* **B161,** 275 (1964).
49. Nossal, G. J. V., *Aust. J. Exp. Biol. Med. Sci.* **36,** 235 (1958).
50. Basten, A., Miller, J. F. A. P., Warner, N. L., and Pye, J., *Nature New Biol.* **231,** 104 (1971).
51. Roelants, G. E., and Askonas, B., *Eur. J. Immunol.* **2,** 151 (1971).
52. Weigle, W. O., Chiller, J. M., and Habicht, G. S., *Transplant. Rev.* **8,** 3 (1972).
53. Katz, D. H., Hamaoka, T., and Benacerraf, B., *J. Exp. Med.* **136,** 1404 (1972).
54. Mitchison, N. A., *Cold Spring Harbor Symp. Quant. Biol.* **32,** 431 (1967).
55. Wigzell, H., and Mäkelä, O., *J. Exp. Med.* **132,** 110 (1970).

56. Vitetta, E. S., Bianco, C., Nussensweig, V., and Uhr, J. W., *J. Exp. Med.* **136,** 81 (1972).

57. Greaves, M. F., *Transplant. Rev.* **5,** 45 (1970).

58. Bankhurst, A. D., Warner, N. L., and Sprent, J., *J. Exp. Med.* **134,** 1005 (1971).

59. Nossal, G. J. V., Warner, N. L., Lewis, H., and Sprent, J., *J. Exp. Med.* **135,** 405 (1972).

60. Hammerling, U., and Rajewsky, K., *Eur. J. Immunol.* **1,** 447 (1971).

61. Perkins, W. D., Marnovsky, M. J., and Unanue, E. R., *J. Exp. Med.* **135,** 27 (1972).

62. Greaves, M. F., and Hogg, N. M., *Prog. Immunol.* **1,** 111 (1971).

63. Mason, S., and Warner, N. L., *J. Immunol.* **104,** 462 (1970).

64. Crone, M., Koch, C., and Simonsen, M., *Transplant. Rev.* **10,** 36 (1972).

65. Uhr, J. W., and Landy, M., eds., *Immunologic Intervention*, Academic Press, New York, 1971.

66. Cone, R. E., Marchalonis, J. J., and Rolley, R. T., *J. Exp. Med.* **134,** 1373 (1971).

67. Marchalonis, J. J., Cone, R. E., and Atwell, J. L., *J. Exp. Med.* **135,** 956 (1972).

68. Feldmann, M., Cone, R. E., and Marchalonis, J. J., *Cell. Immunol.,* **9,** 1 (1973).

69. Cone, R. E., Sprent, J., and Marchalonis, J. J., *Proc. Natl. Acad. Sci. U.S.* **69,** 2356 (1972).

70. De Weck, A. L., *Transplant. Rev.* **10,** 3 (1972).

71. Dunham, E. K., Unanue, E. R., and Benacerraf, B., *J. Exp. Med.* **136,** 403 (1972).

72. Taylor, R. B., Duffus, P. H., Raff, M. C., and de Petris, S., *Nature New Biol.* **233,** 225 (1971).

73. Wilson, J. D., Nossal, G. J. V., and Lewis, H., *Eur. J. Immunol.* **2,** 225 (1972).

74. Loor, F., Forni, L., and Pernis, B., *Eur. J. Immunol.* **2,** 203 (1972).

75. Diener, E., and Paetkau, V. H., *Proc. Natl. Acad. Sci. U.S.* **69,** 2364 (1972).

76. Hadden, J. W., Hadden, E. M., and Middleton, E., *Cell. Immunol.* **1,** 583 (1970).

77. Parker, C. W., Smith, J. W., Steiner, A. L., *Int. Arch. Allergy Appl. Immunol.* **41,** 40 (1971).

78. Basten, A., and Howard, J. H., in *Contemporary Topics in Immunobiology*, Davies, A. J. S. and Carter, R. L., ed. Plenum Publishing Corp., New York, 2, 265 (1973).

79. Sela, M., Mozes, E., and Shearer, G. M., *Proc. Natl. Acad. Sci. U.S.* **69,** 2696 (1972).

80. Dukor, P., and Hartmann, K. U., *Cell. Immunol.* **7,** 349 (1973).

81. Claman, H. N., Chaperon, E. A., and Triplett, R. F., *J. Immunol.* **97,** 828 (1966).

82. Claman, H. N., and Chaperon, E. A., *Transplant. Rev.* **1,** 43 (1969).

83. Davies, A. J. S., Leuchars, E., Wallis, V., and Koller, P. C., *Transplantation* **4,** 338 (1966).

84. Davies, A. J. S., Leuchars, E., Wallis, V., Marchant, R., and Elliot, E. V., *Transplantation* **5,** 22 (1967).

85. Chaperon, E. A., and Claman, H. N., *Fed. Proc.* **26,** 640 (1967).

86. Miller, J. F. A. P., Doak, S. M. A., and Cross, A. M., *Proc. Soc. Exp. Biol. Med.* **112,** 785 (1963).

87. Mitchell, G. F., and Miller, J. F. A. P., *J. Exp. Med.* **128,** 821 (1968).

88. Mishell, R. I., and Dutton, R. W., *J. Exp. Med.* **126,** 423 (1967).

89. Marbrook, J., *Lancet* **2,** 1279 (1967).

90. Mosier, D. W., and Coppleson, L. W., *Proc. Natl. Acad. Sci. U.S.* **61,** 542 (1968).

91. Hartmann, K. U., *J. Exp. Med.* **132,** 1267 (1970).

92. Haskill, J. S., Byrt, P., and Marbrook, J., *J. Exp. Med.* **131,** 57 (1970).

93. Johnson, J. M., and Wilson, D. B., *Cell. Immunol.* **1,** 430 (1970).

94. Scott, D., and Howard, J. C., *Cell. Immunol.* **3,** 421 (1972).

95. Abdou, N. I., and Richter, M., *Adv. Immunol.* **12,** 201 (1970).

96. Ozer, H., and Waksman, B. H., *J. Immunol.* **109,** 410 (1972).

97. Atkins, R. C., Robinson, W. A., Eiseman, B., *J. Exp. Med.* **131,** 833 (1970).

98. Rouse, B., and Warner, N. L., *Nature New Biol.* **236,** 79 (1972).

99. Landsteiner, K., *Biochemistry* **119,** 294 (1921).

100. Weigle, W. O., *J. Exp. Med.* **116,** 913 (1962).

101. Benacerraf, B., Green, I., and Paul, W. E., *Cold Spring Harbor Symp. Quant. Biol.* **32,** 569 (1967).

102. Salvin, S. B., and Smith, R. F., *J. Exp. Med.* **111,** 465 (1960).

103. Benacerraf, B., and Levine, B. B., *J. Exp. Med.* **115**, 1023 (1962).

104. Ovary, Z., and Benacerraf, B., *Proc. Soc. Exp. Biol. Med.* **114**, 72 (1963).

105. Rajewsky, K., and Rottlander, E., *Cold Spring Harbor Symp. Quant. Biol.* **32**, 547 (1967).

106. Plescia, O. J., Palzuk, N. C., Braun, W., and Cora-Figueroa, E., *Science* **148**, 1102 (1965).

107. Senyk, G., Williams, E. B., Nitecki, D. E., and Goodman, J. W., *J. Exp. Med.* **133**, 1294 (1971).

108. Singer, S. J., *Immunochemistry* **1**, 15 (1964).

109. Levine, B. B., *J. Exp. Med.* **121**, 873 (1965).

110. Mitchison, N. A., *Eur. J. Immunol.* **1**, 10 (1971).

111. Brownstone, A., Mitchison, N. A., and Pitt-Rivers, R., *Immunology* **10**, 465 (1966).

112. Mitchison, N. A., Taylor, R. B., and Rajewsky, K., in *Developmental Aspects of Antibody Formation and Structure*, Sterzl, J., ed., Czechoslovak Academy of Science, Prague, 1970, p. 547.

113. Katz, D. H., Paul, W. E., Goidl, E. A., and Benacerraf, B., *J. Exp. Med.* **132**, 261 (1970).

114. Mitchison, N. A., *Eur. J. Immunol.* **1**, 18 (1971).

115. Cheers, C., Breitner, J., Little, M., and Miller, J. F. A. P., *Nature New Biol.* **232**, 248 (1971).

116. Segal, S., Globerson, A., and Feldman, M., *Cell. Immunol.* **2**, 205 (1971).

117. Mitchell, G. F., and Miller, J. F. A. P., *Proc. Natl. Acad. Sci. U.S.* **59**, 296 (1968).

118. Raff, M. C., *Nature* **226**, 1257 (1970).

119. Lawton, A. R., Asofsky, R., Hytton, M. B., and Cooper, M. D., *J. Exp. Med.* **135**, 277 (1972).

120. Miller, J. F. A. P., Sprent, J., Basten, A., Warner, N. L., Breitner, J. C. S., Rowland, G., Hamilton, J., and Martin, W. J., *J. Exp. Med.* **134**, 1266 (1971).

121. Feldmann, M., and Basten, A., *Eur. J. Immunol.* **2**, 213 (1972).

122. Anderson, R. B., Sprent, J., and Miller, J. F. A. P., *J. Exp. Med.* **135**, 711 (1972).

123. Möller, G., and Michael, G., *Cell. Immunol.* **2**, 309 (1971).

124. Roelants, G. E., and Goodman, J. W., *Nature* **227**, 175 (1970).

125. Valentine, F. T., and Lawrence, H. S., *Science* **165**, 1014 (1969).

126. Lawrence, H. S., and Landy, M., *Mediators of Cellular Immunity*, Academic Press, New York, 1969.

127. Kreth, H. W., and Williamson, A. R., *Nature New Biol.* **234**, 454 (1971).

128. Feldmann, M., and Basten, A., *Nature New Biol.* **237**, 13 (1972).

129. Feldmann, M., and Basten, A., *J. Exp. Med.* **136**, 49 (1972).

130. Feldmann, M., and Basten, A., *J. Exp. Med.* **136**, 722 (1972).

131. Basten, A., and Feldmann, M., in *Microenvironmental Aspects of Immunity*, Jankovic, B. D. and Isakovic, K., Plenum Publishing Corp., New York, p. 171 (1973).

132. Feldmann, M., *J. Exp. Med.* **136**, 737 (1972).

133. Hornick, C. L., and Karush, F., in *Topics in Basic Immunology*, Sela, M., and Prywes, M., eds., Academic Press, New York, 1969, p. 29.

134. Feldmann, M., *J. Exp. Med.* **135**, 735 (1972).

135. Klinman, N. R., *J. Exp. Med.* **133**, 963 (1971).

136. Del Guercio, P., and Leuchars, E., *J. Immunol.* **109**, 951 (1972).

137. Feldmann, M., and Basten, A., *J. Exp. Med.* **134**, 103 (1971).

138. Mitchell, G. F., Humphrey, J. H., and Williamson, A. R., *Eur. J. Immunol.* **2**, 460 (1972).

139. Schmidtke, J., and Unanue, E. R., *Nature New Biol.* **233**, 84 (1971).

140. Mosier, D. E., *J. Exp. Med.* **129**, 351 (1969).

141. Sulitzeanu, D., Kleinman, R., Benezra, D., and Gery, I., *Nature New Biol.* **229**, 254 (1971).

142. Feldmann, M., *J. Exp. Med.* **136**, 737 (1972).

143. Shortman, K., Williams, N., Jackson, H., Russell, P., Byrt, P., and Diener, E., *J. Cell. Biol.* **48**, 566 (1971).

144. Cone, R. E., Feldmann, M., Marchalonis, J. J., and Nossal, G. J. V., *Immunology*, **26**, 49 (1974).

     J. J., and Nossal, G. J. V., *Immunology*, submitted, 1973.

145. Lachmann, P. J., *Proc. Roy. Soc.* (*London*) **176**, 425 (1971).

146. Gutman, G., and Weissman, I. L., *Immunology* **23**, 465 (1972).

147. Rieber, E. P., and Riethmuller, G., personal communication, 1972.

148. Gorczynski, R. M., Miller, R. G., and Phillips, R. A., *J. Immunol.* **108**, 547 (1972).

149. Gorczynski, R. M., and Phillips, R. A., personal communication, 1972.

150. Katz, D. H., Paul, W. E., Goidl, E., and Benacerraf, B., *J. Exp. Med.* **133**, 169 (1971).

151. Dutton, R. W., Falkoff, R., Hurst, J. A., Hoffman, M., Kappler, J. W., Kettman, J. R., Lesley, J. F., and Vann, D., *Prog. Immunol.* **1**, 535 (1971).

152. Schimpl, A., and Wecker, E., *Eur. J. Immunol.* **1**, 304 (1971).

153. Schimpl, A., and Wecker, E., *Nature New Biol.* **237**, 15 (1972).

154. Katz, D., *Transplant. Rev.* **12**, 141 (1972).

155. Sjöberg, O., Andersson, J., and Möller, G., *Eur. J. Immunol.* **2**, 99 (1972).

156. Waldmann, H., Munro, A., and Hunter, P., personal communication, 1972.

157. Remold, H. G., Ward, P. A., and David, J. R., in *Cell Interactions and Receptor Antibodies in Immune Responses*, Mäkela, O., Cross, A., and Kosunen, T. U., eds., Academic Press, London, 1971, p. 411.

158. Mackaness, G. B., *J. Exp. Med.* **129**, 973 (1969).

159. Mitchell, G. F., in *Contemporary Topics in Immunobiology*, Vol. 3, in press, 1974, N. L. Warner, ed. Plenum, New York.

160. Pepys, M. B., *Nature New Biol.* **237**, 157 (1972).

161. Dukor, P., Schumann, G. ,Gisler, R. H., Dierich, M., Wolfgang, K., Hadding, U., and Bitter-Suermann, D. *J Exp. Med.* **139**, 337 (1974).

162. Hirschhorn, R., Brittinger, G., Hirschhorn, K., and Weissman, G., *J. Cell Biol.* **37**, 412 (1968).

163. Brittinger, G., Hirschhorn, R., Douglas, S. D., and Weissman, G., *J. Cell Biol.* **37**, 394 (1968).

164. Hill, J. H., and Ward, P. A., *J. Exp. Med.* **130**, 505 (1969).

165. Taubman, S. B., Goldschmidt, P. R., and Lepow, I. H., *Fed. Proc.* **29**, 434 (1970).

166. Weissmann, G., Zurier, R. B., Spieler, P. J., and Goldstein, I. M., *J. Exp. Med.* **134**, 149s (1971).

167. Götze, O., and Müller-Eberhard, H. J., *J. Exp. Med.* **134**, 905 (1971).

168. Bitter-Suermann, D., *Klin. Wochenschr.* **50**, 277 (1972).

169. Parish, C. R., personal communication, 1972.

170. Allison, A. C., Denman, A. M., and Barnes, R. D., *Lancet* **2**, 135 (1971).

171. Katz, D. H., and Benacerraf, B., *Adv. Immunol.* **15**, 1 (1972).

172. Okumura, K., and Tada, T., *J. Immunol.* **106**, 1019 (1971).

173. Gershon, R. K., and Kondo, K., *Immunology* **21**, 903 (1971).

174. Kerbel, R. S., and Eidinger, D., *J. Immunol.* **106**, 917 (1971).

175. Jacobson, E. B., Herzenberg, L. A., Riblet, R., and Herzenberg, L. A., *J. Exp. Med.* **135**, 1163 (1972).

176. Ada, G. L., personal communication, 1972.

177. Basten, A., Miller, J. F. A. P., Sprent, J., and Cheers, C., *J. Exp. Med.* in press, 1974.

178. Feldmann, M., and Nossal, G. J. V., *Quaft. Rev. Biol.* **47**, 269 (1972).

179. Taussig, M. J., and Lachmann, P. J., *Immunology* **22**, 185 (1972).

180. Uhr, J. W., and Möller, G., *Adv. Immunol.* **8**, 81 (1968).

181. Bystryn, J.-C., Schenkein, I., and Uhr, J. W., *Prog. Immunol.* **1**, 628 (1971).

182. Diener, E., and Feldmann, M., *Transplant. Rev.* **8**, 76 (1972).

183. Henry, C., and Jerne, N., *J. Exp. Med.* **128**, 133 (1968).

184. Humphrey, J. H., and Dourmashkin, R. R., in *Ciba Foundation Symposium "Complement,"* Wolstenholme, G. E. W., and Knight, J., eds., Churchill, London, 1968, p. 175.

185. Cheers, C., and Miller, J. F. A. P., *J. Exp. Med.* **136**, 1661 (1972).

186. Ramseier, H., and Lindenmann, L., *Transplant. Rev.* **10**, 57 (1972).

187. Cosenza, H., and Kohler, H., *Proc. Natl. Acad. Sci. U.S.* **69**, 2701 (1972).

188. Barchilon, J., and Gershon, R. K., *Nature* **227**, 71 (1970).

189. Tubergen, D. G., and Feldman, J. D., *J. Exp. Med.* **134**, 1144 (1971).

190. Cerottini, J. C., Nordin, A. A., and Brunner, K. T., *Nature* **227**, 72 (1970).

191. Sprent, J., and Miller, J. F. A. P., *Cell. Immunol.* **3**, 361 (1972).

192. Asofsky, R., Cantor, H., and Tigelaar, R. E., *Prog. Immunol.* **1**, 369 (1971).

193. Tigelaar, R. E., and Asofsky, R., *J. Exp. Med.* **135**, 1059 (1972).

194. Sprent, J., and Miller, J. F. A. P., *Cell. Immunol.*, **3**, 213 (1972).

195. Playfair, J. H. L., and Purves, E. C., *Nature New Biol.* **231,** 149 (1971).

196. Andersson, B., and Blomgren, H., *Cell. Immunol.* **2,** 411 (1971).

197. Wagner, H., Feldmann, M., Boyle, W., and Schrader, J. W., *J. Exp. Med.* **136,** 331 (1972).

198. Gery, I., Gershon, R. K., and Waksman, B. H., *J. Exp. Med.* **136,** 128 (1972).

199. Remold, H. G., *Transplant. Rev.* **10,** 152 (1972).

200. Turk, J. L., *Immunology* **5,** 478 (1962).

201. Blanden, R. V., *J. Exp. Med.* **133,** 1074 (1971).

202. Blanden, R. V., and Langman, *Scand. J. Immunol.* **1,** 379 (1972).

203. McGregor, D. D., and Koster, F. T., *Cell. Immunol.* **2,** 317 (1971).

204. Giroud, J. P., Spector, W. G., and Willoughby, D. A., *Immunology* **19,** 857 (1970).

205. Evans, R., and Alexander, P., *Nature* **228,** 620 (1970).

206. Basten, A., and Beeson, P. B., *J. Exp. Med.* **131,** 1288 (1970).

207. Walls, R. S., Basten, A., Leuchars, E., and Davies, A. J. S., *Br. Med. J.* **3,** 157 (1971).

208. Rothwell, T., and Love, R., unpublished observations, 1972.

209. Dvorak, H. F., and Mihm, M. C., *J. Exp. Med.* **135,** 235 (1972).

210. Lafferty, K. J., Walker, K. Z., Scollay, R. G., and Killby, V. A. A., *Transplant. Rev.* **12,** 198 (1972).

211. Bach, F. H., Bach, M. L., Sondel, P. M., and Sundharadas, G., *Transplant. Rev.* **12,** 30 (1972).

212. McCullagh, P. J., *Transplant. Rev.* **12,** 180 (1972).

213. Sprent, J., Miller, J. F. A. P., and Mitchell, G. F., *Cell. Immunol.* **2,** 172 (1971).

214. Ford, W. L., and Atkins, R. C., *Nature New Biol.* **234,** 178 (1972).

215. Crewther, P., and Warner, N. L., *Aust. J. Biol. Med. Sci.,* **50,** 625 (1972).

216. Good, R. A., Biggar, W. D., and Park, B. H., *Prog. Immunol.* **1,** 699 (1971).

217. Dresser, D. W., *Eur. J. Immunol.* **2,** 50 (1972).

218. Mitchell, G. F., Mishell, R. I., and Herzenberg, L. A., *Prog. Immunol.* **1,** 324 (1971).

219. Cone, R. E., and Johnson, A. G., *J. Exp. Med.* **133,** 665 (1971).

220. Gershon, R. K., and Paul, W. E., *J. Immunol.* **106,** 872 (1971).

221. Gershon, R. K., and Kondo, K., *Immunology* **23,** 335 (1972).

222. Iverson, G. M., *Nature* **227,** 273 (1970).

223. Parish, C. R., *J. Exp. Med.* **134,** 1 (1971).

224. Parish, C. R., *J. Exp. Med.* **134,** 21 (1971).

225. Bloom, E. T., and Hildemann, W. H., *Transplantation* **10,** 321 (1970).

226. Benjamin, D. C., and Weigle, W. O., *J. Exp. Med.* **132,** 66 (1970).

The review of the literature pertaining to this chapter was concluded in December 1972.

CHAPTER TEN

Uptake of Lysosomal Enzymes by Cultured Fibroblasts

STUDIES OF MUCOPOLYSACCHARIDOSES AND I-CELL DISEASE

ELIZABETH F. NEUFELD

National Institute of Arthritis,
Metabolism and Digestive Diseases
National Institutes of Health
Bethesda, Maryland

THE MUCOPOLYSACCHARIDOSES STUDIED IN CELL CULTURE

The genetic mucopolysaccharidoses have amply shown the value of studying the abnormal situation in order to understand the normal. The intensive studies of the last few years have been rewarded by an appreciation of the importance of certain hydrolases and by some understanding of cell interaction in the packaging of lysosomal enzymes.

The mucopolysaccharidoses comprise a group of rare disorders that were classified in 1965 by Maroteaux and Lamy (1) and by McKusick and colleagues (2) on the basis of certain consistent clinical differences, of the mode of inheritance, and of the nature of urinary mucopolysaccharides. Although these classifications had to be modified in the light of recent biochemical studies, they were essentially correct. I shall use in this chapter the 1966 nomenclature of McKusick (3) even though it has been revised in the light of recent biochemical findings (3a): mucopolysaccharidosis I, the Hurler syndrome; II, the Hunter syndrome (the only X-linked disorder in the group, all others being inherited in autosomal recessive manner); III, the Sanfilippo syndrome; IV, the Morquio syndrome; V, the Scheie syndrome; and VI, the Maroteaux–Lamy syndrome. All the disorders but IV involve the urinary excretion of dermatan sulfate and/or heparan sulfate, and are therefore biochemically related.

These are lysosomal storage diseases. Electron microscopy of most cells of a Hurler patient reveals large vacuoles filled with amorphous material. These structures were first observed in hepatocytes by van Hoof and Hers (4), who considered them lysosomes filled with undigested mucopolysaccharide and suggested that the Hurler syndrome (the term being used at the time to include any mucopolysaccharidosis) might be analogous to Pompe disease—a glycogen storage disease due to the deficiency of a lysosomal glucosidase. However, studies of mucopolysaccharidoses to prove or disprove the hypothesis of lysosomal enzyme deficiency were not feasible at the time, in part because of ignorance of the chemistry and metabolism of dermatan sulfate and heparan sulfate, and in part because material from patients was limited to samples of blood and urine, and to occasional biopsy or autopsy specimens.

This limitation was soon removed by the finding that fibroblasts cultured from the skin of mucopolysaccharidosis patients also store sulfated mucopolysaccharide, as shown by metachromatic staining with a histochemical dye (5,6) or by chemical analysis (7). The use of cell culture has greatly accelerated the pace of research in the mucopolysaccharidoses, as in other genetic disorders, by making available significant quantities of genetically marked material without imposing any hardships on the patient other than the taking of one skin biopsy.

It was possible to ask experimentally whether the fibroblasts accumulate excess mucopolysaccharide because of overproduction or decreased secretion of catabolism. This was accomplished by labeling the culture medium with $^{35}SO_4^{2-}$ and comparing the fate of the radioactive mucopolysaccharide that was formed in fibroblasts from normal individuals and from mucopolysaccharidosis patients (8). These experiments and their interpretation have been discussed at length elsewhere (9), and the conclusions can be summarized as follows:

1. The metabolism of sulfated mucopolysaccharide in skin fibroblasts, normal or not, is complex. There are several chemically and metabolically distinct pools: an extracellular pool of proteoglycan in the culture medium, a pericellular pool, presumably lysosomal, devoted to degradation. A very small intracellular pool of mucopolysaccharide in the process of biosynthesis or secretion has been inferred, but not measured.

2. It is the degradative pool that is enlarged and turns over more slowly in fibroblasts from mucopolysaccharidosis patients. Entry of mucopolysaccharide into that pool (either by pinocytic ingestion or intracellular transfer through some unknown mechanism) proceeds normally, and it is therefore the exit (i.e., degradation) that is blocked. The block is usually not absolute; whereas in a few cases the kinetics are consistent with enzyme activity that is markedly reduced or altogether consistent with enzyme activity that is markedly reduced or altogether absent, in other cases an increased K_m value may be the more likely explanation.

CORRECTION

In 1967 Danes and Bearn (10) showed that cells of Hunter heterozygotes obeyed the Lyon hypothesis (11). That is, in any given cell, and in all the progeny of that cell, only one X-chromosome is active, the other being present but not functional. This was ascertained by isolating two types of clonal fibroblast cultures from mothers of Hunter patients; by the criterion of metachromatic staining, some clones were normal, whereas others were indistinguishable from cultures of the affected sons.

The Hunter mother is thus a mosaic of normal and abnormal cells.* We found it surprising that such women should have a completely normal phenotype and postulated that the normal cells were somehow assisting the abnormal ones in degrading mucopolysaccharide. This hypothesis was readily verified by mixing normal and Hunter cells; these "synthetic heterozygotes" had a normal pattern of mucopolysaccharide accumulation (13).

An even more dramatic demonstration of cell interaction was observed when fibroblasts of a Hunter patient were mixed with those of a Hurler patient. Though these two genetically distinct lines of fibroblasts accumulated ^{35}S-mucopolysaccharide in identical manner, the mixture of cells showed a pattern that was normal (Fig. 1).

It proved unnecessary to mix the cells, since the interaction could be observed by using medium preincubated with cells of one genotype to make normal the cells of the other (13,14). Clearly, fibroblasts were elaborating

some substance into the medium, which we designated a "corrective factor" qualified by the name of the cell line that it corrects; for example, the Hunter corrective factor is that which reduces mucopolysaccharide accumulation in Hunter cells. The Hunter corrective factor was found to be present in medium preincubated with fibroblasts of any genotype other than Hunter (14).

Cross-correction by cell mixing or by the use of secretions from used medium was immediately put to use for classification and diagnosis. Early tests revealed some interesting deviations from the McKusick classification; the clinically distinct Hurler and Scheie syndromes did not show cross-correction and were therefore deficient in the same factor, the "Hurler corrective factor" (15), whereas the Sanfilippo syndrome turned out to be biochemically heterogeneous (16). Cell lines derived from Sanfilippo patients could be sorted into two groups, arbitrarily designated A and B, which showed cross-correction and hence the deficiency of a different corrective factor.

Cross-correction retains its theoretical interest, but as a diagnostic tool, it has been superseded in most cases by the use of purified factors of known specificity or by enzymatic tests. Nevertheless, cross-correction must still be used in special situations: if purified factors are not available, if the basic enzyme defect is not known or is difficult to assay, or if a novel disorder is under consideration. Some cautionary remarks on the technique are therefore in order. It is clear that a disparate rate of growth of two cell lines in the course of a mixing experiment would render the results uninterpretable. The recently discovered importance for mucopolysaccharide degradation of the pH of medium means that experiments must be conducted at or below pH 7.0; if the medium is somewhat alkaline, one encounters an environmen-

---

* It may be expected that occasionally, inactivation would occur in a nonrandom fashion or that one population of cells would have a selective advantage over the other. In that case the Hunter heterozygote would effectively become a hemizygote for normal or Hunter cells. The latter probably occurred in the case of a girl of normal 46XX karyotype with a clinically fully expressed and biochemically characterized Hunter syndrome (12).

tally generated mucopolysaccharidosis (17). Cultures that are not healthy—for example, infected with mycoplasma or senescent—accumulate mucopolysaccharide poorly and fail to secrete corrective factors. Experiments using Hurler and Hunter fibroblasts, in which the accumulation of mucopolysaccharide and the effect of correction are great, are easier to perform than experiments on Sanfilippo fibroblasts. In the latter the relatively lower level of mucopolysaccharide accumulation and the consequently smaller effect of correction call for particularly careful technique.

Finally, several conditions must be fulfilled if cross-correction experiments are to be conducted by the use of preincubated medium. To demonstrate reciprocal correction of cell lines a and b, it is necessary that line a secrete the b corrective factor and line b secrete the a corrective factor in adequate amounts; that both factors be stable in the medium and through the manipulations used for collection and concentration; and that each factor be taken up by the other cell line. In experiments that include Maroteaux–Lamy cells, at least one of these conditions must not be met, since fibroblasts from Maroteaux–Lamy patients correct the defect in cells of other genotype, whereas the reciprocal correction is not reproducibly observed (18).

It is implicit in this discussion that the cross-correction test is only valid for cells deficient in a single factor. A special situation arises in fibroblasts with multiple defects, as discussed in the section on I-cell disease.

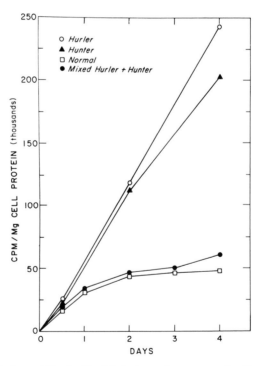

Fig. 1. Interaction of Hurler and Hunter fibroblasts, to produce a normal pattern of mucopolysaccharide metabolism. Reprinted, by permission, from E. F. Neufeld and J. C. Fratantoni, *Science* **169**, 141 (1970).

PURIFICATION OF CORRECTIVE FACTORS

Purification of the factors was the obvious next step. Preliminary experiments indicated that the factors were proteins (19), and the purification steps were selected on that basis. The difficulties of obtaining adequate amounts of culture fluid prompted the search for a better source. Of several sources tested—tissue homogenates, plasma, and urine—normal human urine proved to have the factors in highest specific activity; for this reason, as well as for its obvious availability, it was used for subsequent work.

A quantitative assay was developed, based on the dose–response curve shown in Fig. 2. The corrective effect is a hyperbolic function of the quantity of factor added; a straight line may be obtained by plotting the double reciprocal of dose versus that of correction (correction being the difference between the control sample and that with a given amount of corrective factor added). This relationship, presented in Fig. 2 for Hurler corrective factor activity in crude concentrates from human urine, holds for other corrective factors and for any degree of purification. It is applicable whether correction is measured as reduction of ^{35}S-mucopolysaccharide accumulated (as the test is usually performed) or as reduction in ^{35}S-mucopolysaccharide retained during a chase. The analogy to Michaelis–Menten kinetics (of which the significance will be discussed later) led to a definition of a corrective unit in terms analogous to the Michaelis constant: one unit is that dose of corrective factor that gives half-maximal corrective activity.

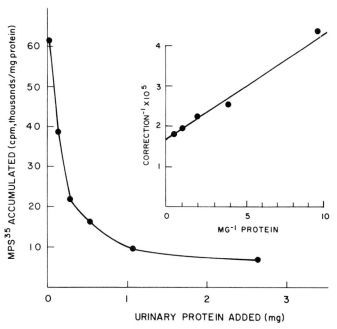

Fig. 2. Correction of the defect of Hurler fibroblasts as a function of the amount of corrective-factor preparation added. Reprinted, by permission, from E. F. Neufeld and M. J. Cantz, *Ann. N.Y. Acad. Sci.* **179.** 580 (1971).

Rigorously standardized conditions are essential for reproducible measurement of corrective factor activity. It is well to remember, however, that the bioassay is not as precise as an enzyme assay and that the inherent variability of the system must be taken into account in the experimental design.

Several corrective factors have been purified to date, though none yet to homogeneity: the Hurler factor, 1000-fold (20); the Hunter factor, over 100-fold (21); the Sanfilippo A factor, 800-fold (22); and the Sanfilippo B factor, 150-fold (23). Further purification has been limited by technical difficulties; presumably elucidation of the enzymatic nature of the corrective factors will lead to the development of appropriate systems for affinity chromatography.

The specificity of the purified corrective factors confirmed the results obtained by cross-correction tests. The most purified Hurler corrective factor has a corrective effect on Hurler and Scheie fibroblasts, but not on fibroblasts of other genotypes; the Hunter corrective factor affects only fibroblasts from Hunter patients, whether of the mild or severe variety [adult or juvenile, respectively, in the classification of Spranger (24)]. There is a unique, though as yet unpurified, corrective factor for the Maroteaux–Lamy genotype (18); and the corrective factors for the two subtypes of the Sanfilippo syndrome are distinct proteins with distinct functions.

FUNCTION OF THE CORRECTIVE FACTORS

Studies in Whole Cells

Since the defect of the mucopolysaccharidoses is a block in the degradation of dermtan sulfate and/or heparan sulfate, it was to be expected that the corrective factors would restore degradation of these polymers by the recipient fibroblasts. This can be experimentally demonstrated in a number of ways: (1) the disappearance of preformed stores of ^{35}S-mucopolysaccharide, accompanied by the appearance of an equivalent amount of inorganic sulfate in the medium (14); (2) the disappearance of mucopolysaccharide species of specific size in fibroblasts of mucopolysaccharidosis patients and the appearance of the normal, polydisperse distribution (20,22); (3) enhanced rate of degradation of proteodermatan sulfate ingested by pinocytosis (21)—the latter is observed in Hurler and Hunter fibroblasts, but not in fibroblasts from Sanfilippo patients, which are blocked only in the degradation of heparan sulfate. It was postulated, therefore, that each corrective factor might be a lysosomal enzyme, and that the ^{35}S mucopolysaccharide that accumulates in its absence would be its substrate.

The Sanfilippo Corrective Factors

On the above hypothesis the ^{35}S-mucopolysaccharide, of which the degradation was blocked, was isolated from Sanfilippo A fibroblasts. It proved to be heparan sulfate by the criteria of electrophoresis, digestibility with heparanase, and content of ^{3}H-glucosamine rather than galactosamine demonstrated after incubation of the cells in the presence of ^{3}H-glucosamine (22,25).

The Sanfilippo A corrective factor releases inorganic sulfate from this mucopolysaccharide, and can therefore be considered a sulfatase for heparan sulfate (22). It was further suggested that it is specific for N-sulfated residues, but the evidence on this point was preliminary. Confirmation will probably require preparation of heparan sulfate specifically labeled either in the O-sulfate or the N-sulfate groups.

The defect in the syndrome of the B subtype seems firmly established as a deficiency of N-acetyl-α-D-glucosaminidase. Not only is the activity of this enzyme missing in fibroblasts and viscera of patients and reduced in fibroblasts of their parents (26), but the Sanfilippo B corrective factor has been shown to be chromatographically and electrophoretically inseparable from N-acetyl-α-D-glucosaminidase (23).

Since heparan sulfate is a polymer of sulfated α-N-acetyl or N-sulfated glucosaminide residues alternating with uronic acid, the two factors must act in tandem to help degrade the molecule. One might suspect that the mucopolysaccharide fragments stored and excreted by Sanfilippo A and B patients would be, respectively, oversulfated and undersulfated. Such heterogeneity in heparan sulfate excretion by Sanfilippo patients has been observed, but not yet correlated with the enzymatic defect (27). Interestingly, the clinical consequences of the two defects are the same.

Two other metabolic errors specific to the degradation of heparan sulfate should be anticipated, as seen from the structure of this mucopolysaccharide (Fig. 3). If the Sanfilippo A factor is indeed a sulfamino hydrolase, its action should leave an α-glucosaminide (not acetylated) linkage that must be cleaved. In addition, there should be a sulfatase specific to the O-sulfated residues. Errors involving uronic acid hydrolases would not be specific to heparan sulfate since, contrary to previous belief that glucuronide linkages in heparan sulfate are characteristically of the α-configuration, such linkages are probably β-D-glucuronosyl or the analogous α-L-iduronosyl, just as in dermatan sulfate (28,29).*

The Hurler Corrective Factor

The Hurler corrective factor has been identified as the enzyme α-L-iduronidase by the following criteria (31):

1. The ratio of enzymatic to corrective factor activity is similar in crude and highly purified preparations.

2. Two isozyme peaks of α-L-iduronidase activity, separable on hydroxylapatite gel, correspond to two "isofactor" peaks.

3. All cell lines deficient in, and correctible, by the Hurler corrective factor likewise have no detectable α-L-iduronidase activity; these include cultures derived from patients with the Hurler syndrome, the Scheie syndrome, and a newly described group with a phenotype intermediate between Hurler and Scheie (32).†

---

* The chemistry of heparan sulfate is relatively obscure, and the discussion is based on data obtained for heparin, which is chemically closely related. Indeed, one might ask why the Sanfilippo syndrome or other mucopolysaccharidoses do not affect heparin. Our speculation is that they do [heparin-like fractions have been found in Sanfilippo urine (30)] but after partial degradation by those enzymes that remain functional the fragments of heparin and heparan sulfate are not readily distinguished from each other.

† To explain the phenotypic differences between the very severe Hurler syndrome and the relatively benign Scheie syndrome, it has been suggested that the Hurler and Scheie mutations are allelic, the latter allowing some residual iduronidase activity *in vivo* (32). By this hypothesis, matings of a Hurler heterozygote with a Scheie heterozygote would produce individuals carrying one Hurler and one Scheie gene. Such "genetic compounds" would lack α-L-iduronidase activity in cultural fibroblasts and have a phenotype intermediate between that of Hurler and Scheie patients. Several patients fulfilling these criteria have been described.

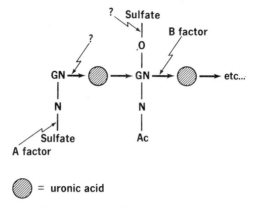

Fig. 3. Structure of heparan sulfate. Arrows indicate the linkages hydrolyzed by Sanfilippo corrective factors A and B, and possible additional factors that might be anticipated.

The finding of identity of Hurler corrective factor and α-L-iduronidase was preceded by the demonstration of the lysosomal and specific nature of that hydrolase (33) and of its deficiency in Hurler fibroblasts and tissues (34,35). Fibroblasts from Hurler heterozygotes show approximately half the normal level of enzyme (36). The role of the enzyme in the degradation of dermatan sulfate is obvious, since L-iduronic acid comprises nearly half the carbohydrate residues of that polymer (37); in the case of heparan sulfate the number of L-iduronic acid residues is not known, but may be small (38). However, a single such residue, located strategically near the nonreducing end, could block all further degradation by exoglycosidases (see footnote on p. 223).

β-Glucuronidase

A novel disorder, β-glucuronidase deficiency, has recently been described (39, 40). The affected child in the first report resembled a Hurler patient, though careful observation revealed some differences. The total absence of β-glucuronidase in his leukocytes and an intermedi-

ate level in those of both parents and several relatives attest to an autosomal recessive mode of inheritance.

Fibroblasts derived from the skin of this patient accumulated excess ^{35}S-mucopolysaccharide; this defect was corrected by the addition of β-glucuronidase from bovine liver (41). As the β-glucuronidase used was not pure, its identity with the corrective factor was verified by coelectrophoresis of enzymatic and corrective (see footnote on p. 223).

The Hunter Corrective Factor

In an approach analogous to that used for the Sanfilippo A system, isotope-labeled mucopolysaccharide stored in Hunter fibroblasts was used as a potential substrate for purified Hunter corrective factor. The factor could release up to 2% of $^{35}SO_4{}^{2-}$ from the polymer; a similar release was catalyzed by acetone powder extracts from fibroblasts of genotype other than Hunter (42). The release of inorganic sulfate was accompanied by a decrease in the proportion of "oversulfated" areas (i.e., dimers containing sulfate on the iduronic acid as well as on the hexosamine residue) of the polymer. The Hunter factor is therefore thought to be a specific sulfatase, probably for the sulfated iduronic acid residues. In addition, these residues may have to be appropriately exposed (e.g., at nonreducing termini).

The number and location of sulfated iduronic acid groups in dermatan and heparan sulfates are not known, but presumably these groups occur at infrequent and variable intervals (43,44). This will probably explain one diagnostic clinical difference between the Hunter syndrome and the α-L-iduronidase deficiencies (Hurler, Scheie, and intermediate phenotypes). In the latter, corneas are always cloudy; in the Hunter syndrome, however severely the patient may be affected, corneas remain clear (3,24, 32). The Hunter corrective factor, in contrast to α-L-iduronidase, must be deemed not essential for the degradation of corneal and scleral mucopolysaccharides, which, by the same argument, would be expected to be devoid of sulfated iduronic residues.

Speculation Concerning Other Mucopolysaccharidoses

Dermatan sulfate contains several linkages for which there should exist as yet unknown hydrolases and, most probably, inborn errors. As seen in Fig. 4, the cleavage of O-sulfate from N-acetylgalactosamine and the removal of the β-linked N-acetylgalactosamine (or, alternatively, the removal of N-acetylgalactosamine sulfate in its entirety) are reactions in search of a disease. The Maroteaux–Lamy corrective factor, whose function has yet to be studied, may catalyze one of these postulated reactions.

All the disorders discussed above are caused by faulty metabolism of dermatan sulfate or heparan sulfate, with the possible exception of β-glucuronidase deficiency, which may involve the chondroitin sulfates as well. Malfunction of the metabolism of other mucopolysaccharides also occurs. The Morquio syndrome, for example, is associated with the urinary excretion of keratan sulfate (3). In our experience the disorder is not manifested in cultured skin fibroblasts, perhaps because these cells do not produce keratan sulfate. It would be interesting to determine whether exogenously supplied skeletal or corneal keratan sulfate is metabolized differently by fibroblasts from normal individuals and from Morquio patients, and if there is a difference, whether it can be corrected by keratan sulfate–degrading enzymes.

CORRECTION OF OTHER LYSOSOMAL ENZYME DEFICIENCIES

Correction—that is, replacement of a lysosomal enzyme—is probably feasible in any lysosomal disorder, provided the fibroblasts manifest a defect whose correction can be observed. This is the case in fibroblasts from patients with metachromatic leukodystrophy, that lack aryl sulfatase A. These fibroblasts accumulate excessive quantities of sulfatide when the glycolipid is introduced into the medium, and the accumulation is reduced to a normal level by addition of human urinary aryl sulfatase A (45,46). The endogenous accumulation of ceramide trihexoside by fibroblasts from patients with Fabry disease disappears on the addition of an α-galactosidase of plant origin (47). In these instances the enzymatic defect was known before correction experiments were attempted; the situation was different, therefore, from that of the mucopolysaccharidoses, where correction was discovered first and led to the eventual identification of the lysosomal enzyme defects.

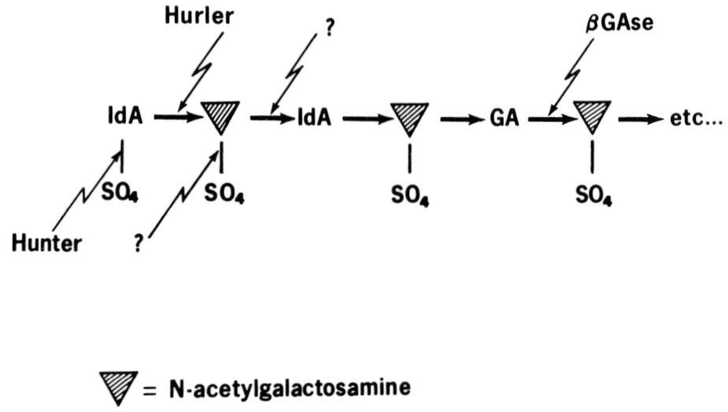

Fig. 4. Structure of dermatan sulfate. Arrows indicate the linkages hydrolyzed by α-L-iduronidase (Hurler factor), β-glucuronidase, and sulfatase (Hunter corrective factor). Linkages for which hydrolases are not yet known are indicated by question marks.

Preliminary experiments suggest that correction may be possible for fibroblasts of patients with Wolman's disease—a deficiency of acid lipase that results in the storage of triglycerides and cholesterol esters (48).

SIGNIFICANCE OF THE KINETICS OF CORRECTION

The possibility that saturation of some transport system for the corrective factor would account for the Michaelis–Menten kinetics has been tested and excluded in two systems. Uptake of α-L-iduronidase by the cell continued to a significant extent considerably beyond the point of 90% correction (31); in the case of β-glucuronidase there was no leveling of uptake until after 90 times the amount necessary for 90% correction had been taken in from the medium (41).

It is probable, therefore, that the ever-diminishing effect of additional corrective factor is due to the saturation of subsequent steps in the degradative pathway by the product of the factor (e.g., in the case of α-L-iduronidase by the polymer from which the terminal iduronic acid has been removed), as well as by the limiting amounts of substrate that become available to the factor after the excess storage material has been degraded.

The rate-limiting steps in the multi-enzyme pathway of mucopolysaccharide degradation by normal fibroblasts, or by fully corrected fibroblasts from mucopolysaccharidosis patients, are not known. It is generally assumed, though without experimental evidence, that lysosomal enzymes do not possess allosteric control mechanisms and that the overall rate of degradation is a function of enzyme and substrate concentration and affinity of the enzymes for their substrates.

SPECIFIC UPTAKE OF LYSOSOMAL ENZYMES: STUDIES OF I-CELL DISEASE

We had heretofore assumed that the uptake of enzymes by the cell occurred by endocytosis, a relatively nonselective process for the entry of macromolecules into cells and their ultimate sequestration into secondary lysosomes (49). It was therefore surprising to discover that Hurler fibroblasts could take up nearly half the activity of α-L-iduronidase supplied in the medium, but only traces of ^{131}I-albumin (31).

The suspicion that uptake of lysosomal enzyme might involve some specific recognition received experimental confirmation in studies of fibroblasts from patients with I-cell disease. The I-cells, or inclusion cells—as these fibroblasts are called because their lysosomes are so large and so full of debris as to be visible in the form of granular inclusions under phase microscopy—have a multiplicity of lysosomal enzyme defects (50–52). Several hydrolases are barely detectable (β-galactosidase, α-L-fucosidase, and aryl sulfatase A); others are present at only 10–30% of the normal level (N-acetyl-β-hexosaminidase, β-glucuronidase, α-mannosidase, and α-L-iduronidase). Only a few hydrolytic activities, such as acid phosphatase and β-glucosidase, are an exception to the general pattern of deficiency. The fibroblasts consequently accumulate both mucopolysaccharides and glycolipids, for which reason the disease has been classified as a "mucolipidosis" (53). Several of the enzymes deficient within fibroblasts are present in considerable excess over the normal in cell culture medium and in the body fluids of I-cell patients (54,55).

The I-cells accumulate ^{35}S-mucopolysaccharide to the same extent as fibroblasts from Hurler or Hunter patients, but do not cross-correct with either one (56). It is easy to see why the accumulation within I-cells would not be reduced: because of the multiplicity of enzyme deficiencies, restoration of all but one activity is not effective in reconstituting the catabolic pathway; for example, I-cells mixed with Hurler cells are still deficient in α-L-iduronidase. On the other hand, I-cells secrete α-L-iduronidase into the medium, so that their inability to correct Hurler cells suggested that something might be wrong with the enzyme itself. Indeed, careful measurements showed that the α-L-iduronidase released by I-cells had but one-fifth the corrective activity of its normal

counterpart, when the two were compared on the basis of their ability to hydrolyze phenyl α-L-iduronide (57). Likewise, on the basis of catalytic activity toward p-nitrophenylglucuronide, the β-glucuronidase secreted by I-cells had less than one-tenth the corrective activity of that secreted by normal fibroblasts.

If correction requires both catalytic activity of a lysosomal enzyme and its ability to enter the cell, the latter might be defective in I-cell enzymes. This was verified directly by the finding that N-acetyl-β-glucosaminidase of I-cell medium entered fibroblasts to only one-tenth the extent of the enzyme from normal medium (for logistic reasons, fibroblasts from a patient with Tay–Sachs O-variant, lacking nearly all N-acetyl-β-hexosaminidase (58), had to be used as recipient cells). By contrast, I-cells were able to take up and retain α-L-iduronidase from normal urine to the same extent as Hurler cells.

The efficient uptake of lysosomal enzymes must require that these have a specific marker. I-cell hydrolases, lacking such markers, enter only by a nonspecific and therefore less efficient mechanism. The result is that they remain outside the cell in culture medium or extracellular body fluid and, if stable, accumulate there in excess. This hypothesis implies that fibroblasts have no intracellular mechanism for channeling hydrolases from the site of synthesis to their eventual lysosomal location [i.e., that the Golgi–endoplasmic reticulum–lysosome pathway postulated by Novikoff et al. (59) does not operate in these cells.] Lysosomal enzymes of fibroblasts are secretory proteins with provision for immediate recapture by the cell of origin or by neighboring cells. The etiology of I-cell disease would be the deficiency of some enzyme required for making the recognition marker on one group of

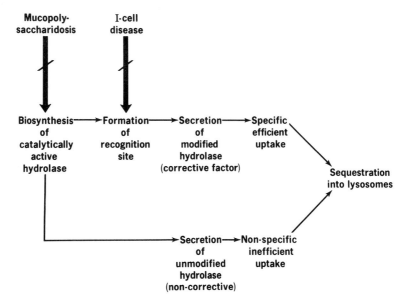

Fig. 5. Schematic representation of the pathway of secretion and uptake of lysosomal hydrolases.

hydrolases. As the disorder is not systemic, but limited to cells of connective tissue origin and lymphocytes [as demonstrated by enzyme analysis or electron microscopy (51,60)], the postulated recognition marker must be specific to those cells.

Though this proposal, shown schematically in Fig. 5, goes considerably beyond available evidence, it can be readily tested experimentally. It is attractive because it can explain not only the facts of I-cell disease but also such previously puzzling phenomena as the constant presence of lysosomal enzymes in the culture medium surrounding healthy cells and the remarkably efficient uptake of enzyme that accompanies correction. This proposal should suggest some caution in the purification of lysosomal enzymes for eventual replacement therapy. A hydrolase purified as a corrective factor is well taken up by fibroblasts; a hydrolase purified on the basis of its catalytic activity alone may not possess the requisite marker for uptake.

The nature of the recognition marker

is at present speculative; it might be a glycoside [by analogy with the role of terminal galactose residues on plasma glycoproteins for their recognition by liver cell membranes and eventual sequestration into lysosomes (61,62)]. Obvious alternatives include a particular amino acid sequence or special subunit. It is apparent from the scheme in Fig. 5 that we should be alert for mutations affecting hydrolase uptake because of alteration of the complementary site at the cell membrane rather than of the hydrolase itself.

Finally, if markers for lysosomal hydrolases differ in different tissues, one might hope to modify proteins in order to guide them to those sites where pathological storage occurs.

SUMMARY

Correction of the biochemical defect of fibroblasts from mucopolysaccharidosis patients is the restoration of the enzymatic pathways for the degradation of dermatan sulfate and/or heparan sulfate. Each corrective factor has been found to be a hydrolase corresponding to the one that is deficient in the cells to be corrected. The Hurler corrective factor (which is equally effective on fibroblasts from Scheie patients) is α-L-iduronidase; the Hunter corrective factor is a sulfatase acting on oversulfated regions of the polymers. The factors corrective for the defect of Sanfilippo A and B fibroblasts are, respectively, a heparan sulfate sulfatase (perhaps an N-sulfatase) and an N-acetyl-α-glucosaminidase. β-Glucuronidase serves as the corrective factor to normalize the mucopolysaccharide catabolism of fibroblasts with β-glucuronidase deficiency.

The secretion and efficient uptake of hydrolases, of which we had taken advantage to elucidate the defect of the mucopolysaccharidoses, are probably part of the normal mechanism for channeling these enzymes into lysosomes. Studies of I-cell disease suggest that fibroblasts have no intracellular mechanism for sequestering lysosomal enzymes. The uptake of lysosomal hydrolases probably requires a specific structural feature that is defective in the hydrolases elaborated by I-cells.

ADDENDUM

The enzyme defects in the Hunter and Sanfilippo A syndromes have been firmly established as sulfoiduronate sulfatase (63–65) and heparan/heparin N-sulfatase (66,67), respectively. Cells and tissues of patients with the Maroteaux-Lamy syndrome are deficient in aryl sulfatase B (68,69), but the relationship of that enzyme to the Maroteaux-Lamy corrective factor (18) and to dermatan sulfate catabolism has not yet been clarified.

Pseudo Hurler polydystrophy (mucolipidosis III) has been found biochem-

ically related to I-cell disease (mucolipidosis II), in spite of the great clinical difference between the two disorders (3a). In both conditions, lysosomal enzymes accumulate extracellularly (70–72) and are not captured by fibroblasts of other genotype (57,71,73).

The selective uptake of certain lysosomal enzymes is now well established (57,74–77). Uptake can be reduced by treatment with periodate under conditions that do not affect catalytic activity, suggesting that the recognition marker is a carbohydrate attached to the enzyme protein (74).

REFERENCES

1. Maroteaux, P., and Lamy, M., *J. Pediatr.* **67,** 312 (1965).

2. McKusick, F. A., Kaplan, D., Wise, D., Hanley, W. B., Suddarth, S. B., Sevick, M. E., and Maumenee, A. E., *Medicine* **44,** 445 (1965).

3. McKusick, V. A., *Heritable Disorders of Connective Tissue*, 3rd ed., C. V. Mosby, St. Louis, 1966, pp. 325–399.

3a. McKusick, V. A., *Hertiable Disorders of Connective Tissue*, 4th ed., C. V. Mosby, St. Louis, 1972, pp. 521–686.

4. van Hoof, F., and Hers, H. G., *C. R. Acad. Sci. (Paris)* **259,** 1281 (1964).

5. Danes, B. S., and Bearn, A. G., *Science* **149,** 987 (1965).

6. Danes, B. S., and Bearn, A. G., *J. Exp. Med.* **123,** 1 (1966).

7. Matalon, R., and Dorfman, A., *Proc. Natl. Acad. Sci. U.S.* **56,** 1310 (1966).

8. Fratantoni, J. C., Hall, C. W., and Neufeld, E. F., *Proc. Natl. Acad. Sci. U.S.* **60,** 699 (1968).

9. Neufeld, E. F., and Cantz, M., in *Lysosomes and Storage Disease*, Hers, H. G., and van Hoof, F., eds., Academic Press, New York, 1973, p. 272.

10. Danes, B. S., and Bearn, A. G., *J. Exp. Med.* **126,** 509 (1967).

11. Lyon, M. F., *Am. J. Human Genet.* **14,** 135 (1962).

12. Milunsky, A., and Neufeld, E. F., *N. Engl. J. Med.*, **288**, 106 (1973).
13. Fratantoni, J. C., Hall, C. W., and Neufeld, E. F., *Science* **162**, 570 (1968).
14. Fratantoni, J. C., Hall, C. W., and Neufeld, E. F., *Proc. Natl. Acad. Sci. U.S.* **64**, 360 (1969).
15. Wiesmann, U., and Neufeld, E. F., *Science* **169**, 62 (1970).
16. Kresse, H., Wiesmann, U., Cantz, M., Hall, C. W., and Neufeld, E. F., *Biochem. Biophys. Res. Commun.* **42**, 892 (1971).
17. Lie, S. O., McKusick, V. A., and Neufeld, E. F., *Proc. Natl. Acad. Sci. U.S.* **69**, 2361 (1972).
18. Barton, R. W., and Neufeld, E. F., *J. Pediatr.* **80**, 114 (1972).
19. Cantz, M., Chrambach, A., and Neufeld, E. F., *Biochem. Biophys. Res. Commun.* **39**, 936 (1970).
20. Barton, R. W., and Neufeld, E. F., *J. Biol. Chem.* **246**, 7773 (1971).
21. Cantz, M., Chrambach, A., Bach, G., and Neufeld, E. F., *J. Biol. Chem.* **247**, 5456 (1972).
22. Kresse, H., and Neufeld, E. F., *J. Biol. Chem.* **247**, 2164 (1972).
23. von Figura, K., and Kresse, H., *Biochem. Biophys. Res. Commun.* **48**, 262 (1972).
24. Spranger, J., *Ergeb. Inn. Med. Kinderheilkd.* **32**, 166 (1972).
25. Kresse, H., and Bach, G., unpublished experiments, 1972.
26. O'Brien, J., *Proc. Natl. Acad. Sci. U.S.* **69**, 1720 (1972).
27. Dean, M., Benson, P. F., and Muir, H., *Biochem. Soc. Trans.* **1**, 54 (1973).
28. Perlin, A. S., Casu, B., Sanderson, G. R., and Johnson, L. F., *Can. J. Chem.* **48**, 2260 (1970).
29. Helting, T., and Lindahl, U., *J. Biol. Chem.* **246**, 5442 (1971).
30. Stone, A. L., Constantopoulos, G., Sotsky, S. M., and Debakan, A., *Biochim. Biophys. Acta* **222**, 79 (1970).
31. Bach, G., Friedman, R., Weissmann, B., and Neufeld, E. F., *Proc. Natl. Acad. Sci. U.S.* **69**, 2048 (1972).
32. McKusick, V. A., Hussels, I. E., Howell, R. R., Neufeld, E. F., and Stevenson, R. E., *Lancet* **1**, 993 (1972).
33. Weissmann, B., and Santiago, R., *Biochem. Biophys. Res. Commun.* **46**, 1430 (1972).
34. Matalon, R., Cifonelli, J. A., and Dorfman, A., *Biochem. Biophys. Res. Commun.* **42**, 340 (1971).
35. Matalon, R., and Dorfman, A., *Biochem. Biophys. Res. Commun.* **47**, 959 (1972).
36. Bach, G., Cantz, M., Hall, C. W., and Neufeld, E. F., *Biochem. Soc. Trans.* **1**, 231 (1973).
37. Cifonelli, J. A., Ludowieg, J., and Dorfman, A., *J. Biol. Chem.* **233**, 541 (1958).
38. Cifonelli, J. A., and Dorfman, A., *Biochem. Biophys. Res. Commun.* **7**, 41 (1962).
39. Sly, W. S., Quinton, B. A., McAlister, W. H., and Rimoin, D. L., *J. Pediatr.* **82**, 249 (1973).
40. Baudet, A. L., DiFerrante, N., Nichols, B., and Ferry, G. D., *Am. J. Human Genet.* **24**, 25a (1972).
41. Hall, C. W., Cantz, M., and Neufeld, E. F., *Arch. Biochem. Biophys.*, **158**, 817 (1973).
42. Bach, G., Cantz, M., Okada, S., and Neufeld, E. F., *Fed. Proc.*, **32**, 483 (1973).
43. Suzuki, S., Saito, H., Yamagata, T., Anno, K., Seno, N., Kawai, Y., and Furuhashi, T., *J. Biol. Chem.* **243**, 1543 (1968).
44. Lindahl, U., and Axelsson, O., *J. Biol. Chem.* **246**, 74 (1971).
45. Wiesmann, U. N., Rossi, E. E., and Herschkowitz, N. N., *N. Engl. J. Med.* **284**, 672 (1971).
46. Porter, M. T., Fluharty, A. L., and Kihara, H., *Science* **172**, 1263 (1971).
47. Matalon, R., Dawson, G., and Li, Y. T., *Pediatr. Res.* **6**, 394 (1972).
48. Kyriakides, E. C., Paul, B., and Balint, J. A., *J. Lab. Clin. Med.* **80**, 810 (1972).
49. DeDuve, C., and Wattiaux, R., *Ann. Rev. Physiol.* **28**, 435 (1966).
50. Leroy, J. G., Spranger, J. W., Feingold, M., Opitz, J. M., and Crocker, A. C., *J. Pediatr.* **79**, 360 (1971).
51. Tondeur, M., Vamos-Hurwitz, E., Mockel-Pohl, S., Dereume, J. P., Cremer, N., and Loeb, H., *J. Pediatr.* **79**, 366 (1971).
52. Leroy, J. G., Ho, M. W., MacBrinn, M. C., Zielke, K., Jacob, J., and O'Brien, J. S., *Pediatr. Res.* **6**, 752 (1972).
53. Spranger, J. W., and Wiedemann, H. R., *Humangenetik* **9**, 113 (1970).
54. Wiesmann, U. N., Lightbody, J., and Herschkowitz, N. N., *N. Engl. J. Med.* **284**, 109 (1971).
55. Wiesmann, U., Vassella, F., and Herschkowitz, N., *N. Engl. J. Med.* **285**, 1090 (1971).

56. Wiesmann, U., and Neufeld, E. F., unpublished observations.

57. Hickman, S., and Neufeld, E. F., *Biochem. Biophys. Res. Commun.* **49**, 992 (1972).

58. Sandhoff, K., and Jatzkewitz, H., in *Sphingolipids, Sphingolipidoses and Allied Diseases*, Volk, B. W., and Aronson, S. M., eds., Plenum Press, New York, 1972, p. 305.

59. Novikoff, A. B., Esner, E., and Quinbara, N., *Fed. Proc.* **23**, 1010 (1964).

60. Kenyon, K. R., and Sensenbrenner, J. A., *Inv. Ophthalmol.* **10**, 555 (1971).

61. Morell, A. G., Gregoriadis, G., Scheinberg, I. H., Hickman, J., and Ashwell, G., *J. Biol. Chem.* **246**, 1461 (1971).

62. Gregoriadis, G., Morell, A. G., Sternlieb, I., and Steinberg, I. H., *J. Biol. Chem.* **245**, 5833 (1970).

63. Bach, G., Eisenberg, F., Cantz, M., and Neufeld, E. F., *Proc. Natl. Acad. Sci. U.S.* **70**, 2134 (1973).

64. Coppa, G. V., Singh, J., Nichols, B. L., and DiFerrante, N., *Anal. Let.* **6**, 225 (1973).

65. Sjöberg, L., Fransson, L. A., Matalon, R., and Dorfman, A., *Biochem. Biophys. Res. Commun.* **54**, 1125 (1973).

66. Matalon, R., and Dorfman, A., *Ped. Res.* **7**, 384 (1973).

67. Kresse, H., *Biochem. Biophys. Res. Commun* **54**, 111 (1973).

68. Stumpf. D. A., and Austin, J. H., *Trans. Amer. Neurol. Assoc.* **97**, 29 (1972).

69. Kihara, H., Fluharty, A. L., and Stevens, R. L., *Am. J. Hum. Gen.* **25**, 41a (1973).

70. Thomas, G. J., Taylor, H. A., Reynolds, L. W., and Miller, C. S., *Ped. Res.* **7**, 751 (1973).

71. Glaser, J. H., McAlister, W. H., and Sly, W. S., *J. Ped.*, in press.

72. Berman, E. R., Kohn, G., Yatsiv, S., and Stein, H., *Clin. Chim. Acta*, in press.

73. Hickman, S., Shapiro, L. J., and Neufeld, E. F., *Biochem. Biophys. Res. Commun.* **57**, 55 (1973).

74. Lagunoff, D., Nicol, S. M., and Pritzl. P., *Lab. Invest.* **29**, 449 (1973).

75. O'Brien, J. S., Miller, A. L., Loverde, A. W., and Veath, M. L., *Science* **181**, 753 (1973).

76. von Figura, K., and Kresse, H., *J. Clin. Invest.* **53**, 85 (1974).

77. Brot. F. E., Glaser, J. H., Roozen, K. J., and Sly, W. S., *Biochem. Biophys. Res. Commun.* **57**, 1 (1974).

CHAPTER ELEVEN

DNA-Mediated Transformation in Mammalian Cells*

ELENA OTTOLENGHI-NIGHTINGALE

National
Academy
of Sciences
Washington, D. C.

Transformation is the transfer of genetic information from one cell to another by means of isolated or of released DNA. In mammalian systems transformation could provide a valuable approach to the understanding of many biological problems. Among such problems are mapping, gene-dosage effects, regulation of gene action, differentiation, aging, and carcinogenesis, as well as the prematurely popularized application of controlled gene transfer in man—correction of genetic defects.

The main advantages of using DNA, a soluble, chemically purified substance, as one of the participants in genetic transfer reactions are the same for mammalian cells as they are for microbial systems. Isolated DNA can be manipulated in the test tube in ways that would damage intact cells. Thus chemicals, radiation, heat, and other agents can be used to alter the genetic material in specific ways and the biological effects of these known alterations can be determined.

* The word "transformation" also refers to the change of a cell from the normal pattern of growth to the unregulated growth of tumor cells. This type of transformation can be brought about by the infection of nonpermissive cells with oncogenic viruses or their nucleic acids. This chapter deals not with oncogenic transformation but with the genetic transformation of mammalian cells by nonviral DNA.

233

DNA can be treated enzymatically or by heat to change its length or configurational properties in very precise ways. Methods are now available for adding or removing desired segments from these macromolecules. The *in vitro* synthesis of specific genes is also technically possible.

The development of feasible and reproducible methods for enabling mammalian cells to accept, integrate, and express foreign DNA would stimulate the growth of mammalian genetics to parallel that of microbial genetics and molecular biology that has occurred over the last 25 years.

The progress that has been made toward the alteration of the genome of mammalian cells by exogenous nonviral DNA and the directions such research may take in the future will be reviewed.

HISTORICAL BACKGROUND

In 1928 Griffith (1) reported the transformation of pneumococcal types. The classic work of Avery, MacLeod, and McCarty (2) established that the pneumococcal transforming principle was deoxyribonucleate (DNA). More than a decade elapsed before DNA was widely accepted as the genetic material for all cellular forms of life. This final acceptance was based on correlating evidence from several different disciplines. More than a century ago, Strasburger, Hertwig, and Boveri found that the nuclei of eggs and sperm were the bearers of heredity. Mendel later defined the precise manner by which hereditary traits were transmitted from one generation to another. Finally, Sutton and Boveri independently proposed that genes were located in visible physical structures, the chromosomes. Miescher was the first to isolate nucleic acids from the nuclei of pus cells. Kossel and Feulgen showed

that chromosomes were composed of nucleic acids and proteins. Later Chargaff, Mirsky, and others documented the constancy in the amount of DNA per cell and the direct relationship of this amount to ploidy (3). Nevertheless, the conclusion of Avery's group that DNA was the genetic material was not accepted until Hershey and Chase (4) showed that only the DNA of bacteriophage enters the host bacterium, with the protein remaining behind, and until a model for the double-helix structure of DNA was proposed by Watson and Crick (5). With this model it was possible to comprehend that DNA contained the necessary features of genetic material, a great potential for variability, and a means of exact self-replication. The possibility that mammalian cells could also be heritably altered by exposure to foreign DNA now seemed real.

Before discussing the attempts at DNA-mediated transformation of mammalian cells, the main features of bacterial transformations will be reviewed briefly since most of the experiments with mammalian cells were patterned after them.

COMPARISON OF TRANSFORMATION IN BACTERIA TO TRANSFORMATION IN MAMMALIAN CELLS

Successful transformation of bacteria depends on the interaction of competent recipient cells with undegraded DNA of appropriate molecular size. Competence includes the factors that enable a bacterium to accept DNA, and these factors vary from genus to genus. Some bacteria require specific constituents of the medium, such as calcium, magnesium, or serum albumin. Others become competent only for brief periods during the division cycle (6). Some bacteria elaborate a small protein, the competence factor, which, when added to physiologically incompetent cells, renders them competent. A genetic basis for competence exists; certain lines of bacteria are genetically unable to accept and to integrate foreign DNA under any circumstances. Bacterial transformation and competence have been reviewed (7,8).

It is significant that the prerequisites for transformation vary from one genus of bacteria to another. Usually the conditions for optimal transformation of a given genus are found only empirically. In addition, only a fraction of any recipient population will be transformed. For some bacteria, such as *Pneumococcus,* this fraction may be fairly large and even approach unity. For other bacteria, such as *Bacillus subtilis,* transformation is a much rarer event (7,8). To date the majority of microbial species have not been transformed. Thus, in trying to design transformation experiments for mammalian cells, it might be useful to keep in mind that the difficulties encountered with microbial systems are even more likely to be found with mammalian cells. For example, the conditions favorable to transformation might, as for bacteria, be specific to a particular cell line grown under particular conditions. Genetic competence might be a property of only a small minority of mammalian cell lines. Given a population of cells capable of achieving competence, only a small fraction might be transformed to the trait in question. As opposed to the situation in bacteria, it is difficult to handle very large numbers of mammalian cells, thus transformations at low frequencies might go undetected. Extremely stable markers that can be selected from a high background of untransformed cells are only now becoming available. Thus detection of transformants and separation of transformation from backmutation are difficult (see p. 237). Also, if the DNA is obtained from

cells similar to the recipients, an enormous amount of irrelevant material is likely to be present (9,10).

Once genetically active DNA has entered a cell, permanent integration of foreign DNA into such highly organized structures as mammalian chromosomes might be much more difficult than the integration of DNA into the simpler and more loosely organized bacterial chromosome. The latter is essentially one large DNA molecule not segregated from the rest of the cell by a nuclear membrane. During interkinesis, however, recombination with foreign DNA might occur in mammalian cells. The diploid nature of mammalian cells would also decrease the chance of detecting the phenotypic effects of recessive genes integrated in a single dose.

In spite of the difficulties outlined here, transformation of mammalian cells seems to be possible. Whether it will become as useful a tool in the study of the genetics of mammalian cells as it has been for microbial systems depends on whether conditions can be found for transforming cells *in vitro* with reproducible and predictable results of relatively high efficiency. Nonetheless, the demonstration that mammalian cells can be altered genetically by exogenous DNA at all is of biological significance. Cells in intact organisms are constantly "at risk" of exposure to DNA released from other mammalian cells, whether viable or dying. DNA of viral, bacterial, or mycotic origin is also present from time to time. Even a rare event could conceivably lead to the acquisition of genetic information with, for example, oncogenic potential (11). Thus, on some occasions, transformation might augment the variability already imposed on somatic or germ cells by mutation or by chromosomal crossing over.

Because of the obvious potential biological signicance of DNA-mediated transformations of mammalian cells, many studies of this nature are now in progress. It is possible to distinguish steps in the interaction of DNA and the recipient cell that are essential for genetic transformation: binding of DNA to the cell surface; penetration of DNA to the nuclear area; integration of the foreign DNA into the resident genome; replication; and expression of the newly acquired trait by the transformed cell and its descendants.

The remainder of this chapter will review the current knowledge of each of these steps in the interaction of mammalian cells with foreign DNA.

UPTAKE OF DNA BY MAMMALIAN CELLS

Numerous studies have demonstrated that a variety of mammalian cells are capable of binding both homologous and heterologous DNA (for reviews, see refs. 12–15).

In most of the *in vitro* studies the donor DNA, which was extracted from homologous tissues or from a variety of other sources, including bacteria and viruses, was labeled with a radioactive isotope. The uptake of such DNA by recipient cells was then estimated by autoradiography or by measuring radioactivity in appropriate cell fractions, such as acid precipitates of cell lysates, isolated nuclei, or isolated DNA. Autoradiographic studies were particularly useful in localizing the DNA within the cells.

The recipient cells used *in vitro* included normal human, mouse, or hamster cells as well as tumor cells such as Ehrlich ascites cells, HeLa cells, leukemic cells, and melanoma cells.

As in bacterial studies, experimental conditions varied, but usually physio-logical conditions (pH 7.4, 37°C, standard buffers, and tissue culture media) were used.

Szybalska and Szybalski (16) used spermine to facilitate DNA uptake by mammalian cells. More recently other polybasic amines, especially DEAE-dextran, have been found to greatly increase the uptake of nucleic acids by mammalian cells. The enhancement is found either by pretreating the recipient cells with DEAE-dextran for a brief period before exposure to DNA or by adding DNA and DEAE-dextran together to the incubation mixture. The mechanism by which polyamines increase the penetration of nucleic acids into mammalian cells is not known, but both stimulation of pinocytosis and increase in membrane permeability are considered to be possibilities (17,18).

A few examples of the different systems used to study DNA uptake by mammalian cells are presented in Table 1. A more extensive list can be found in Bhargava and Shanmugam (15).

Labeling methods alone cannot indicate whether the DNA taken up by host cells was either degraded and reutilized or was present in polymeric form. Kay (19), using DNA with known proportions of ^{14}C-adenine and ^{14}C-thymine, showed by chromatographic analysis of the base content that polymerized DNA penetrated ascites tumor cells. Recently Leavitt and associates (20) studied the uptake of ^{32}P-labeled bacteriophage T_7 DNA by baby hamster cells and found that the DNA was localized in the nuclear region of the hamster cells. Hybridization and sedimentation studies on reextracted DNA showed that the T_7 DNA component was not degraded. These investigators estimated that some of the hamster cells contained 7–10 copies of the T_7 genome per cell.

The conclusion of more than 30 studies published during the last 15 years (15) is that normal mammalian cells or mammalian cells of neoplastic origin can take up undegraded macromolecular DNA of homologous or heterologous origin.

The binding of DNA to the cell surface is at first reversible and sensitive to DNase; subsequently about 5–20% of the bound DNA penetrates the cell and becomes resistant to DNase, as in bacterial cell systems (8,21,22). The DNA can be offered to the cells in macromolecular soluble form or packaged into sperm (23), whole chromosomes (24), or viral protein coats (25). The exogenous DNA can usually be found in the nuclear region of the host cells (see Table 1). The uptake of polymeric DNA has been demonstrated mostly in cell culture systems, but also in intact animals (see, for example, refs. 11 and 26). Thus it seems that the nucleases found in plasma and in body fluids do not preclude the transport of polymerized DNA from the blood or intraperitoneal region to various organs and tissues (15). The possible biological implications of these findings will be discussed in a later section.

EFFECTS OF INCORPORATED DNA ON MAMMALIAN CELLS

There is little doubt that mammalian cells can incorporate foreign DNA. This DNA may exert on the host cell physiological or perhaps genetic effects that do not involve genetic recombination between the resident and the incoming DNA. For example, large amounts of foreign DNA (250 μg/ml) have been found to have toxic effects on L cells (27,28). Other examples of the inhibition of cell viability by DNA have been reviewed by Glick (14). Excess DNA may

become integrated and, when replicated, produce various mutations, since the base sequences may differ from the normal ones of the host DNA. Genetic information in the DNA may also have lethal potential for specific cells. Although no rigorous proof of this premise exists for mammalian cells, several investigators found DNA to be mutagenic for *Drosophila* (29–31). It is evident that, when the induction of genetic changes in mammalian cells by added DNA is being studied, the possible nonspecific effects of DNA should not be overlooked in the interpretation of the results.

The necessity for rigorous controls and the ability to distinguish the gene product of the presumably altered gene from the parental gene products is illustrated by the work of Shin and associates (32). These investigators observed that phenotypically normal cells could be isolated from mouse cell mutants lacking hypoxanthine guanine phosphoribosyl transferase (HGPRT) that had been treated with DNA prepared from phenotypically normal human cells. The human HGPRT isozyme can be electrophoretically distinguished from the mouse isozyme. The enzyme made by the mouse cells was found to be only of the mouse type and therefore presumably due to the expression of mouse genes already present. This system had been thought to be ideal for transformation studies because reversions at the locus in question had not been observed in these cell lines. In fact, the DNA was not involved at all since reversions to normal phenotype were also observed in its absence. The conclusion of these studies was not successful transformation, but that the original mutations that led to HGPRT deficiency could not have been deletions of the structural gene for HGPRT and that the experimental conditions increased the frequency of reversions.

Table 1. Uptake of Exogenous DNA by Mammalian Cells

Recipient Cell Type	Source of DNA	Method of Labeling DNA	Result	Ref.
Mouse L cells	Mouse L cells	^{3}H-Thymidine; bromodeoxyuridine	Radioactivity in nuclei	99
Human bone marrow making hemoglobin S	Human bone marrow making hemoglobin A	^{3}H-Thymidine DNA	Autoradiography revealed ^{3}H in cytoplasm of erythrocytes and in nuclei	45
Mouse gametes	Mouse thymus	^{3}H-Thymidine DNA	^{3}H label in 1/10^4 of cells screened	100
Rat bone marrow	Rat liver, spleen, bone marrow	^{3}H- and ^{32}P-DNA	DNA $>6 \times 10^6$ daltons absorbed to and penetrated cells	101
HeLa cells (human neoplastic origin); mouse L cells	Isolated metaphase chromosomes of L cells	^{3}H-Thymidine	Autoradiography revealed DNA in host nuclear region in random distribution	24
Human and mouse embryo cells	Baby mouse kidney	^{3}H-Thymidine mouse DNA in polyoma pseudovirions	7–24% of label in nuclear fraction; DNA characterized by sedimentation and hybridization	25
HeLa cells	Pneumococci; human leukocytes	^{3}H-Thymidine DNA	Autoradiography revealed ^{3}H localization in nuclear region	102
Murine lymphoma	Murine lymphoma	^{14}C-DNA; bromodeoxyuridine-DNA	Nuclear uptake up to 5% of host cell complement but not chromosomally incorporated	33

	Pneumococci	^{32}P-DNA	^{32}P found in cells; linear uptake in first 5 minutes	103
	Murine thymus	^{3}H-Thymidine DNA	Linear uptake in first 5 minutes, then rate of uptake decreased	104
Lymphoma P388F resistant to iododeoxyuridine	Lymphoma P388F	^{3}H-Thymidine- or ^{3}H-iododeoxyuridine-labeled DNA	Labeled DNA taken up linearly	34
Mouse embryos	E. coli	^{3}H-Thymidine DNA i.v. into pregnant mice	Only embryos younger than 15 days took up DNA	26
Mouse fibroblasts (NCTC)	E. coli	^{3}H- and ^{14}C-DNA; estimated radioactivity of DNA extracted from exposed cells	Uptake of 0.03 μg exogenous DNA for 120 μg of cellular DNA	105
Syrian hamster embryo	Bacteriophage T$_7$	^{32}P-DNA	Autoradiography revealed label in 90% of nuclei; hybridization and sedimentation revealed T$_7$ DNA not degraded	20
Sheep macrophages	Intact rat sperm	Phagocytosis; column chromatography of extracted DNA	Altered patterns of saline elution showed incorporation of sperm DNA in polymerized form by macrophage nuclei	23
Landschütz ascites tumor cells	E. coli	Deuterated, ^{3}H-thymidine DNA injected into animal	Autoradiography revealed radioactivity in cell nuclei	22

These studies emphasize the necessity for caution in selecting an experimental system and in interpreting the results when looking for DNA-mediated transformations in mammalian cells.

INTEGRATION OF DNA INCORPORATED BY MAMMALIAN CELLS

As discussed above, many kinds of mammalian cells will bind macromolecular DNA from almost any source. Penetration of a fraction of the DNA follows. The still fairly intact DNA then seems to become associated with the nuclei of the recipient cells. Some studies indicate that at least some of the nucleus-associated foreign DNA can become truly integrated into the DNA of the recipient genome.

Robins and Taylor (33) used cultures of lymphoma cells and homologous DNA labeled with ^{14}C and 5-bromodeoxyuridine (BU) to distinguish between simple uptake of DNA by a cell and chromosomal incorporation. The recipient cells were exposed to large amounts of DNA (1–400 μg for 2×10^7 cells) for 30 minutes at 37°C. Samples of donor BU-DNA and of host cell DNA after exposure of the cells to exogenous ^{14}C-BU-DNA were subjected to density-gradient centrifugation in cesium chloride. Optical densities and radioactivity of aliquots were measured. The average specific activities of host-cell DNA at the optical density peak were compared with estimates of specific activities calculated from that of the donor DNA, the amount of incorporation, and the proportion of non-BU-substituted DNA. The specific activity in the host cell that originated from the donor DNA and moved with it on centrifugation (presumably because it was tightly bound or integrated) was taken to be the difference between the observed and calculated values in specific activities of host-cell DNA. These differences were always very small ($< 0.3\%$). The authors concluded that, although nuclear uptake of macromolecular DNA by lymphoma cells is significant (up to 5% of host-cell complement), chromosomal incorporation is very small or negligible. The possibility remains, however, that even minute amounts of truly incorporated DNA may have significant genetic effects.

Ayad and Fox (34) have evidence that a mutant line of P388F lymphoma genetically resistant to 5-iodo-2′-deoxyuridine takes up DNA labeled with the ^{3}H-nucleotide that has been extracted from sensitive cells. They conclude that true integration of DNA into the recipient genome has taken place because, after uptake, DNA fractions from the host cells showed a distinct peak well separated from those given by donor or recipient DNA when analyzed by buoyant density distribution in caesium chloride. The reutilization of fragments of degraded DNA was controlled. This "integrated" DNA was calculated to be about 1.5% of the recipient genome, more than that estimated by Robins and Taylor (33). However, since many of the experimental conditions, such as the duration of exposure to DNA and the concentration of cells and DNA, differed in these two studies, a direct comparison is not possible.

Evidence that integration of foreign DNA can occur in nonmammalian systems is accumulating. Recently Ledoux, Huart, and Jacobs (35) reported that germinating seeds of Thalia cress, *Arabidopsis thaliana,* will take up ^{3}H-DNA from *Escherichia coli* or *Streptomyces,* and traces of this DNA can be found even in the F_1 progeny. These results are still preliminary, and the state of the DNA has not yet been determined because the radioactivity is not sufficient for sonication and alkaline hydrolysis procedures to be performed. Hill and Huppert (36) reported that mouse DNA can be integrated into the genome of chick fibroblasts in amounts equivalent to 2% of the host genome. This amount is in the same range as that found by Ayad and Fox for lymphoma cells (34). The replicated strands of host-cell DNA seemed to have exchanged pieces with the mouse DNA (37). Thus recombination has apparently taken place, but again the evidence is biochemical rather than genetic.

That viral DNA can enter into stable, long-term associations with host-cell DNA has been known for some time, but the exact nature of this association is not clear. Recently Lavi and Winocour (38) showed that at high multiplicities of infection true recombination occurs between the DNA of SV-40 virus and the host cell. As a result, viral particles containing host DNA sequences are produced, and presumably some viral DNA sequences are integrated into the DNA of the cell. It now seems possible that foreign nonviral DNA also becomes integrated into the host-cell genome in an amount that may seem small in biochemical terms (0.3–2% of the host genome) but may be significant biologically. Reports of genetic expression of foreign DNA incorporated by mammalian cells and presumably integrated into the host DNA are numerous; some of these are described in the next section.

GENETIC TRANSFORMATIONS OF MAMMALIAN CELLS BY EXOGENOUS DNA

As discussed in the preceding sections, mammalian cells seem to be capable of absorbing, incorporating, and integrating exogenous DNA. It is certainly possible, therefore, that genetic transforma-

tions can occur in such cells.

The earlier reports on attempts to transform mammalian cells by DNA in the environment have been reviewed by Ledoux (12) and more recently by Olenov (13), Glick (14), and Bhargava and Shanmugam (15). Representative experiments with results reported as positive are presented in Table 2.

In most of the transformation systems the difference between the donor cells and the recipient population is of a quantitative, rather than a qualitative, nature. Such traits as the level of resistance to a drug (e.g., 8-azaguanine) or the number of cells producing melanin were used. Recipient cell populations showed a significant increase in the degree of expression of the trait after exposure to DNA from cells that had a greater level of expression. In all cases the significance of the differences disappeared in control experiments using degraded DNA or DNA from the recipient strain itself. In these experiments, however, it is difficult to exclude totally the influence of the environmental conditions on the outcome. Many mammalian cell mutants show a certain degree of reversion, and many cell lines a degree of spontaneous mutation which, when coupled with the low efficiency of transformation, can interfere with the rigorous interpretation of the results (see, for example, Refs. 32 and 39–41). For example, in the experiments of Roosa and Bailey (41), 100 of 10^6 colonies were resistant to 10^{-6} M 8-azaguanine in the presence of DNase-digested DNA from the resistant line and 213 of 10^6 resistant colonies resulted when undigested DNA was used. This twofold difference is significant ($P < .001$). Thus, although transformation probably did occur, its demonstration is not as clear as if the recipient cells had acquired a property that was totally absent or different previously. It is also difficult, with

Table 2. Transformation of Mammalian Cells by Nonviral DNA

Recipient Cells	Source of DNA	Genetic Marker	Result	Ref.
Human bone marrow producing hemoglobin S	Human bone marrow producing hemoglobin A	Ability to synthesize β^A chain of hemoglobin[a]	After 10 days' incubation with 900 μg DNA, synthesis of hemoglobin A in recipients was detected by electrophoresis	45
Human sternal marrow line deficient in HGPRT[b]	Same as recipient but produced HGPRT	Production of HGPRT permitting survival in HAT[c] medium[d]	4×10^{-4} transformants per recipient cell at DNA dose level of 20 μg/ml	16
Mouse liver macrophages resistant to mouse hepatitis virus	Crude extract of macrophages sensitive to virus	Mouse hepatitis virus sensitivity[d]	Conversion of macrophages from virus resistance to sensitivity after exposure to heat-stable, DNase-labile extract from sensitive cells	106
Murine lymphoma in culture sensitive to 8-azaguanine	Same as recipient but resistant to 8-azaguanine	8-Azaguanine resistance (1000-fold increase)[d]	Doubling in frequency of resistant colonies after exposure to DNA from resistant line (20 μg DNA for 2×10^6 cells)	41,107
Rat sarcoma sensitive to sarcolysin	Rat sarcoma resistant to sarcolysin	Resistance to sarcolysin[d]	Twentyfold increase in sarcolysin-resistant tumors	108
Amelanotic melanoma	Melanotic melanoma	Melanin production[d]	Threefold increase in pigment-producing cells	109

Human uvular carcinoma cell line sensitive to 6-azathymine	Same as recipient but resistant to 40 mM/ml 6-azathymine	6-Azathymine resistance[d]	Increase in number of resistant cells after exposure to DNA significant at 1% level when compared to spontaneous mutation	40
Lymphoma cells sensitive to 5-iodo-2′-deoxyuridine	Lymphoma cells resistant to 5-iodo-2′-deoxyuridine	Resistance to 5-iodo-2′-deoxyuridine[d]	Increase in number of resistant colonies significant at 5% level	110
Embryonic albino (tyrosinase$^-$) mouse melanoblasts	Soft tissues of pigmented mice (tyrosinase$^+$)	Melanin production as a result of tyrosinase production[a]	Melanin observed in 10/49 trials with DNA and in 0/37 trials without intact DNA	47
Chinese hamster cells	Mouse Ehrlich ascites tumor cells	Mouse antigen synthesis[a]; oncogenic potential	10^{-4} to 10^{-5} hamster cells per recipient cell produced mouse antigens (immunofluorescence) after exposure to 40 µg mouse DNA; a few of the above produced sarcomas in Syrian hamsters	48

[a] Qualitative difference between donor and recipient.

[b] HGPRT is hypoxanthine guanine phosphoribosyl transferase, previously called inosinic acid pyrophosphorylase.

[c] HAT medium contains 0.1 µg/ml aminopterin to inhibit *de novo* purine synthesis; 5 µg/ml hypoxanthine as sole purine source; 5 µg/ml thymidine to counteract accompanying block in thymidylate synthesis.

[d] Quantitative difference between donor and recipient.

quantitative differences, to analyze specific progeny for the segregation of the acquired trait. Such an analysis is of obvious significance for a genetic study.

The experiments of Szybalska and Szybalski (16) included features necessary for the demonstration of genetic transformation. The D98S human cell line employed as recipient showed a low ($<$ 10^{-7}) rate of spontaneous mutation to production of HGPRT (previously called inosinic acid pyrophosphorylase), whereas transformation frequencies of greater than 10^{-4} were observed, a differential of 1000-fold. Cells lacking this enzyme were totally unable to grow in the selective medium (see footnote c of Table 2). Most importantly, a linear relationship was observed between the concentration of DNA and the number of transformants. The dose response was linear until about 20 μg of DNA per milliliter. Although for pneumococci saturation is reached at 1 μg of DNA per milliliter (6), when the experimental differences (e.g., ploidy, molecular size of DNA) are considered, the results with these mammalian cells clearly parallel those with bacteria. The specific role of the DNA in the transformation process was established by showing that DNase-treated DNA or DNA from the recipient cell line produced no transformants, whereas RNase-treated DNA had full activity. Furthermore, DNA isolated from a transformed clone was fully capable of transforming the enzyme-deficient recipients. It certainly seems as though DNA-mediated transformation took place in this mammalian cell system. The possibility remains that the results were due to an increased reversion rate. Since homologous DNA was used, this question cannot be answered unequivocally. The recipient cell line showed a high degree of transformability. Competence is a heritable property of only a minority of bacteria (8), and its nature

is still understood poorly. This may explain why it has been difficult to obtain results as clear-cut as Szybalski's since 1962 and with other cell lines. A similar situation arose with *E. coli* in the early days of bacterial transformation. The DNA-mediated transformation of capsular polysaccharide in *E. coli* was first reported in 1945 by Boivin. The results and controls in every way paralleled those of pneumococcal transformations. Unfortunately the competent strain of *E. coli* was lost after Boivin's death in 1949 (42). Although unique systems for genetic intervention in *E. coli* by use of protoplasts as recipients (43) or by use of circular phage or plasmid donor DNA have been reported in the intervening years, transformation of chromosomal loci in an auxotrophic strain of intact *E. coli* cells by DNA isolated from a prototrophic strain has only just been reported, a lapse of 27 years since the original transformation report (44). The frequency of transformation is low (10^{-6} per recipient cell) but nonetheless real. In order for transformation to occur in *E. coli,* the cells had to be treated with calcium chloride. Most important, this strain of *E. coli* was chosen as the recipient because it is a mutant of *E. coli* K-12 that lacks ATP-dependent DNase. This enzyme degrades linear DNA, such as transforming DNA, but does not attack double-stranded circular DNA, such as plasmid DNA. The success of these transformations was due to the proper choice of recipients.

Thus, with increased knowledge about the physiology and genetics of mammalian cells, it may become possible to select lines of cells more amenable to genetic intervention.

The ability of a cell to acquire a totally new property after exposure to DNA from cells having that property would provide unambiguous demonstration that a transformation-like event had taken place. Kraus (45) reported in 1961 that human bone marrow cells that produced only hemoglobin S were able to synthesize the β^A-chain of hemoglobin after exposure to a DNA from human marrow cells producing hemoglobin A. Electrophoresis was used to demonstrate the β^A-polypeptide chain in hemolysates of the recipient cells. Unfortunately no controls with DNA treated with DNase, RNase, or proteolytic enzymes were reported in this paper. Weisberger (46) was also able to induce altered globin production in human immature erythrocytes by nucleoprotein preparations from genetically different marrow cells. In these experiments, however, the altered globin production depended on the presence of some intact RNA and protein in the preparation since both RNase and trypsin prevented its occurrence. It is possible that in Krau's work the DNA was the active component, but sufficient information is lacking to reach this conclusion.

Ottolenghi-Nightingale (47) and Borenfreund et al. (48) also reported transformations in which the trait transferred by the DNA is clearly distinguishable from any property previously existing in the recipient cells. In the first of the two studies nucleoprotein was extracted from the soft tissues of pigmented C_{57} black mice. Cell suspensions of skin, including melanoblasts, from albino embryos at 14 days' gestation were exposed to the DNA preparation *in vitro*. The cells were then implanted subcutaneously into the neck of young adult albino hosts. After 14 days the skin-cell colonies were examined for melanin pigment by light and by electron microscopy. The recipient cells were from animals whose total albinism resulted from a lack of the tyrosinase activity necessary to convert dihydroxyphenylalanine (DOPA) to DOPA-quinone (49). This enzyme is not found in the livers and spleens of the pigmented (tyrosinase+) animals. Thus the DNA was extracted from differentiated tissues in which the genetic potential for tyrosinase synthesis is usually unexpressed. In 10 of 49 trials the albino recipient melanoblasts exposed to DNA from pigmented mice were able to synthesize melanin. In equivalent numbers of control trials in which either saline or DNase-digested DNA no pigmented cells were found. Thus transformation seems to have occurred. The advantages of this system are that contamination of cell lines is excluded and the albino mice used have never been observed to revert to pigment production; thus backmutation is not a problem. In mice melanin production behaves as a Mendelian dominant trait, so that substitution at one locus should be sufficient for expression of the newly inserted gene. The melanin can be visualized by light and by electron microscopy. The presence of any melanin is significant. The active agent was most probably DNA since treatment by DNase completely prevented transformation.

Although the regulation of melanin synthesis in mammalian cells is not clear, the number of pigment cell genomes in a cellular hybrid affects the expression of pigmentation. In hamster–mouse hybrids two pigmented genomes to one unpigmented were necessary for melanin synthesis to occur (50,51), whereas in the intact animal melanin production is transmitted as a Mendelian dominant. Although it seems unlikely that Ottolenghi-Nightingale's results were due to release of the host DNA from "extinction" (52), either a genetic or an epigenetic phenomenon may be involved. The main disadvantages of this system are that only small numbers of recipient cells can be screened at any one time, so that many experiments have to be done to obtain a positive result. Also, within an experiment all cells have to be screened since there is no method of selection for the transformed cells. Thus the experiments are cumbersome and tedious to perform and, as a consequence, are of limited practicality. Questions have been raised about these experiments that a control with DNA from unpigmented cells was not included and that the DNA was not treated with RNase (53). Although these points are valid, controls of the nonspecific effects of intact DNA are present in the many experimental trials in this series that had negative results. No effects other than induction of pigment were observed under these experimental conditions. Digestion with RNase was not used to avoid loss of yield since small amounts of nucleoprotein were being processed. The results of the DNase controls were considered to be sufficient to implicate DNA as the active material in the transformations.

Borenfreund and associates (48) studied the acquisition of the ability to synthesize mouse antigens and the expression of oncogenic potential by Chinese hamster cells after exposure either to DNA from Ehrlich ascites (mouse) tumor cells or to the cells themselves. In these experiments controls with Pronase, with RNase- and DNase-treated DNA, and with homologous DNA were included. Antigens were detected by immunofluorescence. A synthetic polyamine (DEAE-dextran) was used to increase DNA uptake by the hamster cells. The frequency of immunofluorescent hamster cells after exposure to mouse DNA was 10^{-4}–10^{-5}, a frequency consistent with that observed by others working in this field [e.g., Szybalski (16)]. A total of 17 mouse-antigen-producing clones were isolated and propagated. The DNA extracted from these clones had transforming activity. A few sarcomas were observed in conditioned weanling Syrian hamsters injected with cells from the

cloned lines. As in other studies, many experiments were totally negative, another indication of our lack of knowledge about the factors involved in the development of competence for transformation by mammalian cells.

Although the expression of oncogenic potential is a very complex genetic marker, the acquisition by hamster cells of the ability to synthesize mouse antigens after exposure to mouse DNA is a good indication that transformation of some cells to a totally new property did, in fact, take place.

Several failures to transform mammalian cells have been reported. Among these are attempts to transform albino rat skin to pigment production (54); murine lymphoblasts to amethopterin resistance (55), and murine lymphoma cells to purine pyrophosphorylase production (56). In this area the reporting of negative results is valuable to inform interested investigators of the trials that have already been made, but unfortunately such reporting is rare.

Because of the difficulties encountered in obtaining stable genetic markers, sufficiently high efficiencies of transformation, and/or reliably competent cell lines, DNA-mediated transformation of mammalian cells is still not a useful tool for the study of cell genetics or for the correction of genetic defects in mammalian cells. Although the potential for each step necessary for transformation exists, not enough is known about competence to approach the problem in a logical, rather than an empirical, manner.

Totally new genetic properties have been acquired by mammalian cells in different experimental approaches. For example, hybrids formed after the fusion of cells from different species (52) have some properties of each parental line. Schwartz, Cook, and Harris (57) enabled deficient mouse cells to produce HGPRT by incorporating small amounts of chick erythrocyte DNA as a consequence of cell fusion followed by "chromosome pulverization." Fragmentation or pulverization occurs if, after fusion, one nucleus is ahead of the other and sets the pace for mitosis in the heterokaryon. In this case the chromosomes of the nucleated chick erythrocyte in the heterokaryon with a mouse nucleus condense prematurely and fragmentation of chick chromosomes occurs, liberating small fragments of genetic material that may be integrated into the functional mouse chromosomes. This technique, in effect, results in the production of minute pieces of chromosomes of one cell in intimate contact with the intact chromosomes of another and provides a delivery system for foreign DNA. Although genetic repair was effected by this method, the repair was not permanent since in a nonselective medium the chick genes were frequently lost from the dividing cells.

McBride and Ozer (111) reported recently that mouse cells deficient in HGPRT production could make this enzyme after incubation with purified metaphase chromosomes from wild-type Chinese hamster fibroblasts. Although the frequency of correction of the enzymatic defect was low (10^{-6} to 10^{-7}) the enzyme produced by the mouse cells was not of the murine type and was indistinguishable from that of Chinese hamster cells by DEAE-cellulose chromatography and gel electrophoresis.

Aside from experiments involving the fusion of whole cells of distinct genotype or the ingestion of whole chromosomes, attempts have been made to introduce new genes using virus coats as delivery systems for mammalian DNA (pseudovirions), but expression of these genes has not been reported (25,58–60). Correction of genetic defects in mammalian cells by the use of viruses has also been tried. Some of these attempts are summarized in Table 3.

Merril and colleagues (61,62) achieved correction of the enzymatic defect found in individuals with galactosemia, a condition transmitted as an autosomal recessive, by the use of a transducing bacteriophage. Human fibroblasts from galactosemic patients lack the enzyme α-D-galactose-1-phosphate uridyl transferase. The transducing bacteriophage λ, which has acquired the galactose operon from the bacterium *E. coli,* is a viral particle whose DNA carries the information for the synthesis of the transferase lacking in the cells of galactosemics. Ninety-six hours after galactosemic fibroblasts (transferase$^-$, or T$^-$) were exposed to this phage (λ pgal T$^+$) or to small amounts of DNA isolated from it (1 μg DNA for 5×10^5 cells), the transferase activity of the culture as a whole increased by a factor of 10–80, well within the range expected for normal human fibroblasts. When a transferase$^-$ mutant of the phage (λ pgal T$^-$) was used, the transferase activity of the galactosemic culture remained the same. The mammalian cells appeared capable of transcribing microbial DNA into RNA and of translating it into protein. In fact, λ-specific RNA was found by RNA–DNA hybridization. Since enzyme was not produced in the presence of cycloheximide (which blocks protein synthesis in mammalian cells but not in bacteria), translation was necessary for transferase activity to appear.

These experiments support the universality of the genetic code. They also provide a specific genetic correction to defective human cells. From these data it is not possible to determine how the fibroblast cultures were able to maintain the foreign genes, but transferase activity did persist undiminished for well over a month. Evidence of the true integration of the *E. coli* genes into the host genome is lacking, and the new DNA

Table 3. Virus-Mediated Correction of Genetic Defects in Mammalian Cells

Recipient Cell	Genetic Defect	Virus and/or Source of Correcting Genes	Result	Ref.
Human fibroblasts	Deficiency of α-D-galactose-1-phosphate uridyl transferase	1. Bacteriophage λ carrying *E. coli* galactose operon (includes transferase) 2. Naked DNA from above λ pgal T⁺	1. Fiftyfold increase in transferase activity after 52 days 2. Seventy-five-fold increase in transferase activity after 96 hours	61,62
L Cells (mouse)	Deficiency of thymidine kinase	Herpes simplex virus, UV-inactivated, carrying thymidine kinase gene	10^{-5}–10^{-3} of infected cells produced 7–24 times more thymidine kinase than uninfected cells (reversion rate of $<10^{-8}$); transformed cells stable at least 10 months	65,66
Two children with arginase deficiency	Deficiency of arginase	Shope papilloma virus, which induces arginase in man	No effects 2 years after infection with papilloma virus (laboratory workers have low serum arginine levels)	67,112

may well be replicating as a plasmid. For permanent correction of a genetic defect, stable integration of the new genes would be necessary. Also the gene product was not identified as corresponding to cellular DNA or added DNA. Electrophoretic analysis of the transferase enzyme could characterize it as mammalian or bacterial. However, the hybridization of 0.2% of the RNA to phage DNA would be evidence of a bacterial enzyme. The possibility of contamination of the serum with coliphages (63) and the contention that rapidly growing cells from galactosemic lines can make up to 85% of the normal amount of enzyme (64) have been raised as objections to the interpretation of the results. Enzyme production as a whole was studied, whereas for genetic analysis the production of enzyme by individual clones should be evaluated. Merril and colleagues are investigating these possibilities. As already mentioned, in any genetic experiments with cell lines in tissue culture the positive identification of the cell lines and the verification of lack of contamination and of genetic reversion are important.

In spite of objections, the experimental approach of correcting genetic defects by the introduction of specific genes from bacteria or viruses is a sound one. The vast amount of irrelevant DNA present in homologous wild-type DNA would be eliminated. However, it is possible that restriction nucleases might defend a genome against invasion by heterologous DNA. Preferential digestion of foreign DNA might interfere with the usefulness of transformation by DNA from an unrelated source.

A conceptually similar approach to Merril's for the correction of genetic defects in mammalian cells was taken by Munyon (65). His group used a mutant line of mouse L cells that lack the enzyme thymidine kinase (Tk⁻). Herpes

simplex, an animal virus that usually induces thymidine kinase activity in the cells it infects, was used as the source of "correcting" DNA. The virus was inactivated by ultraviolet irradiation so that it would not kill the recipient cells. After exposure to irradiated herpes simplex virus, about 0.1% of the TK⁻ L cells were able to produce thymidine kinase. The newly acquired property remained stable for many months. Cells infected with herpes simplex mutants that have lost the ability to induce thymidine kinase and uninfected cells did not have enzyme activity. Again the mechanism for the correction of the genetic defect is not clear: a viral gene might have been integrated by the L cell's genome, or a viral gene product might have induced the stable expression of a mouse gene. The kinetics of thermal inactivation of Tk⁺ L cell enzyme and of enzyme produced by infection with herpes virus are different. The interpretation of these results depends on the identification of the gene product as an enzyme of mouse or of viral origin. Recent studies on the electrophoretic migration of thymidine kinase in polyacrylamide gels indicate that the enzyme activity in the virus-transformed cells is different from that in unselected TK⁺ L cells, and thus is presumably of viral origin (66).

GENE THERAPY FOR HUMAN GENETIC DISEASE

One attempt to correct a genetic defect in man has been made, but so far without success. It has been observed that laboratory workers who handled the Shope papilloma virus, a virus oncogenic for several mammalian species in the wild, had low serum levels of arginine for periods as long as 20 years. No other effects of handling the virus were observed. The low arginine levels resulted from virus-induced arginase synthesis (67). Terheggen and associates (68) described the occurrence of argininemia due to a deficiency of arginase in two German sisters. This genetic lesion results in profound mental and physical impairment. The two children were infected with Shope papilloma virus in the hope of inducing arginase synthesis, thereby correcting the argininemia. Two years after infection, no effects of the virus were demonstrable in the children, and the disease continued to progress. Moral and ethical objections were raised since a virus with known oncogenic potential was used in attempting to correct a disease in man. The defense of the experiments was that there was no therapy for the disease in question and that no deleterious effects had been observed in laboratory workers exposed to this virus. Nonetheless, the experiment has so far been unsuccessful. In contrast, a recent article (112) indicates that fibroblasts cultured from skin biopsies taken from these same patients were induced to produce arginase after inoculation with Shope papilloma virus. Perhaps the Shope virus may yet prove useful in the treatment of argininemia.

The possibility of genetic intervention in mammalian cells by the introduction of viral genes or of mammalian genes carried by viral genomes or particles is an intriguing one. So far only attempts at genetic intervention in somatic cells have been described. Recently Bracket and associates (69) reported that germ cells could transmit viral DNA during fertilization. After rabbit sperm were exposed to SV-40, a small DNA virus that is oncogenic for some mammals, the DNA of the virus was found to be located in the postacrosomal area of the sperm. These sperm were used for the uterine insemination of rabbits. Ova fertilized by sperm carrying the SV-40 DNA were then cocultivated with a line of monkey kidney cells that was susceptible to SV-40 infection and was permissive for viral replication. Infectious SV-40 virus was recovered from the cocultured cells. Thus a heterologous genome was acquired by the sperm and carried into the ovum by fertilization. If the SV-40 genome could be constructed so as to include a gene lacking in either parent, it is conceivable that the defect might be corrected in all the cells of the new individual by manipulation of the sperm and by the use of artificial insemination.

In fact, techniques are becoming available for the chemical synthesis of particular genes. Khorana and colleagues (70) have achieved the synthesis of the total DNA duplex comprising the structural gene for an alanine t-RNA from yeast. It will be possible to introduce specific changes into such sequences or to add the synthetic gene to other genomes, such as those of transducing phages. Harvey, Wright, and Nussbaum (71) have already reported the successful covalent joining of a chemically synthesized oligonucleotide to a naturally occurring DNA molecule (λ DNA). Jackson, Symons, and Berg (72) have also developed methods for "welding" DNA from different sources. By using the appropriate restriction endonuclease, terminal DNA transferase, and DNA ligase, they were able to covalently integrate the bacteriophage λ DNA containing the galactose operon from E. coli into the circular DNA of the SV-40 tumor virus. The method is generally applicable to joining DNA molecules together. The aim of the work was to "develop a method by which new, functionally defined segments of genetic information can be introduced into mammalian cells" (72). Once this aim has been fulfilled, the new sequences must be integrated and replicated. Recent work from the laboratories of Nathans (73–76) and of Berg (77,78) on the specific cleavage of SV-40 DNA by restriction enzymes provides a means of identifying specific regions of the SV-40 DNA to be included in an artificially constructed molecule. For example, if a segment of SV-40 DNA that includes the initiation region is welded to other desired nucleotide sequences, it would ensure replication of the particle to be introduced into a recipient cell. Lavi and Winocour (38) report that, at high multiplicities of infection in monkey cells, SV-40 particles that have acquired sequences homologous to host DNA can be recovered. The use of such particles might increase the chance of recombination between the SV-40 DNA (and whatever sequences had been attached to it) and recipient mammalian DNA, and facilitate true integration of the new genes.

Many articles have been written both in scientific and in lay publications on the feasibility, need, justification, and moral and ethical implications of manipulating the primary genetic material of mammalian cells and, eventually, of man (see, for example, refs. 79–86). It seems that, although techniques are becoming available that will eventually make genetic manipulation feasible on a cellular level (see above), permanent correction of genetic defects in man by transformation or transduction-like phenomena is still remote. Moreover, the unwanted effects of such manipulations must be kept in mind. As pointed out in *Nature* (87), if a transducing phage of *E. coli,* such as λ, is joined to the genetic material of a virus, such as SV-40, with oncogenic potential for human cells, the phage might infect *E. coli.* This bacterium is a normal inhabitant of the human gastrointestinal tract and could be a vehicle for the introduction of SV-40 genes into human populations by infective means, with possibly dire consequences.

Aside from these more dramatic consequences of induced genetic changes in mammalian cells, other interactions with genetic effects might be considered. It has been shown for pneumococci (88) and for some other bacteria (89–91) that release of genetically active DNA occurs as a normal concomitant of population growth. If competent recipient cells are present in the environment, DNA-mediated transformations of bacteria occur in the test tube, in the mouse, and in the human host in the absence of added DNA (92–94).

Under certain conditions, human lymphocytes also excrete DNA into the environment (95,96). Hill (97) found that lymphocytes take up polymerized DNA released from damaged mouse L cells. Polymerized DNA of cellular, bacterial (98), or viral origin, if present in the environment and absorbed by cells susceptible to transformation, may play a role in the acquisition of new traits or of oncogenic potential by somatic and even germ cells. Bendich (11) discussed the possibility that circulating DNA might be a possible factor in oncogenesis. Perhaps environmental DNA plays a role in introducing other genetic changes in animal as well as bacterial cells.

CONCLUSIONS

Mammalian cells have the capacity to absorb polymeric DNA from a wide variety of sources. Environmental DNA can penetrate mammalian cells, but the mechanism of penetration is not yet understood. Once inside the cell, undegraded DNA becomes localized and associated with nuclei. In some instances there is biochemical evidence of the integration of exogenous DNA into the genome of the recipient cell. Expression of new traits, such as drug resistance or synthesis of specific enzymes, by the host animal cell after uptake of exogenous DNA and replication of the new DNA has been reported in several experimental systems. These events would represent DNA-mediated transformations. At present transformations are of only limited usefulness in the study of mammalian cell genetics or in the correction of genetic defects in mammalian cells. Some of the problems encountered are the low frequency of transformation; the lack of knowledge about the factors involved in competence of animal cells for transformations; the scarcity of stable markers amenable to selection in mammalian cell lines; and the problems of unambiguous characterization of mammalian cell lines in culture. Nonetheless, the rapidly advancing knowledge of techniques for constructing DNA duplexes with desired specific nucleotide sequences and with high potential for infectivity, integration, and self-replication may make the insertion of genes into mammalian cells easily attainable in the near future.

When gene insertion into mammalian cells is easily and reproducibly performed, it will be extremely valuable for the study of mammalian cell genetics and the control of both normal and abnormal development. As for any large-scale correction of human genetic defects, even then many difficulties will still remain, such as the problem of correcting polygenic traits, diseases with dominant patterns of inheritance, or abnormalities involving whole chromosomes. Therefore prevention rather than therapy will probably remain the most effective approach to the problem of genetic disease in man.

REFERENCES

1. Griffith, F., J. Hyg. 27, 113 (1928).
2. Avery, O. T., MacLeod, C. M., and McCarty, M., J. Exp. Med. 79, 137 (1944).
3. Stent, G., Molecular Genetics, An Introductory Narrative, W. H. Freeman, San Francisco, 1971.
4. Hershey, A. D., and Chase, M., J. Gen. Physiol. 36, 39 (1952).
5. Watson, J. D., and Crick, F. H. C., Nature 171, 737 (1953).
6. Hotchkiss, R. D., in The Chemical Basis of Heredity, McElroy, W. D., and Glass, B., eds., Johns Hopkins Press, Baltimore, 1957, p. 321.
7. Hotchkiss, R. D., and Gabor, M., Ann. Rev. Genet. 4, 193 (1970).
8. Tomasz, A., Ann. Rev. Genet. 3, 217 (1969).
9. Siniscalco, M., Adv. Biosci. 8, 307 (1972).
10. Britten, R. J., and Kohne, D. E., Science 161, 529 (1968).
11. Bendich, A., Wilczok, T., and Borenfreund, E., Science 148, 374 (1965).
12. Ledoux, L., Prog. Nucleic Acid Res. Mol. Biol. 4, 231 (1965).
13. Olenov, J. M., Int. Rev. Cytol. 23, 1 (1968).
14. Glick, J. L., Axenic Mammalian Cell Reactions, Dekker, New York and London, 1969, p. 117.
15. Bhargava, P. M., and Shanmugam, G., Prog. Nucleic Acid Res. Mol. Biol. 11, 103 (1971).
16. Szybalska, E. H., and Szybalski, W., Proc. Natl. Acad. Sci. U.S. 48, 2026 (1962).
17. McCutchan, J. H., and Pagano, J. S., J. Natl. Cancer Inst. 41, 351 (1968).
18. Howard, B. V., Estes, M. K., and Pagano, J. S., Biochim. Biophys. Acta 228, 105 (1971).
19. Kay, E. R. M., Nature 191, 387 (1961).
20. Leavitt, J. C., Schechtman, L., Ts'O, P. O. P., Borenfreund, E., and Bendich, A., World Symposium on Model Studies in Chemical Carcinogenesis, Baltimore, October 31–November 3, 1972.
21. Schell, P. L., Biochim. Biophys. Acta 166, 156 (1968).
22. Charles, P., Remy, J., and Ledoux, L., in Informative Molecules in Biological Systems, Ledoux, L. G. H., ed., North-Holland, Amsterdam–London, 1971, p. 88.
23. Reid B. L., and Blackwell, P. M., Aust. J. Exp. Biol. Med. Sci. 45, 323 (1967).
24. Burkholder, G. D., and Mukherjee, B. B., Exp. Cell Res. 61, 413 (1970).
25. Qasba, P. K., and Aposhian, H. V., Proc. Natl. Acad. Sci. U.S. 68, 2345 (1971).

26. Ledoux, L., and Charles, P., *Exp. Cell Res.* **45,** 498 (1967).

27. Smith, A. G., and Cress, H. R., *Lab. Invest.* **10,** 898 (1967).

28. Smith, A. G., *Cancer Res.* **24,** 603 (1964).

29. Mathew, C., *Genet. Res.* **6,** 163 (1965).

30. Fahmy, O. G., and Fahmy, M. J., *Nature* **191,** 776 (1961).

31. Gershenson, S. M., *Genet. Res.* **6,** 157 (1965).

32. Shin, S., Caneva, R., Schildkraut, C. L., Klinger, H. P., and Siniscalco, M., *Nature New Biol.* **241,** 194 (1973).

33. Robins, A. B., and Taylor, D. M., *Nature* **217,** 1228 (1968).

34. Ayad, S. R., and Fox, M., *Nature* **220,** 35 (1968).

35. Ledoux, L., Huart, R., and Jacobs, M., in *Informative Molecules in Biological Systems*, Ledoux, L. G. H., ed., North-Holland, Amsterdam–London, 1971, p. 159.

36. Hill, M., and Huppert, J., *Biochim. Biophys. Acta* **213,** 26 (1970).

37. Hill, M., and Hillova, J., in *Informative Molecules in Biological Systems*, Ledoux, L. G. H., ed., North-Holland, Amsterdam–London, 1971, p. 113.

38. Lavi, S., and Winocour, E., *J. Virol.* **9,** 309 (1972).

39. Szybalski, W., and Szybalska, E. H., *Univ. Michigan Bull.* **28,** 277 (1962).

40. Majumdar, A., and Bose, S. K., *Br. J. Cancer* **22,** 603 (1968).

41. Roosa, R. A., and Bailey, E., *J. Cell. Physiol.* **75,** 137 (1970).

42. Ravin, A. W., *Adv. Genet.* **10,** 61 (1961).

43. Chargaff, E., Schulman, H. M., and Shapiro, H. S., *Nature* **180,** 851 (1957).

44. Cosloy, S., and Oishi, M., *Proc. Natl. Acad. Sci. U.S.* **70,** 84 (1973).

45. Kraus, L. M., *Nature* **192,** 1055 (1961).

46. Weisberger, A. S., *Proc. Natl. Acad. Sci. U.S.* **48,** 68 (1962).

47. Ottolenghi-Nightingale, E., *Proc. Natl. Acad. Sci. U.S.* **64,** 184 (1969).

48. Borenfreund, E., Honda, Y., Steinglass, M., and Bendich, A., *J. Exp. Med.* **132,** 1071 (1970).

49. Gaudin, D., and Fellman, J. H., *Biochim. Biophys. Acta* **141,** 64 (1967).

50. Fougère, C., Ruiz, F., and Ephrussi, B., *Proc. Natl. Acad. Sci. U.S.* **69,** 330 (1972).

51. Davidson, R. L., *Proc. Natl. Acad. Sci. U.S.* **69,** 951 (1972).

52. Ephrussi, B., *Hybridization of Somatic Cells*, Princeton University Press, Princeton, N.J., 1973.

53. Bendich, A., in *Informative Molecules in Biological Systems*, Ledoux, L. G. H., ed., North-Holland, Amsterdam–London, 1971, p. 3.

54. Bearn, J. G., and Kirby, K. S., *Exp. Cell Res.* **17,** 547 (1959).

55. Mathias, A. P., and Fischer, G. A., *Biochem. Pharmacol.* **11,** 69 (1962).

56. Ozer, H. L., *J. Cell. Physiol.* **68,** 61 (1966).

57. Schwartz, A. G., Cook, P. R., and Harris, H., *Nature New Biol.* **230,** 5 (1971).

58. Osterman, J. V., Waddell, A., and Aposhian, H. V., *Proc. Natl. Acad. Sci. U.S.* **67,** 37 (1970).

59. Grady, L., Axelrod, D., and Trilling, D., *Proc. Natl. Acad. Sci. U.S.* **67,** 1886 (1970).

60. Yelton, D. B., and Aposhian, H. V., *J. Virol.* **10,** 340 (1972).

61. Merril, C. R., Geier, M. R., and Petricciani, J. C., *Nature* **233,** 398 (1971).

62. Geier, M. R., and Merril, C. R., *Virology* **47,** 638 (1972).

63. Merril, C. R., Friedman, T. B., Attallah, A. F. M., Geier, M. R., Krell, K., and Yarkin, R., *In Vitro* **8,** 91 (1972).

64. Russell, J. D., and De Mars, R., *Biochem. Genet.* **1,** 11 (1967).

65. Munyon, W., Kraiselburd, E., Davis, D., and Mann, J., *J. Virol.* **7,** 813 (1971).

66. Munyon, W., Buchsbaum, R., Paoletti, E., Mann, J., Kraiselburd, E., and Davis, D., *Virology* **49,** 683 (1972).

67. Rogers, S., *Res. Commun. Chem. Pathol. Pharmacol.* **2,** 587 (1971).

68. Terheggen, H. G., Schwenk, A., Lowenthal, M., Van Sande, M., and Columbo, J. P., *Z. Kinderheilkd.* **107,** 298 (1970).

69. Bracket, B. G., Baranska, W., Sawicki, W., and Koprowski, H., *Proc. Natl. Acad. Sci. U.S.* **68,** 353 (1971).

70. Caruthers, M. H., Kleppe, K., Van de Sande, J. H., Sgaramella, V., Agarwal, K. L., Büchi, H., Gupta, N. K., Kumar, A., Ohtsuka, E., Raj Bhandary, U. L., Terao, T., Weber, H., Yamada, T., and Khorana, H. G., *J. Mol. Biol.* **72,** 475 (1972).

71. Harvey, C. L., Wright, R., and Nussbaum, A. L., *Science* **179**, 291 (1973).

72. Jackson, D. A., Symons, R. H., and Berg, P., *Proc. Natl. Acad. Sci. U.S.* **69**, 2904 (1972).

73. Danna, K., and Nathans, D., *Proc. Natl. Acad. Sci. U.S.* **68**, 2913 (1971).

74. Danna, K., and Nathans, D., *Proc. Natl. Acad. Sci. U.S.* **69**, 3097 (1972).

75. Nathans, D., and Danna, K. J., *Nature New Biol.* **236**, 200 (1972).

76. Sack, G. H., and Nathans, D., *Virology*, **51**, 517 (1973).

77. Mertz, J. E., and Davis, R. W., *Proc. Natl. Acad. Sci. U.S.* **69**, 3370 (1972).

78. Morrow, J. F., and Berg, P., *Proc. Natl. Acad. Sci. U.S.* **69**, 3365 (1972).

79. Hotchkiss, R. D., *J. Heredity* **56**, 197 (1965).

80. Tatum, E. L., *Perspect. Biol. Med.* **10**, 19 (1966).

81. Lederberg, J., *Bioscience* **20**, 1307 (1970).

82. Davis, B. D., *Science* **170**, 1279 (1970).

83. Aposhian, H. V., *Perspect. Biol. Med.* **14**, 98 (1970–1971).

84. Fletcher, J., *N. Engl. J. Med.* **285**, 776 (1971).

85. Freese, E., *Science* **175**, 1024 (1972).

86. Friedmann, T., and Roblin, R., *Science* **175**, 949 (1972).

87. Anon., Genetics: A two-edged sword, *Nature* **240**, 73 (1972).

88. Ottolenghi, E., and Hotchkiss, R. D., *J. Exp. Med.* **116**, 491 (1962).

89. Catlin, B. W., *J. Bacteriol.* **79**, 579 (1960).

90. Takahashi, I., *Biochem. Biophys. Res. Commun.* **7**, 467 (1962).

91. Ephrati-Elizur, E., *Genet. Res.* **11**, 83 (1968).

92. Ottolenghi, E., and MacLeod, C. M., *Proc. Natl. Acad. Sci. U.S.* **50**, 417 (1963).

93. Auerbach-Rubin, F., and Ottolenghi-Nightingale, E., *Infect. Immun.* **3**, 688 (1971).

94. Ottolenghi-Nightingale, E., *Infect. Immun.* **6**, 785 (1972).

95. Rogers, J. C., Boldt, D., Kornfeld, S., Skinner, A., and Valeri, C. R., *Proc. Natl. Acad. Sci. U.S.* **69**, 1685 (1972).

96. Sarma, D. S. R., and Rutman, J., *Fed. Proc.* **31**, 607 (1972).

97. Hill, M., *Exp. Cell Res.* **45**, 533 (1967).

98. Anker, P., and Stroun, M., *Science* **178**, 621 (1972).

99. Gartler, S. M., *Nature* **184**, 1505 (1959).

100. Yoon, C. H., and Sabo, J., *Exp. Cell Res.* **34**, 599 (1964).

101. Rusinova, G. G., Rogacheva, S. A., and Libinzon, R. E., *Vop. Med. Khim.* (*Moscow*) **13**, 485 (1967).

102. Borenfreund, E., and Bendich, A., *J. Biophys. Biochem. Cytol.* **9**, 81 (1961).

103. Sirotnak, F. M., and Hutchison, D. J., *Biochim. Biophys. Acta* **36**, 246 (1959).

104. Glick, J. L., *Cancer Res.* **27**, 2350 (1967).

105. Laval, F., Malaise, E., and Laval, J., in *Informative Molecules in Biological Systems*, Ledoux, L. G. H., ed., North-Holland, Amsterdam–London, 1971, p. 124.

106. Kantoch, M., and Bang, F. B., *Proc. Natl. Acad. Sci. U.S.* **48**, 1553 (1962).

107. Bradley, T. R., Roosa, R. A., and Law, L. L., *J. Cell. Comp. Physiol.* **60**, 127 (1962).

108. Podgajetskaja, D. J., Bresler, V. M., Surikov, I. M., Ignatova, T. N., and Olenov, J. M., *Biochim. Biophys. Acta* **80**, 110 (1964).

109. Glick, J. L., and Salim, A. P., *J. Cell Biol.* **33**, 209 (1967).

110. Fox, M., Fox, B. W., and Ayad, S. R., *Nature* **222**, 1086 (1969).

111. McBride, O. W., ank Ozer, H. L., *Proc. Nactl. Aad. Sci. U. S.* **70**, 1258 (1973).

112. Rodgers, S., Lowenthal, A., Terheggen, H. G., and Columbo, J. P., *J. Exp. Med.* **137**, 1091 (1973.)

Index

Acetylcholine, 51, 52, 61, 62
Acetylcholine receptors, 37, 59
Acetylcholinesterase, 56, 60
Actinomycin, 100
Action potentials, 54
Active transport, 81
Adaptive immunity, 188
Adenine phosphoribosyltransferase (APRT), 68
Adenosine 3′, 5′-monophosphate, 71
Adenyl cyclase, 161
Adhesions, cell-substratum, 157
  intercullular, 157
Alignment, fibroblasts, 173
α-bungarotoxin, 17, 58
Amino Acids, 81
Amoeboid locomotion, 147
Anchorage dependence, 179
Antibody, 195
Antibody-dependent killing, 190
Antibody molecule, 191
Antibody production, 193
Antigen-lymphocyte interaction, 191
Antigen presentation, 199
Antigen recognition, 191, 192, 195
Antigens, 73, 188

Baby hamster kidney cells, 70
  ARPT-deficient hamster cells, 71
  $BHK_{21}C_{13}$, 70
  TK-deficient hamster cells, 71
  wild-type BHK cells, 71
Basal epidermal layer, 114
B-cell activation, 189
Blebbing activity of cell, 154
B-lymphocytes, 189, 190, 191
Botulinum toxin, 56
BSC-1 cell, 73
Burnett's hypothesis, 191
Bursa of Fabricis, 189

Calcium, 47, 49, 57, 81
Cancer, 152
Cancer cells, 156
Cap formation, 192
Capping of cells, 166
Cation transport, 89
Cell aggregation, 117
Cell apposition, 94
Cell associations, 73
Cell-cell adhesive force, 121
Cell-cell recognition, 188
Cell-to-cell transfer of macromolecular RNA, 100
Cell-collagen adhesion, 121
Cell communication, 67
  pigment donation, 122
  pigment transfer, 123
Cell confluence, 118
Cell contact, 67, 68, 72
Cell cycles, 116
Cell-glass adhesion, 121
Cell interactions, 70
Cell junctions, 31, 35, 109
  desmosomes, 117
  gap junction, 32
  hemidesmosomes, 114
  septate junction, 32
  tight junction, 32, 119
Cell membranes, apposition of, 72
Cell migration, 44, 114, 116
Cell mixture, 73
Cell monolayer, 70
Cell nuclei, 68
Cells involved in immune response, accessory cells, 188
  B lymphocytes, 189
  dendritic reticulum cell, 188
  granulocytes, 188
  lymphocytes, 188
  macrophages, 188

255